# ■ GEOLOGY IN THE FIELD

Robert R. Compton
Stanford University

# ■ GEOLOGY
# IN THE FIELD

John Wiley & Sons
New York ■ Chichester ■ Brisbane ■ Toronto ■ Singapore

*To my students.*

Design: Candace Compton

Produced by: Alphabetics/Design With Type

Illustration: Robert R. Compton

Cover: The circles show the limiting diameters of pebble and cobble size-classes (p. 49). To classify a clast, estimate where its median diameter (the diameter of a sphere of equal volume) would fall in the array of circles. The centimeter grid on the inside back cover can be used to obtain specific dimensions of clasts and other objects.

*Library of Congress Cataloging in Publication Data:*

Compton, Robert R.
  Geology in the field.

  Includes index.
  1. Geology — Field work. I. Title.
QE45.C63   1985      551'.0723      85-2325
ISBN 0-471-82902-1

Printed in the United States of America
10

**Printed and bound by Courier Companies, Inc.**

# ▪ Preface

Geology has evolved greatly since I wrote the *Manual of Field Geology* in the 1950s. Advances in theory have transformed the formation mapping of that time into increasingly broader and more interpretive studies. We have much to seek at the outcrop, and the rising costs of field work compel us to recognize key features the first time around. In addition, mapping and data collecting must be more consistent and accurate than ever. This book is intended as a guide for these modern studies. Its form is compact so that it can be carried in the field, yet most procedures are spelled out completely. Half the book consists of brief descriptions of textures and structures helpful in interpreting depositional environments, kinds of volcanic activity, and plutonic events and conditions. To encourage full interpretation during the field season, procedures are included that are often reserved for the laboratory or office: staining rocks, correcting orientations of current indicators, constructing profile sections of folds, measuring strains, making photogeologic interpretations, and so on.

Broadly, the book proceeds from pre-field considerations to methods of observation and measurement, and then to recognition of key geologic features, and finally to preparation of a report. Chapter 1 presents the general philosophy of field geology together with the steps in a typical field project. Chapter 2 describes field equipment and its uses, and Chapter 3 observation, interpretation, and accumulation of information. Hand-lens identification of textures and rocks is the subject of Chapter 4, and Chapter 5 describes how to recognize and trace rock units and structures, including the details of a pace and compass traverse. Geologic mapping on a topographic base is covered in Chapter 6 and mapping on aerial photographs and other remote-sensed imagery in Chapter 7, the latter including sections on photogeologic interpretation and on compilation of photo data. Chapter 8 covers mapping with the plane table and alidade, including surveys of control systems. Structures and depositional environments of marine and nonmarine deposits are described in Chapters 9 and 10, and measurement and description of stratigraphic sections in Chapter 11. Chapter 12 presents summaries of tectonic structures and mélanges, and Chapters 13, 14, and 15 primary features of volcanic rocks, of plutons, and of metamorphic rocks. The final chapter describes ways of planning and preparing geologic illustrations and reports. Fourteen appendixes provide systematized data and procedures.

In order to make the book as compact as possible, involved or special methods that can be anticipated before the field season were omitted. Underground mapping, for example, requires many safety precautions and specific instructions, and is well described elsewhere. Trigonometric tables were kept to a minimum because of the general availability of pocket calculators. Many of the drawings are generalized or composite in order to save the space that would have been needed for more specific examples.

In 1980 I corresponded with many instructors who had used my manual, asking for criticisms and suggestions. The following persons provided useful ideas and materials: A. K. Baird, S. S. Beus, W. A. Braddock, M. L. Bregman, P. W. G. Brock, E. R. Brooks, V. M. Brown, G. H. Davis, J. Deen, J. A. Dorr, Jr., P. L. Ehlig, V. Fischer, D. R. Foutz, R. E. Garrison, C. A. Hall, Jr., C. W. Harper, E. A. Hay, H. E. Hendriks, J. F. Karlo, M. Kay, S. R. Kirkpatrick, S. A. Kirsch, R. G. Lawrence, W. E. LeMasurier, K. A. McDonald, D. F. McGeary, R. G. McWilliams, M. E. Maddock, W. D. Martin, M. A. Murphy, T. L. Péwé, K. J. Schulz, K. Servos, R. L. Shreve, P. Snavely, III, L. A. Standlee, C. A. Suczek, A. G. Sylvester, A. N. Ward, Jr. and H. Zantop.

A number of persons and organizations answered questions, offered valuable advice, or supplied copies of papers or other materials during the writing stage: T. H. van Andel, W. R. Dickinson, L. G. Duran S., J. R. Dyer, Earth Sciences Associates, W. R. Evitt, J. H. Fink, R. V. Fisher, R. W. Galster, R. E. Garrison, the Geological Survey of Canada, R. T. Holcomb, C. M. Isaacs, D. L. Jones, J. R. Keaton, D. K. Keefer, V. A. M. Langenheim, R. V. Laniz, D. K. Larue, C. McCloy, G. Mahood, D. M. Miller, E. L. Miller, T. H. Nilsen, B. M. Page, T. L. Péwé, J. B. Pinkerton, M. C. Powers, E. I. Rich, the Shell Oil Company, M. F. Sheridan, D. A. Swanson, A. G. Sylvester, R. L. Threet, P.-J. Uebel, and the U.S. Geological Survey.

Persons who read completed chapters or parts of chapters and provided many valuable criticisms and suggestions are: E. R. Brooks (Chapters 2, 4, and 15), Lindee Glick (Chapter 12), H. E. Hendriks (Chapter 7), R. T. Holcomb (first half of Chapter 13), J. R. Keaton (the tables and engineering geology section of Chapter 5), G. Mahood (Chapter 13), N. McLeod (first half of Chapter 13), C. Meyer (the hydrothermal alteration part of Chapter 15), D. M. Miller (Chapter 12 and Appendix 7), E. L. Miller (Chapter 12), R. J. Newberry (part of Chapter 15), T. H. Nilsen (Chapter 9 and part of Chapter 10), B. M. Page (the mélange section of Chapter 12), T. L. Péwé (Chapter 10 and part of Chapter 7), A. G. Sylvester (Chapters 1, 3, 5, 6, and part of 7), and V. R. Todd (Chapter 14, part of Chapter 12 and Appendix 7). In addition, the manuscript was reviewed for John Wiley & Sons by F. R. Ettensohn, R. L.

Kaesler, and S. H. Wood, who each suggested a number of improvements.

The staffs of the Branner Library and the Department of Geology at Stanford were very helpful. Lyn Dearborn and Dave Olson typed and computerized the manuscript, making many corrections in it. Last and far from least, the staff at Alphabetics/Design With Type, especially my daughter Candace, made many improvements during design and composition.

Robert R. Compton

# ■ Contents

## Chapter 6 Geologic Mapping on a Topographic Base

## Chapter 7 Use of Aerial Photographs and Other Remote Imagery

## Chapter 8 Mapping with the Plane Table and Alidade

## Chapter 9 Primary Features of Marine Sedimentary Rocks

## Chapter 10 Surficial Sediments; Continental Environments

## Chapter 16 Preparing Illustrations and Writing Reports

## Appendixes

# ■ GEOLOGY IN THE FIELD

# 1

# ■ Philosophy and Organization of a Field Study

## 1-1. Field Geology in General

Field studies offer several unique advantages in solving geologic problems. For one, earth materials and structures can be identified most easily when seen with related features. For another, interpretations made in the field can be checked immediately against relations they predict. And for a third, studying actual associations of materials and structures can lead to discovery of new kinds of features or relations, and thus to new ideas.

Field studies are founded on three kinds of information, the simplest being the straightforward facts arising from direct observation and measurement. The texture of a rock, the strike and dip of beds, and the geographic relations between two rock bodies are examples. These are objective data. They constitute the main information on geologic maps, and ideas drawn from them, alone, are truly inductive. In this book, methods of gathering and organizing objective data are the main topics of Chapters 2 through 8, and 11.

The second kind of information is basically interpretive. Certain associations of rocks and primary structures, for example, are thought to imply specific genetic conditions or environments. Geologists "see" environments of the past by means of these implications. This approach has great power because genetic insight can clarify a host of interrelated data. Interpretations, however, depend on the state of geologic theory and on the geologist's perception and experience, and thus may be incorrect. They are nonetheless the most valued basis of many modern field studies, and the descriptions, drawings, and references in Chapters 9, 10, and 12 through 15 are intended to help in making them.

The third kind of information consists of age relations. These relations are partly objective and partly interpretive but form a particular category because they put geologic events in order. Ordering of events is essential in geology because the science is basically historical. Every geologic event is unique and each rock and structure thus has intrinsic uniqueness, a dependency differing radically from those of broadly abstract sciences. In physics, for example, all atomic particles of one kind are considered identities, each represented by a numeral representing charge or mass; the history of a given particle in no way distinguishes it. In geology, on the contrary, each rock unit is unique because it represents history. Each lava flow, for example, represents a unique eruption from a specific volcano. Lava flows can thus be used to work out locations and eruptive histories of groups of volcanoes and thereby the tectonic and thermal development (history) of that part of the earth.

Age relations among events are determined in part by cross-cutting or superimposed structural relations in rocks, as described in parts of Chapters 9, 10, and 12 through 15. Stratigraphic sequence, a second means of ordering events, is treated in parts of Chapters 9, 10, 11, and 13 through 15. Determining geologic ages and numerical ages of rocks is especially valuable for ordering events, and those topics are treated briefly in several chapters.

To summarize, field studies are both operational and philosophical. Factual field data are at the heart of the science because they force us to see the earth as it is and permit us to discover new geologic relations. Field interpretations are based on theory and thus may be incorrect, but they enrich and enlarge studies in many ways. Firm age relations are the basis for ordering facts and interpretations into actual histories. They provide the strongest test of broad geologic theory.

## 1-2. Geologic Maps and Mapping

Three conditions tend to make geologic field studies intriguing and sometimes difficult: (1) the great length of time and the many events that may be represented in a given area; (2) the typically incomplete record at any one place; and (3) the thinness of the earth's exposed skin. Geologic mapping deals uniquely with each of these conditions. Geologic time is set in order by a carefully constructed geologic map showing the geometric relations among the different rocks and structures. The incompleteness of the local record may be compensated by correlating the map's features with more complete records mapped elsewhere. The thinness of the exposed skin may be resolved by using the map to project structures to depth in vertical cross sections.

A geologic map is a precisely oriented, scaled-down diagram of the earth's surface, or perhaps of an underground level in a mine. Its position relative to other parts of the earth is shown by lines of latitude and longitude, as well as by other cadastral lines or geographic boundaries. Sizes, orientations, and positions of geologic features can thus be compared exactly with those on other maps. Rock units and structures are identified in an explanation, which also shows the age sequence of the rock units. Contours and symbols for waterways, roads, and buildings make it possible for anyone to take the map to the field and find the geologic features shown.

Geologic maps are thus of great value for others to use; they are, in addition, absolutely essential to the geologist who makes one. Many geologic features are too large, too complex, or too diffuse to be recognized outright on the ground. Mapping is a means of discovering such features and finally of seeing them exactly in three dimensions. Mapping may also develop patterns of specific kinds of data that imply genetic relations; for example, the zonal patterns of metamorphic minerals or small-scale structures around igneous intrusions. As already noted, the geologist uses mapping to resolve age relations and thus to work out geologic history.

No matter how great the geologist's experience and capacity for memory, mapping always leads to discoveries and often to surprises. It commonly sends him or her back to certain localities to look again.

### 1-3. Selecting a Field Study

Although many thorough and valuable field-based studies have been made, geologic theory has advanced so rapidly over the past few decades that the science needs almost countless additional studies. Over the same period, the potential of field work has been increased by: (1) new means of biostrati-graphic zonation, (2) increased availability of numerical dating, (3) new or improved laboratory methods, (4) new remote imagery and improved base maps, (5) new field techniques, and (6) digital analysis of data. Many areas and problems studied more than 20 years ago might thus be restudied.

Selection of a study can therefore be made from a wide range of possibili-ties. A primary need is that the study truly interest the person undertaking it. It should constitute a scientific challenge and at the same time promise enjoyment. Ideas for studies are usually gained or refined by reading and by discussions with other geologists. Pertinent literature is indexed by subject, area, and author in these standard references: *Bibliography of North American Geology* by the U.S. Geological Survey (up to 1970); *Bibliography and Index of Geology Exclusive of North America* by the Geological Society of America (for 1933-1968); *Bibliography and Index of Geology* by the Geological Society of America (1969-1978) and by the American Geological Institute (since 1978). These reference sets and many of the items to which they refer are available in most university libraries and in libraries of U.S. Geological Survey regional offices.

Some state agencies (e.g., the California Division of Mines and Geology) have also compiled topics and locations of theses, dissertations, and other academic work, and may be able to supply information on studies in pro-gress. Trautmann and Kulhawy (1983) have described data sources for engi-neering geologic studies which can be used for other kinds of geologic studies. Many computer facilities can provide access to search systems that include earth science data bases (e.g., *Dialog*).

Geologic maps are included in the indexes cited above, and most states in the United States and provinces in Canada can supply geologic maps or references to maps published by other agencies. Inquiries regarding geologic maps and information can also be made to specific national surveys (Geologic Inquiries Group, U.S. Geological Survey, 907 National Center, Reston, VA

22092; Geological Survey of Canada, 601 Booth St., Ottawa, Ontario K1A
OE8, etc.).

*The focus of the study* should be a major consideration. Some projects are
directed toward specific questions; some others explore more general questions, and still others examine promising areas to develop questions. Focusing on a specific question can have great value but is sometimes overdone.
Some questions are concocted so carefully that they are answered largely
before the field season. Some topics are such current favorites that they will
be difficult to investigate without bias. Projects should be planned so as to
permit discoveries and innovation.

Generally, however, it should be clear not only what is to be accomplished,
but also who will be interested in the results. How will the study be reported,
and to whom? It may be helpful to read Chapter 16 during the planning
stage.

## 1-4. Reconnaissance

Although a perfect topic may be selected by reading, discussion, and study,
it cannot become a practical project until an area has been reconnoitered.
The reconnaissance has two basic purposes: (1) to make sure the area is
suitable for the topic selected; and (2) to plan the field work in light of time
and funds available. The reconnaissance will be especially effective if copies
of pertinent geologic maps are taken to the field, or if data from maps and
reports are copied onto a map of the area. A preliminary photogeologic study
may prove invaluable (Section 7-3). If possible, a large-scale topographic
map of the area should be carried during the reconnaissance so that geologic
data and ideas can be located precisely. Locations and sets of data or ideas
must be recorded in carefully organized notes (Section 3-3). In addition to
the geology, the reconnaissance must evaluate the terrain and its accessibility. Specific items and questions will vary with the geography, but in
most cases the following will be useful:

1. What are the principal rock units in the area? Do they differ from
descriptions of the same units studied elsewhere? Are they sufficiently exposed
and fresh enough to meet the needs of the project?

2. Are there homoclines of layered rocks that can be used to establish an
age sequence? Might fine-grained sedimentary rocks be collected to see if
they contain usable microfossils?

3. Do major folds, faults, or unconformities mapped elsewhere continue
into the area? What major structures can be seen otherwise?

4. If intrusive bodies are present, are their contacts exposed? Do contacts between adjoining intrusions indicate age relations?

5. Does the area have different structural terrains, perhaps expressed
topographically? How are they joined?

6. For deformed rock units, what structural features appear clear and consistent from place to place?

7. What scale will be needed to map the more important structural details? Will there be time to do this?

8. Are there fine-grained rocks that should be collected for petrographic identifications?

9. How will the topography and vegetation affect field work? Does vegetation preclude use of aerial photographs as a base for mapping (Section 1-5)?

10. Who are the chief landowners? Will their permission be necessary for camping, collecting, and mapping?

11. Will camping during the field season save time and funds? If it is not feasible, what time and funds must be allotted for commuting and living elsewhere?

12. Which of the roads in the area can be driven safely? Are there trails or other easy means of access otherwise? Must parts of the area be mapped from backpack camps?

13. If plane table methods are being considered, does the area meet the special needs described in Chapter 8?

14. Reviewing all the information gained, what field methods and map scale will be optimal to serve the project's purpose? Is there a particularly well exposed area at which mapping should be started? Should a stratigraphic section be studied in moderate detail before beginning the mapping (Sections 5-1 and 11-1)?

A thorough reconnaissance is likely to take about a week for a moderately complex area of 40 to 100 sq km (15 to 40 sq mi). About twice that much time is likely to be needed for complex areas. The time spent will rarely be wasted.

### 1-5. Preparations for the Field

A reconnaissance commonly gets one to reread the literature, typically with much more interest than before. The general idea is to be as well prepared as possible for the actual features in the area, and to see what they may contribute to the topic or question of the study. Another major task is preparing or obtaining field equipment. Appendix 1 provides a checklist; Chapter 2 describes basic equipment; and Chapters 6, 7, and 8 mention additional equipment needed for certain methods or kinds of surveys. Small items, such as pencils, pens, and protractors, are easily lost, and thus several should be obtained. The magnetic declination must be determined for the field area (Section 2-4). It may also be necessary to confirm permissions to map on private land, to arrange for additional persons to help with the mapping, and to arrange for funding and insurance.

*Choice of a base for mapping* depends on the characteristics of available topographic maps and aerial photographs as well as the nature of the area being studied and the scale to be used. Areas covered by forest are very difficult to map on aerial photographs, and areas of monotonous brush or grass will show few features by which to make accurate locations. Otherwise, aerial photographs have a characteristic that makes them virtually invaluable: They may show geologic features that do not appear on topographic maps and would probably go unnoticed on the ground. Another advantage of aerial photographs in many areas is that locations can be made on them accurately and quickly because they show many details, such as individual trees and the smallest of streams.

Otherwise, topographic maps might be preferred for several reasons. They are far less expensive than aerial photographs, especially in mapping quadrangle-size areas. They give a more exact three-dimensional view of large areas, and cross sections can be drawn from them more readily. They can be enlarged inexpensively so as to give more room for plotting features and locality numbers, and although photographs can also be enlarged, they cannot then be used easily to obtain stereographic (three-dimensional) views of the terrain. Finally, the relatively small area covered by each photograph, and the patchy dark-and-light nature of the photo image, make it difficult to get an overview of features mapped on a number of photographs.

In conclusion, mapping in many areas can be done more accurately, completely, and rapidly on aerial photographs, but the data should be transferred frequently to a topographic base. The latter will give a clear view of the geology covered and permit construction of cross sections at any stage of the mapping. Geologists should nonetheless become experienced in mapping directly on topographic maps, because this experience instills an ability to visualize the third dimension as well as skill in making accurate locations on both maps and photographs. In addition, topographic maps are the preferred base for some areas.

### 1-6. Work in the Field

This book is composed of instructions and suggestions for work in the field but does not advise about many specifics, especially the commonsense ones that may be important to the success of a project but generally must be tried and learned in an individual's own way. Hints that will surely help in this regard were brought together in an essay by an expert field geologist, S. W. Muller (1983). One suggestion deserves emphasis: Field geology is learned in the field; therefore one must go there as soon and as frequently as possible.

Geologic mapping is a good way to start field work because it is reasonably simple in concept. A map can be constructed from a blank sheet of paper, but usually a topographic map or an aerial photograph is carried in the field to serve as a base on which to plot geologic data. Mapping should be started in an area where rock units are well exposed and where their sequence has been reconnoitered (Section 1-4). As rock units and other features are encountered, their exact positions are located and marked by a point, a line, or a closed form. The localities are numbered in order to relate them to field notes that describe the features. If structures are measured, such as strike and dip of bedding, a structure symbol is plotted on the map. This is done at once so that the symbol can be checked against the outcrop, and to see structural relations develop as soon as possible. Where a locality exposes a contact surface (interface) between two rock bodies, a line is drawn on the map depicting a scaled-down copy of the contact's trace across the ground. Faults and fold hinges are examples of other structures that are mapped as lines. Localities of collected fossils, rocks, or minerals are plotted, as are sites where photographs are taken or where drawings are entered in the notes to record structural relations too small or too complex to map to scale. In short, the map becomes a picture of the formations and structures that can be drawn to scale as well as a geographic record of all other data and ideas recorded. The map will be easier to read if the mapped units are colored lightly, and the recorded notes will be easier to use if organized, dated, and paginated with strict consistency.

Mapping is likely to be exploratory and deliberate until the principal rocks have become familiar and mapping units have been tested and proven. Visits to nearby mapped areas and to type sections of formations may be needed to complete this selection. When mapping becomes more rapid and systematic, it should be directed toward parts of the area that seem most likely to resolve the project's major questions. As parts of the area are more or less completed, they can be checked by constructing cross sections through them. Recesses may be needed to do laboratory work on critical fossils or rocks.

**Work in the field camp or office** must be done routinely during evenings or on an occasional office day. Data are transferred frequently from separate field sheets or aerial photographs to an overall base map, an *office map*, which is colored lightly to emphasize the distribution of rock units and structures. The transferring is done frequently in order to see the geology develop concurrently with field mapping, and thereby keep the project moving ahead as planned. Other important routines are to read field notes and study the map and rock samples together in order to examine recent discoveries and think over geologic puzzles. These reviews are the basis for planning the field strategies of the next few days or weeks. They also insure that the project will proceed as originally intended (Section 1-3).

As parts of the area are completed, trial cross sections should be constructed from them. Faulted or folded parts of the area may thus be examined to see if they are geometrically consistent in three dimensions. Sense of movement on faults may be estimated from relations described in Section 12-5. Folds will be clearer if constructed in profile view (Section 12-3), and the consistency of foliations and lineations may be checked by compiling these data on equal-area stereographic diagrams (Appendix 14).

An important routine is to write summary descriptions of rock units and large structures as field work progresses, including interpretations that cite evidence pro and con. These summaries indicate what has been resolved and how, as well as which questions are left to be answered. They are an invaluable basis for writing the final report.

Mechanical operations include inking note numbers and structure symbols on field sheets or photographs. Contacts, faults, and fold hinge-lines may also be inked when they seem to be resolved. Faults are commonly inked by thin colored lines rather than the standard thick black lines, which may cover topographic detail. Rock and fossil samples should be checked after each day's work to be sure their labels are legible. If samples are segregated per unit, they can be reexamined easily when writing descriptions of units or considering problems.

### 1-7. Completing a Field Study

Late in the field study, questionable faults, unconformities, and intrusive contacts should be reexamined. Specific kinds of data or samples may also have to be augmented. Commonly, the more intensive parts of the study are expanded during a second field season, or after laboratory or office study of data and samples. Samples for numerical age determinations should not be selected until field work has established the events that will be dated.

While still in the field, the geologist can anticipate the final report by doing the following:

1. Comparing the office map to field sheets and notes to make sure that all data have been compiled and designated correctly on the map.

2. Selecting cross-section lines and preparing complete pencil drafts of the sections. Typically, the data on the map will be insufficient to resolve all parts of the sections, so that traverses must be made along the lines of section to obtain additional data.

3. Reviewing notes and summary descriptions to be sure that (a) all units and structures can be described fully; (b) rock and fossil samples will be adequate to resolve questions of interpretation; and (c) structural data are sufficient to define major structures and principal episodes of deformation.

4. Writing an outline of the geologic history in order to note any age relations that have not been explored adequately.

5. Studying Chapter 16 to be sure that any miscellaneous data needed for a report will be available.

***On completion of field work,*** all notes, maps, photographs, and samples are organized for office and laboratory work. A final geologic map, cross sections, and perhaps columnar sections are prepared and finally copied in ink. Each plate must carry a complete title and explanation. Writing a report (the last major task) should be based largely on the map and sections, and on the summary descriptions written during the field season. The organization and clarity of the report are crucial, because it will have to be understood by persons who have not seen the rocks and structures described. The clarity of the report will always be improved if the report, like the field project, is organized so that the purpose of the study is the central theme.

### References Cited

Muller, S. W., 1983, Some field hints from an old top hand: *Journal of Geological Education*, v. 31, p. 36-37.
Trautmann, C. H., and Kulhawy, F. H., 1983, Data sources for engineering geologic studies: *Association of Engineering Geologists Bulletin*, v. 20, no. 4, p. 439-454.

# 2

# Basic Equipment and Its Uses ■

## 2.1 Equipment for Sampling and Recording

Appendix 1 lists all equipment that is likely to be needed in the field or field office, and the nature and use of the more basic items are described in this chapter. Topographic base maps are described in Section 6-1, aerial photographs in Section 7-1, equipment used in mapping on photographs in Section 7-1 and 7-4, and equipment for plane table mapping in Chapter 8. Suppliers, such as Miners, Inc. (P.O. Box 1301, Riggins, ID 83549) and Forestry Suppliers, Inc. (Box 8397, Jackson, MS 39204-0397) sell most kinds of geologic equipment, and generally send catalogs on request.

A *hammer* with a pick or chisel end is used for cleaning exposures, for digging, for breaking rocks, and for trimming samples. Standard geologists hammers have heads weighing 1.5 to 2 lb (0.68 to 0.9 kg) and are adequate for most geologic work. A small sledge—for example a 2 or 3-lb head on a 14-in. handle—may be needed to collect fresh samples of especially tough rocks. *All* hammers are potentially dangerous, because heavy blows may send off rock spalls or steel flakes at high speeds. It is truly important, therefore: (1) to wear safety goggles (which fit over glasses); (2) not to strike heavy blows when people are nearby; (3) never to strike one hammerhead with another, as when using one as a wedge; and (4) to strike angular rock edges so that spalls will fly to the side rather than toward oneself (Fig. 2-1*A*). The hammerhead is less likely to spall if the corners and edges of the striking face are filed or ground to a bevel.

A *cold chisel* or a *moil* (a piece of tempered drill steel with a pointed end) may be used with a hammer to split rocks parallel to bedding or foliation and

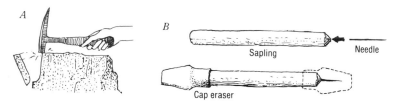

**Fig. 2-1.** *A.* Correct method of breaking a spall from the angular edge of an outcrop. *B.* Mounting a needle to use for scratch or probe tests, or for marking points on aerial photographs. A cap eraser used to cover the needle can be placed over the other end of the sapling when the needle is being used.

to free fossils or specific mineral samples from nonfoliated rocks. A *pocket knife* is used for cutting suitable materials rather than prying and scratching rocks, which will spoil the blade. Scratch tests can be made with a sharp *steel needle*, which may be mounted permanently as follows: (1) cut a 4-in. (10-cm) piece from a green sapling branch as thick as a heavy pencil; (2) cut a needle in half with a pair of pliers; (3) holding it in the pliers, force the cut end about 1 cm into the piece of sapling; and (4) add a cap eraser that will cover the needle and fit snuggly on the sapling (Fig. 2-1*B*).

A *map holder* must be large enough to carry 9 X 9 in. (23 X 23 cm) aerial photographs and should be made of masonite rather than metal (which is uncomfortable to carry) or plastic (which may break when cold) (Fig. 2-2). The same holder can be used for 8.5 X 11 in. (21.6 X 27.9 cm) ruled paper sheets for field notes (which are placed over the aerial photograph or map). Field notes may also be taken in a separate *field notebook*, which can be obtained in sizes small enough (about 4.5 X 7 in.) to fit in a large pocket. Or it may be preferable to take notes in a small loose-leaf notebook which is carried in a leather case on one's belt. Engineers' field books and level books are available with high-rag or waterproof papers, which are essential for note-taking in wet conditions. If separate sheets are used, as in the map holder shown in Fig. 2-2, heavy ledger paper of top quality should be used in wet or very dry climates (the latter eventually cause light, inexpensive papers to disintegrate).

*Pencils* and *pens* are used for plotting data on maps or photographs, for taking field notes, and for marking samples. The one used for plotting on a base map must make fine lines that are easily visible, will not smudge, and are erasable. The choice will vary with the climate and the operator but is generally a well-sharpened pencil of about 2H hardness. Mapping on plastic-base photographs such as those currently supplied by the U.S. Geological Survey (Section 7-1) requires a sharply pointed ink-flow pen (as a Pilot *ultra-fine-point, permanent, SC-UF*). Red or green pens mark black and white photographs more legibly than black or blue pens. Pens should be tested on

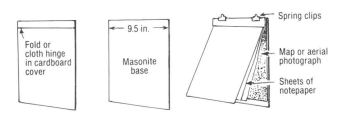

**Fig. 2-2.** Lightweight holder for carrying maps, aerial photographs, and 8.5 X 11 in. notepaper.

photographs before the field season, and should make a fine, even line that can be wiped off the photograph with a moist piece of cloth or a moistened soft eraser. These erasures should not damage the photograph if made within a day, and should damage it only slightly if made at a later time. Ink-flow pens with standard (moderately thick) points are used for marking rock or fossil samples. A pencil or ball-point pen used for note-taking must make easily legible copy that is waterproof and will not smudge. Several extra sets of pencils and pens should be available in the field camp and an extra set should be carried in the knapsack.

A *scale*, used for measuring features or laying off distances on maps and photographs, should have fine, distinct graduation marks that are equivalent to even increments at the map scale used. The engineer's 20-scale, for example, has divisions equal to 100 ft on maps of 1:24,000 (1 in. = 2000 ft), and a millimeter scale corresponds to even units on a metric map. A scale 6 in. (15 cm) long is adequate for most work. Transparent plastic scales with built-in protractors are generally available (as from the C-Thru Ruler Co., Bloomfield, CT 06002). Sets of 6-in. scales graduated in various metric and English units corresponding to specific map scales are available from some suppliers.

A *protractor* is used for plotting bearing lines and structure symbols on maps and photographs and for measuring angles between structures in rocks. In order to plot lines by the method illustrated in Fig. 6-1*A*, the protractor should have a base 4 to 6 in. (10 to 15 cm) across.

A *camera* is carried routinely in most field studies and should thus be compact and strong. Older models of several of the top-grade 35 mm cameras have retractable and exchangeable lenses as well as solid workings that make them ideal, moderately priced field cameras. All 35 mm cameras have a greater depth of focus than cameras with longer focal lengths, and this is a decided advantage in photographing irregular outcrops at close range. A wide-angle lens (as a 2.8 cm lens for a 35 mm camera) gives extreme depth of focus (commonly 1 m to infinity) and a much fuller view of outcrops that must be photographed at close range. A *lightweight tripod* permits longer exposures and thus the reduced apertures required for maximum depth of focus. A flash attachment may prove valuable, as described in Section 3-4.

Of a miscellany of items, *sample bags* of cloth or plastic may be obtained through most suppliers, or bags may be of extra heavy paper, the variety often used as nail bags. A small spring-wound *6-ft tape* can be used to measure bed thicknesses, clast sizes, and so on, and a compact pair of *binoculars* may be helpful in finding outcrops. The use of a precise altimeter is described in Section 6-3. *Hydrochloric acid* will be needed, and should be diluted just to the strength that causes effervescence of calcite but not dolomite (except when powdered). A *shovel, pick,* and *soil auger* are likely to be essential in studies of surficial deposits (Chapter 10) or in mapping rocks through poorly exposed areas. A *gold pan* is useful for concentrating heavy minerals that might otherwise go unnoticed—for example kyanite or staurolite in an area of mica schist.

## 2-2. Selecting and Using a Hand Lens

Of the hand lenses generally available, 10X and 14X lenses are used most widely. With a 14X lens, one can see the 0.01 mm divisions of a micrometer scale, whereas objects only as small as 0.02 mm can be distinguished with a 10X lens. The depth of focus of the 14X lens, however, is only 0.8 mm, whereas that of the 10X lens is 2.5 mm. This is a major consideration because it is tiring to the eyes to raise and lower a lens in order to see all parts of grains sharply. The 14X lens must also be held closer to the object being viewed, which tends to interfere with lighting.

Whatever magnification seems best, the quality of the lens can be tested by examining fine print or a finely graduated scale on a flat surface. The field of view should be sharply in focus out to its edge. Good quality triplet lenses typically give excellent images. In testing a lens, and in all other viewing, the following are important:

1. Hold the sample so that the area being viewed is in full light—in sunlight, if possible.

2. Hold the lens exactly at the distance of sharp focus, with its optical axis perpendicular to the surface being viewed.

3. Bring the eye to the point where the eyelashes are almost touching the lens (this is the only position from which the entire field of view will be sharply and comfortably in focus).

Hand-lens procedures seem awkward at first because of the unnaturally close range of viewing and because of the images seen through the unused eye. With practice, however, the method will simply become an additional way of seeing things. The broken rock surface will then appear as a well-lighted landscape, every part of which can be "visited" and examined in great detail. Because objects only 0.01 to 0.02 mm in diameter are visible, a hand lens has far greater potential in rock study than is usually realized.

Mineral properties that can be used to identify grains with a hand lens differ in some ways from those used on large mineral specimens. Color and luster are readily observable with a hand lens and generally are reliable properties in unaltered rocks. Crystal form or habit, fracture, cleavage, and parting are particularly valuable, although working them out accurately may require study of several grains. The angle at which two sets of cleavages intersect can be estimated by facing the sun and holding the sample so that the line of intersection of the two cleavages is approximately horizontal. If the sample is then rotated on this horizontal axis, the amount of angular rotation between the two sets of cleavage reflections gives a measure of the angle between the cleavages. Cleavages can also be determined directly by examining crushed material under the lens.

Streak is a reliable property that is easily determined where hammer blows have powdered the rock, yet left it coherent; dark feldspar and quartz grains, for example, become white in such areas. Streaks of small grains softer than 5 on the Mohs scale can be made by scratching them with a steel needle.

Hardness of small grains is difficult to test, except by using sharply pointed objects, as a steel needle ($H=5$), a sharpened copper wire ($H=3$), or a set of hardness points. Hardness tests must be made with care so that the grains are not crushed in the process of trying to scratch them. A needle can also be used to determine whether flexible cleavage plates are elastic (a true mica) or brittle (other sheet silicates). With moderate practice, the lens and sample can be held firmly in one hand (the left hand of a right-handed person) while using the other hand to probe grains with a needle or place a drop of HCl on a grain.

## 2-3. Materials and Methods for Staining Rocks

Several important rock-forming minerals may be distinguished in the field by staining procedures similar to those used in the laboratory. Smooth, clean outcrop surfaces have been found to give results almost as good as sawed and lap-ground surfaces, and freshly broken surfaces have given adequate results. The methods described here, based on those of Laniz and others (1964), are a sequential use of cobaltinitrite, which stains potassium feldspar bright yellow, and amaranth dye, which stains plagioclase (except albite) medium crimson and cordierite deep crimson. Albite is stained crimson by prior use of $CaCl_2$ solution. Dolomite, benitoite, celsian, hydrogarnet, pectolite, vesuvianite, witherite, and wollastonite also stain medium to deep crimson, and because calcite stains pale pink, it can be distinguished from dolomite as well as from several minerals common in calcsilicate rocks.

Specific reasons for staining include the following:

1. Making accurate estimates of the mineral composition of quartzofeldspathic and calcsilicate rocks, including fine-grained varieties.

2. Making fabrics more discernible (Sections 4-2, 4-6, and 14-2).

3. Seeing the distribution of phenocrysts, irregular metacrysts, clots, veins, compositional layering, and segregations in granitic rocks, metamorphic rocks, dolomitic limestone, and dolomitic chert.

4. Using a hand lens to see textural details such as intergrowths, zoned crystals, partially resorbed or reacted grains, potassic halos next to veins, and poikiloblasts (irregular metacrysts containing large numbers of other grains).

The reagents needed are: (1) concentrated hydrofluoric acid (52% HF);

(2) a saturated solution of sodium cobaltinitrite; (3) amaranth solution, which is made by dissolving 1 oz of F., D. and C. Red No. 2, 92% pure coal-tar dye, in 2 liters of water; and (4) calcium chloride and barium chloride solutions, both made by dissolving 5 gr of the salt in 100 cc of water. These reagents can be obtained from any chemical supplier. Essential equipment consists of a pair of rubber or plastic gloves, heavy plastic-covered tongs, and several shallow plastic bowls large enough to hold the samples.

Hydrofluoric acid is highly corrosive and poisonous. Its use should be supervised by experienced persons. These rules are essential: Wear gloves; use tongs to handle samples; do not breathe the acid fumes; do the etching as far away from the face as possible; have ample water at hand for washing; use the HF over a small pit dug in soil well away from any stream or vegetation; and cover it with soil when the staining is completed.

The procedure that follows is based on the laboratory method described by Laniz and others (1964):

1. Find clean, smoothly worn outcrop surfaces from which to break samples for staining.

2. If the samples need washing, dry them thoroughly.

3. Pour hydrofluoric acid over the surface to be stained, using the precautions noted above.

4. After 10 to 15 seconds dip the sample in water or pour water gently across the etched surface.

5. To stain for potassium feldspar, immerse the etched surface in sodium cobaltinitrite solution for 1 minute, or, if the sample is large, pour the solution over the surface.

6. Remove excess solution by pouring water gently over the surface.

7. Dry the stained surface.

8. Immerse the surface for 15 seconds in 5% barium chloride solution.

9. Dip surface once quickly in water; blow off excess water and let dry.

10. Immerse the surface for 15 seconds in amaranth solution.

11. Dip the surface once quickly in water.

12. Tilt the surface and gently blow off remaining excess of amaranth solution and let dry.

13. If milky white (etched) areas suggest albite, select another sample and go through steps 1 through 4 with it, then dip it in 5% calcium chloride solution, dry it, and proceed with steps 5 through 12 (or, if potassium feldspar is absent, steps 8 through 12).

An advantage of using this method for staining dolomitic carbonate rocks and cherts is that only one set of reagents needs to be taken to the field. A disadvantage is the poisonous nature of hydrofluoric acid. Thus if *only* carbonate rocks will be stained, other methods should be adapted for field study, as those described by Friedman (1959).

### 2-4. The Compass

 A variety of compasses are available at a considerable range of prices (see, for example, a supplier's catalog). Of these, the Brunton compass (Fig. 2-3) has the advantages of combining a precise sighting-clinometer and hand level with a compass that can be used to measure bearings at either waist-height or eye-height. The less expensive and more compact Silva Ranger Compass is excellent for measuring strike and dip, or trend (bearing) and plunge, but cannot be used as a precise sighting-clinometer or hand level. The Silva compass also lacks a leveling system for sighting bearings and thus is not as precise as the Brunton compass when traversing or making locations by intersection (Section 6-3).

 Brunton compasses, however, may be damaged when handled roughly, and should therefore be checked before use to be sure that: (1) the compass needle swings freely; (2) the hinges of the sighting arm and lid are firm; and (3) the point of the sighting arm touches the axial line of the mirror when the arm and lid are turned together. The clinometer level, which may become misaligned, can be checked by setting the vertical angle index at 0 and placing the compass on a smooth surface that has been leveled with a long carpenter's level or an alidade (the bull's-eye level on the Brunton compass is not precise enough). If the tube bubble does not move to center, the plastic (or glass) cover must be removed and the level-mount rotated gradually until the bubble is centered.

 The compass must also be opened to clean or dry the needle bearing or to level the magnetic needle by moving the small wire coil on it. For mapping south of the equator, the coil must be removed and placed on the north-seeking half of the needle.

 **The magnetic declination** (the local difference between magnetic north and true north) must be added or subtracted from all bearings taken with a magnetic compass. The correction is made beforehand and automatically for most compasses by turning the graduated compass circle the amount of the local declination. The circle of the Brunton compass is turned by a screw

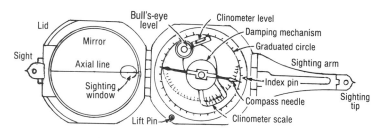

**Fig. 2-3.** View directly down on a Brunton compass with lid and sighting arm fully open.

on the side of the compass box. For example, if the declination is 17° east of true north, the circle is turned until the index pin (at the base of the sighting arm) points to 17 on the side of the graduated circle marked with an E. To check this setting, hold the compass level and oriented so that the white end of the needle points to 0 (thus the sighting arm should be pointing to true north). Now see if the needle is pointing 17° to the *right* of the index pin—the direction you know geographically to be eastward of true north. If it is, the setting was made correctly.

In the rare cases where a modern map cannot be used to obtain the direction and amount of the magnetic declination, the declination can be read before the field season from an isogonic chart, available in most surveying texts and from government agencies (in the United States from the National Ocean Survey, Department of Commerce). These charts also give the amount of change in the declination per year, which is less than 3 minutes per year for all parts of the United States except the southeastern states, where it is as much as 5 minutes per year.

### 2-5. Taking a Compass Bearing

A compass bearing is the geographic direction from one point to another, as N 10° W (or, on an azimuth scale, 350°). A bearing is taken by standing at one point, orienting the sighting arm of the compass toward the other point, and reading the compass graduation indicated by the north-seeking (white) end of the needle. The following detailed procedure for the Brunton compass is used when the point to be sighted is visible from waist height. This is the easiest way to take an accurate bearing.

1. Open the compass as in Fig. 2-4*A*, and cup it firmly in one hand at waist height.
2. Level the compass by centering the bull's-eye level and adjust the lid until the sighting tip and the point sighted appear in the mirror. The sighting arm may be turned upward as necessary.
3. Rotate the leveled compass on a vertical axis until the mirror images of the point sighted and the sighting tip coincide with the axial line of the mirror.
4. Check the bull's-eye level and read the bearing indicated by the white end of the needle. If the needle has not come to rest (it may not in old models not equipped with a damping mechanism), carefully estimate the center of its swing.
5. Record the bearing and repeat the procedure as a check.

If the point sighted can be seen only at eye-level, or by a steep downhill sight, the procedure is as follows:

1. Open the compass as shown in Fig. 2-4*B* and hold it with the sighting

arm pointing toward the eye. Hold it about 1 ft (30 cm) from the eye so that the point sighted and the axial line of the mirror will both be in sharp focus.

2. After leveling the compass by observing the bull's-eye level in the mirror, rotate it on a vertical axis until the point sighted is seen through the window in the lid.

3. Checking to be sure the compass is level, bring the point sighted, the sighting tip, and the axial line of the window into coincidence.

4. Read the bearing in the mirror and transpose it to a foresight before recording it (e.g., N 20° W becomes S 20° E, or 340° becomes 160°). Transposition is necessary because the compass is oriented 180° from its bearing sense. Repeat the procedure as a check.

Both procedures require patience in holding the compass level, which is essential for accurate inclined sights. The needle should be nearly still by the time the sight is completed, so that the bearing can be read to the nearest degree, perhaps the nearest ½ degree. This is the precision of the instrument; the accuracy of the reading can be affected by sighting the wrong point, by tilting the compass even slightly when taking an inclined sight, by reading or recording the bearing incorrectly, or by local magnetic anomalies.

**Magnetic anomalies** may be caused by any nearby iron-bearing objects or by magnetic rocks. A steel belt buckle is likely to affect readings taken at waist height, and a hammer must be dropped to the ground at least a meter away. Wire fences, pipelines, and steel tapes may cause notable anomalies, as may mafic and ultramafic rocks. Any suspect object can be tested by bringing it close to the leveled compass when the needle is at rest. Local magnetic rock bodies are indicated when a foresight and backsight between two widely separated points do not agree. The actual magnetic anomaly at any point can be determined by locating the point precisely on a topographic map and taking a compass bearing to another point that can also be located on the map. The compass bearing is then compared with the true bearing read from the map by the use of a protractor.

In areas where magnetic anomalies are typical, as over large bodies of gabbro, basalt, or ultramafic rocks, it may be necessary to use a *sun compass* (available from Miners, Inc.), which indicates bearings by means of a linear shadow cast by the sun.

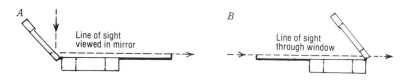

**Fig. 2-4.** Brunton compass with lid opened for taking a bearing at waist height *(A)* and at eye height *(B)*.

## 2-6. The Clinometer; Calculating Vertical Distances

A clinometer is a sighting instrument with a tube level mounted on a movable graduated arc. It is used to measure the vertical angle between the horizontal and any line of sight. A clinometer is built into the box of the Brunton compass (Fig. 2-3), which is used to measure a vertical angle as follows:

1. Open the lid and sighting arm as in Fig. 2-4*B*, but hold the compass so that the box is on edge and the clinometer is upright. The sighting tip should be about 1 ft from the eye.

2. Move the compass until the point to be sighted is visible through the window in the lid and coincides with the axial line and the sighting tip.

3. Holding the compass in this position, rotate the clinometer by the lever on the base of the compass box until the tube bubble, as observed in the mirror, is centered.

4. Recheck the alignment, and finally bring the compass down in order to read the angle indicated by the vertical angle index. Record it at once and repeat the entire procedure as a check.

If a horizontal distance is measured (Section 2-8), or is scaled from a map, a vertical angle can be used to calculate a vertical distance by the relation: *vertical distance = horizontal distance × tangent of the vertical angle* (Appendix 11). If a slope distance is measured, the relation is: *vertical distance = slope distance × sine of the vertical angle*. The height of one's eye above the ground is then added for uphill sights and subtracted for downhill sights (Fig. 2-5).

## 2-7. The Hand Level

A hand level has a tube level set parallel to the sighting axis and is designed so that the level bubble can be observed while sighting. The clinometer of the Brunton compass is converted to a hand level by bringing the clinometer to exactly 0°. The compass is then held as for taking vertical angles (Section 2-6) and is brought to level by observing the level bubble in the mirror. All sighted points coinciding with the axial line and the sighting tip will be at the observer's eye level.

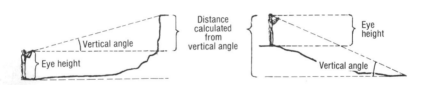

**Fig. 2-5.** Relations among eye height, distance calculated from a measured vertical angle, and actual vertical distance between two ground points.

The vertical distance between two points can be measured with a hand level by: (1) determining the height of one's eye above the ground; (2) standing at the lower of the two points and sighting a level line to select any convenient intermediate point at eye level; (3) walking to that point and repeating the procedure; and (4) continuing similarly until the last measure before the second point, which will generally be less than eye height and can be estimated or measured with a roll-up tape. The difference in elevation is the number of intermediate points multiplied by the height of one's eye, plus the last fractional measure. This measurement can be made rapidly and has an inherent precision of ±2 in. (5 cm) if made with reasonable care. Major errors result from occupying an intermediate position other than the one actually sighted, and from miscounting the intermediate points. Miscounting can be avoided by using a tally counter or making a pencil tally.

A Jacob staff (Section 11-4) provides a solid base against which to hold the hand level and thus obtain precise vertical measures on rough ground.

## 2.8 Taping and Pacing Distances

Fairly accurate measurements of distance are needed occasionally in most field projects and are essential routine in measuring stratigraphic sections and in making unusually large-scale maps where no suitable base is available. Section 5-1 describes a full procedure for detailed traversing with a compass.

Tapes ordinarily used are made of nylon, fiberglass, or steel ribbon, and mounted in a roll-up case or on a reel. Hip or belt chains that have digital recorders operated by a pull-out thread may be precise enough for many purposes (see a supplier's catalog). Steel tapes are heaviest and least comfortable to handle but are easiest to keep clean. A dirty tape may jam the wind-up mechanism as well as become difficult to read. The rule is, always wipe a tape clean before rolling it up. Be sure not to step on any kind of a tape, and not to kink steel tapes (which shortens them).

Distances are taped most rapidly by holding the tape level. One person can tape in a downhill direction by fastening the 0 end of the tape to the ground with a nail or chaining pin and carrying the case or reel ahead until it is at eye level when the tape is within a few degrees of level, as determined with a hand level. The measure can be carried to the ground by dropping a pebble. The point is then marked clearly so that it can be found after bringing the 0 end of the tape ahead, and the measured distance is recorded in a notebook. Using reasonable care, distances of several hundred meters (600-900 ft) can thus be taped to the nearest 0.2 m (8 in.). Taping longer distances with greater precision is described for measuring a base line in Section 8-7.

Pacing is less precise than taping but is much faster and eliminates carrying a tape. The length of one's pace (double step) is determined by walking a taped course of 100 m or so until the results compare within one pace. The gait must be kept natural, and the calibration should be repeated after several days and occasionally thereafter. Precision of pacing decreases with steepness of slope and roughness of ground, and if adverse conditions are typical, a calibration should be made on similar ground. When walking uphill, for example, steps tend to be shorter than when walking downhill. Where trees are in the way, the course must be offset at right angles. The number of normal paces through a bush or across a large stone can usually be estimated by noting where steps would fall if the obstacle were not there. Errors of miscounting can be eliminated by using a *tally counter*, which is advanced one digit by pressing a lever. A *pedometer*, which is advanced by the jolt of each footstep, does not allow for offsets and broken paces.

With practice, the precision of pacing is 1 to 2 parts in 100, which is adequate for most geologic mapping but not for detailed measurement of stratigraphic sections (Chapter 11).

*Rangefinders* might be considered in place of pacing for projects that will require rapid traversing over rough ground or many measurements of moderate distances. A rangefinder about 11 in. (27 cm) long has an advertised precision of 1 part in 100 at distances up to 100 m (see a supplier's catalog).

## References Cited

Friedman, G. M., 1959, Identification of carbonate minerals by staining methods: *Journal of Sedimentary Petrology*, v. 29, p. 87-97.

Laniz, R. V., Stevens, R. E., and Norman, M. B., 1964, *Staining of plagioclase feldspar and other minerals with F. D. and C. red No. 2*: U.S. Geological Survey Professional Paper 501-B, p. 152-153.

# 3

# Basic Procedures at Outcrops ■

### 3-1. Observations in the Field

Outcrops, or exposures, consist of earth materials that can be examined in place. They constitute all of the earth that can be observed directly, and examining them is the most fundamental procedure of field geology. Finding enough of them may be a problem, especially at the outset of a study. Unusually large exposures can generally be located by photogeologic studies (Section 7-3). Topographic maps and aerial photographs may also show the general grain or trend of rock structures, and streams aligned across this trend commonly develop many outcrops. Otherwise, outcrops tend to be located on steep slopes or in road and railway cuts, ditches, landslide scars, and at places where streams have undercut their banks. Where outcrops are scarce or geologic relations complex, a pick and shovel may be needed to enlarge or make them.

The primary physical operation in studying outcrops is looking. Most outcrops have much to observe, and the challenge is to look carefully enough to see as much as possible. The first thorough examination may thus take an hour or more. It must also be as objective as possible and might be organized as follows:

1. Look over the outcrop and nearby outcrops broadly in order to spot the principal bodies of rock or other materials. Walk around or over large outcrops several times and view them from various distances. Do all the bodies of rock appear to have once continued beyond the outcrop, or do some end there against other bodies? Do they end against faults, intrusive contacts, or unconformities? Why does the outcrop itself end where it does? Does it grade into soil or is it overlain by surficial deposits?

2. Continue to study the outcrop from a moderate distance. Are the rock bodies tabular, irregular, lenticular, or with some other distinctive shape? What are their orientations and dimensions? If they are internally layered, are the layers parallel to any of their bounding surfaces? In overall view, do any rock bodies appear variable? Do any contain fragments?

3. Now study the boundaries (contact surfaces) between the bodies, both at a moderate distance and up close. Are they sharp or gradational? Do they cut across grains or structures in either adjacent body? Do any rocks or deposits vary in color or texture near these contacts? The views in Fig. 3-1 may be helpful.

4. Break off, or scoop up, samples of the main materials, and examine

weathered as well as fresh surfaces with a hand lens. Identify the constituent mineral and rock grains, and note their sizes, shapes, and surface features, as well as their part in the overall fabric and porosity of the rock or deposit. To test for the distribution of soft mineral grains, especially carbonates, probe grains with a needle, apply dilute HCl, and examine weathered surfaces for pits and insoluble residues. Estimate the composition of each sample, in percent by volume of each kind of grain, and identify the rock or material (Chapter 4). Even if it seems obscure, give it a provisional name.

5. Now examine the rocks closely for primary fabrics and structures. Look especially for structures that establish tops and bottoms of deposited layers that were once sediments or igneous deposits (Section 9-10). Do relations at contacts support these indications of sequence?

6. Look for all features indicative of depositional current direction or direction of magmatic flow. Measure a number of them in order to judge their consistency (Sections 9-4, 13-2, 13-4, and 14-1 through 14-5).

7. To detect deformation, see if rock layers, veins, or planar structures have been folded (Section 12-3). If no folds are obvious, perhaps foliations, cleavages, or lineations indicate folding (Section 12-4).

8. Whether the rocks are folded or not, examine them for grains or other small bodies that have been deformed into planar or linear shapes that give a measure of deformation (Section 12-2). How are these grains oriented relative to other structures?

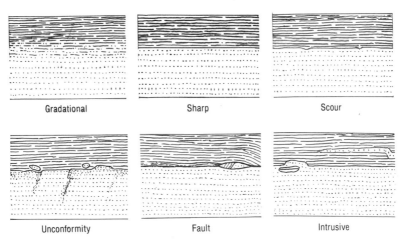

Gradational    Sharp    Scour

Unconformity    Fault    Intrusive

**Fig 3-1.** Six basic kinds of contacts, viewed here at single outcrops of shale overlying sandstone. In addition to the simple geometric relations shown, depositional gradations are discussed in Section 5-2, scour surfaces and major unconformities in Section 9-6, faults in Sections 5-4 and 12-5, and intrusive contacts in Sections 13-7 and 14-1.

9. Examine all faults, even those that displace rocks as little as a centimeter. Are there gouge or breccia along them? Any indications of actual directions of displacement (Section 12-5)? Are the faults younger than other tectonic features?

10. To determine the physical state of the rocks, especially if they are sedimentary, test the degree of compaction and cementation by hefting dry samples (porous rocks are lighter than nonporous ones), by their reaction to breaking in the hands and to hammer blows (Appendix 4), and by the rate they soak up water. Porous fine-grained rocks tend to stick to the tongue. Depth and strength of weathering generally increase with porosity and permeability.

11. What are the typical spacings and orientations of joints in the different rocks? Are there secondary color changes, and are they related to fractures? Fracturing and weathering characteristics may help in identifying the same rock unit in other outcrops (see the subsection that follows).

12. Bring together all observations made thus far in order to identify rocks and structures that were originally obscure, or to refine identifications.

13. Systematically measure and record: (a) the thickness of each layered unit of rock; (b) structural attitudes of all primary structures, as bedding; and (c) attitudes of all secondary features, as folds and faults (Sections 3-5 and 3-6).

14. Make a thorough search for fossils (Section 3-7).

15. Collect rocks that seem particularly useful, either as typical samples or to resolve identifications (Section 3-8).

16. Before interpreting the various rocks and structures, be sure that you have observed all possible indications of relative ages of the rocks in the outcrop.

**Weathering and related secondary characteristics** are seldom described in the literature and almost never treated in geology courses, which traditionally present fresh, solid rock specimens. Secondary characteristics are, however, of great value in field studies. They can be used (1) to recognize specific minerals and rocks, and (2) to recognize individual map units of rocks. They are essential data in engineering geologic studies (Section 5-7).

Some rocks disintegrate into individual mineral grains and others into fracture-bounded pieces that may have distinctive sizes and shapes. Disintegration is often along microscopic fractures that may be cogenetic with the rock and thus highly specific to one rock formation. One kind of granite in a composite pluton, for example, may crumble into mineral grains and another granite in the same pluton into small fragments of rock. Disintegration may also be caused by unstable minerals that swell when altered to OH-bearing minerals of greater specific volume. Alteration of feldspar and mica to clay and of olivine to serpentine are examples.

Discoloration of rocks to shades of yellow, orange, red, and brown is generally due to leaching of $Fe^{++}$ from unstable minerals and its fixation as $Fe^{+++}$ in oxide pigments. The degree of discoloration thus indicates degree of stability of minerals in rocks and may be highly specific for distinguishing between closely similar rocks. One dolomite, for example, may be stained more than another because of containing ferroan carbonates, and two granites may stain differently because the biotite in one is more stable than the same mineral in another.

Exposed rock surfaces may also develop forms characteristic of certain rock units. These forms can typically be classified as one or more of: smoothly rounded, fluted, pitted, cupped, parallel-ribbed, reticulate-ribbed, beaded, knobby, chalky, hackly, irregularly rough, and exfoliated.

### 3-2. Interpretation of the Outcrop

When the objective examination seems complete, the rocks and structures are interpreted as fully as possible. This must be done at the outcrop so that the observer can test interpretations against additional materials or relations they predict. The scientific method is thus fully encompassed: (1) an examination has been made, as suitable to the purpose of the study; (2) the resulting data and ideas are thought through completely to compose *all possible explanations* (hypotheses) for the observed materials and features; and (3) additional relations predicted by each explanation are sought at the same outcrop and elsewhere in order to settle on the most satifactory explanation. In actual practice, steps 2 and 3 are likely to be taken in part during the initial examination. The basic imperative in every case is that all three steps be completed before leaving the outcrop.

The explanations or hypotheses of step 2 are based partly on the specific materials and relations examined. They are also based on resemblances to features, models, or ideas stored in one's memory, in field notes, or deduced from theory. Important prerequisites are research of the literature and an open mind, but most important is a fertile imagination (in the sense of forming mental images—*image-ination*). Some of these images will be largely geometric, as in seeing a feature of the outcrop in its broad structural setting. Others will be genetic explanations that may arise from similarities to rocks and features having genetic significance, as the possibilities described in Chapters 9, 10, and 12 through 15. These images answer the question "How did this particular association of rocks and structures form?"

The most involved and demanding images (and the most exciting and powerful ones) are histories of how observed features came to be, thus including their various attachments through time. The questions asked are "If that is how such-and-such formed, how did that come to happen?" and "What are some consequences of its happening?" The power of these historical images derives from their forcing one to seek causes as well as results. The images can therefore be tested and sharpened by a variety of data and relations. G. K. Gilbert described these interrelations as follows (1886, p. 285):

> Phenomena are arranged in chains of necessary sequence. In such a chain each link is the necessary consequent of that which precedes, and the necessary antecedent of that which follows. The rising of the sun is consequent on the rotation of the earth. It is the logical antecedent of morning light. Morning light is in turn the consequent of sunrise and the antecedent of numerous other phenomena. If we examine any link of the chain, we find that it has more than one antecedent and more than one consequent. The rising of the sun depends on the position of the earth's axis as well as on its rotation, and it causes morning heat as well as morning light. Antecedent and consequent relations are therefore not merely linear, but constitute a plexus; and this plexus pervades nature.

Perhaps a partial example will give a sense of how this "plexus" may be utilized. Fig. 3-2 illustrates an outcrop to be interpreted and notes the main features that were observed and identified in the first part of the examination. The discordant contact and the bend in the shale beds near it suggest a fault, and thus the contact is reexamined for other evidence of faulting. A few millimeters of gouge and some slickensides are found on parts of the contact, but striations on the slickensides are oriented nearly horizontally and therefore do not match the direction of displacement indicated by the bent shale beds. Moreover, faulting does not explain the fragments of shale included well within the sandstone.

Are there other explanations for the contact? For one, it could be an unconformity that has served as a surface for minor to moderate fault displacement. It could also be the margin of a sand body intruded into a fracture (possibly a fault?) in the shale, and subsequently displaced along this same surface. What consequent and antecedent relations might be sought to test the latter explanation? If the sand was injected into the shale from beneath (a relation supported by the upward bend in the shale beds), the shale should be underlain somewhere by more of the same sandstone. In addition, if the

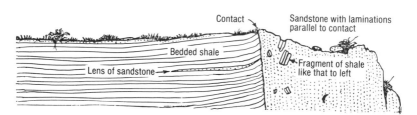

**Fig. 3-2.** Vertical face exposing shale and sandstone with features as noted.

sand was injected forcefully, other intrusions might be found in the shale. The small sandstone lens shown may thus be part of a sill and, if so, should match in detail the sandstone on the right side of the outcrop. Perhaps some of the shale inclusions in the sandstone have small sand injections.

To further test the intrusion hypothesis, the other wall of the body could be sought somewhere to the right of the view, in other outcrops. Similarly, evidence for unconformity (Section 9-6) and additional evidence for faulting (Sections 5-4 and 12-5) should be sought at nearby outcrops in order to test those explanations.

The investigator may well leave the outcrop favoring the sand-intrusion hypothesis, but remain unconvinced because of the incomplete record. Interpretation and testing have nonetheless sharpened several explanations and thereby the field study in general. They have given the investigator some specific things to look for. Crucial but poorly preserved features that would otherwise be overlooked are thus more likely to be found at the next outcrop.

### 3-3. Taking Field Notes

All observations and interpretations are recorded in field notes, which will generally be better organized and less redundant if written after parts of the examination seem more or less complete. Notes can be kept from becoming verbose or illegible by use of engineering-style lettering, telegraphic composition, and abbreviations (Appendix 2, Fig. 3-3). The part of the notes covering the interpretation, however, should probably be composed of well-thought-out sentences rather than cryptic words and phrases; it is easy to forget the exact meaning and relevance of ideas that were clear at the outcrop. Each set of notes should be completed at the outcrop and not modified later unless the changes are identified clearly.

Materials for taking notes are described in Section 2-1. Each page of notes must be numbered consecutively for a given notebook or project, and must be headed by the geologist's name, the date, and preferably by a brief title for the area covered by the notes on that page (Fig. 3-3). The number of the map sheet or aerial photograph must be recorded, and numbers of additional sheets or photographs entered in the left margin as work progresses. Field locations are numbered consecutively (strictly). Chapters 6 and 7 describe how they are located and marked on the front or back of field sheets or aerial photographs. Points of unusual interest, such as specimen locations and critical relations or questions, should be noted in the left margin (Fig. 3-3) or flagged by a colored line or a box. Notes taken on loose-leaf sheets should be stored in order in a safe place at the field camp or office.

The descriptive parts of the notes should present facts and thus be kept as free as possible from terms that are basically genetic. Rocks and struc-

Gold Mt. 7.5' (1:24,000) quad     by R.C. Jamison     10 Oct. 84
Field sheet #3                                          page 1

1.  Starting on first spur NW of Woods Ranch HQs
Beaut. lg. otcp along bluff exposes ~ 55 m ss and
intbdd ls. Distinct bdg + cross-joints make otcp
stepped and blocky. Bdg consistently N 17 W 12 E. More
prominent jts. N 5-15 W 60-90 W; lesser system N 70-90 E
90 (±15). Ss in 4 tabular sets, 4-6 m thick, separated
by somewhat lenticular ls beds .3-.7 m thick. Ss beds
grade from .5-1 m thick at base of each set to .1-.2 m
thick at top.
    Ss is lt. gray (6.5/0), weathering pale brn (10YR 7/2);
it is coarse + moderately sorted in basal beds of sets
+ grades upward to fine + well sorted in upper beds;
shells (up to 6 cm) scattered throughout, frag. in lower
beds + commonly whole in upper beds. Relics of distorted
X-lam indicate ~ 90% bioturbation. All ss is tough calcite-
cemented feld. semiquartzose arenite, with 20% feld., 5%
dark chert grains, 3% bio, + <1% bright green pellets of
glauconite(?). Grain porosity low but fracture porosity
moderate (unit supports dense oak woodland along strike).
    Ls tough, sparsely jointed, v. lt. gray (7.5/0), weathering
nearly white; chiefly sandy shelly calcirudite, mod. to
poorly sorted(?); matrix largely rexlized but locally with
subang. bioclasts and oolites; sand like that of ss 5-25
% (+ up to 50% near ctcs with ss). Matrix most abnt (40-
80%) near base of beds, + fossils best preserved there.
Beds unlam. but local imbrication gives these current
directions: → S15W, S12W; S3W; N14E, ~S20W, N10E,
N4E, S8W, suggesting reversing (tidal) currents.
    Fossils in ls and lower beds of ss sets frag; better
preserved ones (sample A) chiefly pelecypods (both thick
and thin-shelled species), with few echinoids + gastropods.
Sample B of 2 species pelecypods from upper beds of ss
sets, one species quite commonly articulated.

*margin notes:* Current directions · fossil Samples

**Fig. 3-3.** Two pages of field notes, selected to show a variety of entries.

Field sheet # 3, cont.                                    10 Oct. 84, p. 2.

Interpretation: the textures, fossils, and oolites suggest
a shallow-water, moderate-energy environment. The tidal
currents + high % of bioturb. would suit a shallow shelf
off a river or estuary (??).

2. Same ss + ls but with more of ss unbioturb. The latter
collected in oriented samples for study of imbricate fabric

Oriented   (all marked on up-facing bdg planes): A, →N 80W 6 N; B,
rocks      →N 76W 5 N.; C, →N 83W 7N.

3. Rather hurried traverse up spur NW of pts 1 + 2. developed
this sequence:

NE         ┌ Resistant ss ┐         3 yellow-weathering .5-1 m beds
Summit     │ bed, 1m Thick │    Unit of    of pale brn dol., locally lam.
           └              ┘    note 1
                                                          Bdg N30E 4NW
                                                               Pt. 3
Intbdd. bioclastic                                                SW
grainstone + biocl., pelletal          └─────100m─────┘
mdst, the latter being        │ Brown (sapropelic), white-weathering
dominant in upper 1/4         │ diatomaceous lam. shale. Bdg at top N31E 30W
section.

4. Interp.: the laminated sapropelic shale must be Carster's
Marion Shale; if so, my unit of note 1 is probably his "Upper
Sandstone Member" which he assumed to be closely related to
the shale. Judged from the contrast in depositional environments, however, the ss/sh etc is a major disconformity.
I saw no specific evidence, but the etc is poorly exposed.
The ls unit overlying the unit of note 1 is gradational to it
and suggests either deepening shelf conditions or an increasingly protected lagoonal environment.
   The joint systems and bdg attitudes of pts. 1-3 indicate an
anticline plunging ~ 5° to the north (like the folds in the
SW part of Carster's area). The entire sequence should thus
be exposed on the high ridge SW of Woods Ranch — which
looks craggy and a better place to sample for microfossils
(the stuff seen today in the shale unit appears leached).

tures identified with certainty can be given firm names, but other identifications should be queried, noted as problems, or simply stated as unknowns. Interpretations interwoven with descriptions must be identified clearly so that they will not later be read as facts.

Notes covering a number of outcrops will be easier to use later if each set is written in the same order. The order will probably seem most natural if broader aspects are described before more specific ones (Fig. 3-3, Note 1).

The final item, the interpretation, includes a brief description of the origin of each rock and feature. Events implied by structural relations should be summarized together with the evidence on which they are based (Fig. 3-3, Note 1). All events should be recorded as a history if possible, with specific age relations noted. Inconclusive, even highly speculative ideas should be included and identified as such, together with whatever suggests them (Fig. 3-3, Note 4). Sketches or schematic diagrams will often be helpful. The gaps and doubts in the interpreted history are noted so that they can be sought at the next outcrop. In this way, field work and interpretation evolve together.

Before moving on to the next outcrop, however, the history might be reconsidered carefully. If the rocks are as old as they are thought to be, can you imagine *all* that has happened since? Think of additions to the local history implied by relations outside the immediate area. Perhaps a last careful look at the outcrop will be worthwhile.

**Descriptions of rock units.** As the study is extended to additional outcrops, rocks that continue as uniform materials are simply recorded as "the same." This must be done cautiously, however, because it may be easy to gloss over small but important differences. Variations are described per outcrop, including any geographic sense of variation discovered at that point. When a specific rock or deposit has been walked out and examined as much as it is likely to be, a complete description is generally recorded, one that includes all variations as well as the overall nature of the rock unit or deposit—from one contact to the other. The following outline suggests items that might be included and an order for presenting them.

1. Stratigraphic name of the unit, or its relations to named units.
2. Area to which description applies.
3. Nature of the terrain underlain by the unit—its topography, soils, vegetation, and outcrops.
4. Overall shape or structure of the rock unit in this area.
5. Thickness of the unit.
6. Principal kinds of rocks and their distribution in the unit.
7. Unusual rocks and their stratigraphic (or other) position and genetic implications.
8. Primary structures in the unit.
    *a.* How bedding or other layer-structures (as flow structures) are

expressed, as by color, texture, induration, and so on.
   *b.* Range of thicknesses and typical thicknesses of beds or other prim-
      ary layer-structures.
   *c.* Shapes of beds or other layer-structures.
   *d.* Primary structures within beds or other structures, as grading,
      laminations, cross-stratification, channeling, and inclusions.
 9. Fossils.
   *a.* Distribution of fossils, stratigraphically and laterally.
   *b.* Special characteristics of the more fossiliferous rocks.
   *c.* Positions and condition of fossils, as growth position, fragmental,
      rounded, and any signs of reworking (Section 3-7).
10. Description of rocks, most abundant kind first.
   *a.* Color—fresh, weathered, moist, dry (Appendix 6).
   *b.* Firmness of fresh and of weathered rock (Appendix 4).
   *c.* Grain sizes— range and average (or typical) sizes.
   *d.* Degree of sorting or equigranularity.
   *e.* Typical shapes of the principal kinds of grains.
   *f.* Fabrics (orientations) of tabular or linear grains, especially in rela-
      tion to rock structures.
   *g.* Kinds and proportions of mineral cements, matrix, or groundmass.
   *h.* Nature and proportion of pores (porosity) and indications of
      permeability.
   *i.* Kinds of grains and the approximate percent by volume of each.
11. Contacts (Fig. 3-1).
   *a.* Sharp or gradational (describe gradations).
   *b.* Indications of scour, unconformity, intrusive relations, or faulting.
   *c.* Criterion or criteria for locating the contact in the field.
12. Characteristic secondary features, such as cleavage (fissility), concretions, veins or other fillings, presence of hydrocarbons, and deformational structures, including joints.
13. Characteristics that are particularly useful in distinguishing this unit
   from all others in the area.
14. Interpretation of the unit.
   *a.* Geologic environment or conditions under which the unit was orig-
      inally deposited or crystallized.
   *b.* Specific processes contributing to its origin.
   *c.* Genetic relations to associated rocks.
   *d.* Later modifications within the rock at grain-scale, as cementation,
      compaction, autometamorphism, and recrystallization.
   *e.* Tectonic and other structural modifications, as folding (Section
      12-3), fracturing (Sections 12-4, 5, and 6), and homogeneous strains
      (Section 12-2).
   *f.* Geologic age of the unit or age relations to other rock units.

## 3-4. Drawing and Photographing Outcrops

Drawings often save time in note-taking and may be essential for recording complex shapes or relations. Drawing contributes to observation by forcing one to look closely, especially at how specific features come together. One general method of note-taking is to make a page-size drawing of the outcrop or some part of it, and record descriptive notes right on the drawing (Fig. 3-4A). Small cross sections can be used to record stratigraphic sequence and structural relations (Fig. 3-3, Note 3), and a page-size columnar section (Fig. 11-3) can be used in overall descriptions of rock units. All drawings should be accompanied by a bar scale, by indication of geographic directions, and by labels to identify features.

Drawings need not be "artistic" or otherwise attractive, but they must show proportions, angular relations, and shapes of important features correctly. If drawing the parts of an outcrop to proportion proves difficult,

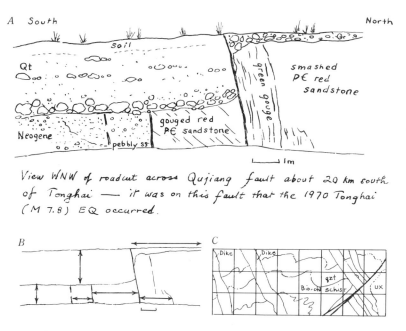

**Fig. 3-4.** *A.* Drawing of a roadcut exposure, copied from a page of a field notebook of A. H. Sylvester, and reduced approximately 30%. Note that the Quaternary terrace sediments *(Qt)* are displaced along the most recently active fault trace, and that older traces cut Neogene sediments and are truncated by the erosion surface beneath the Qt. *B.* Outline of the outcrop of *A,* showing measurements *(arrows)* that could be used to lay out the main lines of a field drawing. *C.* Rectangular grid used as a basis for a drawing (here a very large-scale map) of a horizontal outcrop surface *(ux = unexposed).*

measurements of the outcrop's main divisions can be used to block out the main lines of the drawing. The drawing of Fig. 3-4*A*, for example, could have been started by making the measurements and drawing the light pencil lines shown in Fig. 3-4*B*. Important inclined lines, such as the faults in Fig. 3-4*A*, may be sighted with a clinometer or compass and plotted with a protractor. Irregular lines and curving boundaries should then be drawn a little at a time while observing the outcrop frequently. A method worth practicing is to draw these lines while observing the outcrop rather than the drawing. A given line is started by placing the point of the pencil at the correct place in the drawing, then looking at the outcrop and *feeling that the pencil point is being held against the actual line on the outcrop*. As the eyes follow the outline of the feature, the pencil is moved along it, thus scribing the outline on the drawing. If the line goes somewhat astray, it can be corrected afterward. The method will seem awkward at first but will lead to these advantages: (1) observing more exactly; (2) making drawings more quickly, and (3) gaining a *sense* of proportion in drawing.

If an outcrop is large, complex, and with no clearcut divisions to serve as a framework, a rectilinear grid may be constructed on the outcrop surface to control the drawing. The grid lines might be spaced at 0.5 to 3 m apart, depending on the size of the features to be recorded, and can be made from tape, string, sticks, marker pen, or pieces of rock. A grid is then plotted to a convenient scale on a note sheet, and each square on the outcrop is examined in order to draw its part of the picture (Fig. 3-4*C*). On more or less horizontal outcrop surfaces, the drawing will be a very large-scale map of the outcrop; indeed, this is a mapping method that may be used for small but critical areas.

When the principal lines and forms of the outcrop have been drawn, details and labels can be added as necessary. The general rule is to keep the drawing as simple as possible, so that important features will stand out clearly. Unexposed areas are generally left blank, and incidental features like roots and washovers of soil are omitted. Shading is generally unnecessary in line drawings like Fig. 3-4*A* but might be needed to show three-dimensional shapes, as those of Fig. 4-17.

**Photographs** take little time to complete in the field and show features exactly as they are, thus being the most convincing kind of graphic evidence. Some useful photographs are: (1) overall views of outcrops, showing relations to their surroundings; (2) features showing age relations; (3) close-up views of primary and secondary structures; (4) well-exposed contacts; and (5) compositional or textural variations in a rock, by a series of close-up views.

Taking clear, informative photographs requires thought and practice. Considerations before taking a photograph include direction of the view, direction of lighting, closeness of the subject, and depth of focus. The direction of the view should, if possible, be parallel to bedding or other planar structures, such as faults. The planes will thus appear as lines in the photograph and true bed thicknesses will be shown. For folded rocks, the camera should be pointed parallel to hinge lines, thus recording true shapes of folds and beds. The horizon or some other indication of horizontal or vertical directions should be included wherever possible. Subjects to be illustrated in the round, such as fossils and landforms, should be lighted from one side, and from the upper left, if possible. Local shadows on irregular outcrops can be resolved by taking photographs: (1) on a cloudy day, (2) when the outcrop is in full shadow, or (3) by using flash lighting. Depth of focus is crucial in most close-up views and can be increased by using a tripod, fast film, or flash lighting, all of which permit small lens apertures.

An exposure meter must be held over the part of the outcrop that is of interest. A hammer or scale (*not* a coin) should be included in the view, and a sketch drawn in the notes to help in identifying features in the photograph. When taking color photographs in which exact color will be important, a card used to indicate scale can be composed of color strips that will serve later to check the accuracy of the photograph. A Polaroid camera permits marking prints in the field and thus using them as adjuncts to field notes. Using an ordinary camera, it may be worthwhile returning to outcrops in order to use prints in the same way.

### 3-5. Measuring Strike and Dip

Strike and dip are used to record the structural attitude of beds, layers in igneous and metamorphic rocks, planar fabrics, cleavages, faults, joints, and planar veins. Outcrops must first be checked to be sure that they are in place rather than large half-buried boulders or parts of landslides (Section 10-7). Planar structures in ductile rocks, especially rocks rich in clay, are likely to be distorted by downhill creep, an effect that can be avoided by using hilltop or valley-bottom outcrops. Bedding in some sandstones and mudstones is obscure except for occasional minor layers of coarser grains, or a planar fabric due to flakes of mica, chlorite, or carbonaceous material. These various features and relations should be clear by the time an outcrop has been examined thoroughly (Section 3-1). The measurements are made with a compass and clinometer (Chapter 2). Suggestions for plotting them on maps or aerial photographs are given in Section 5-4.

Strike is the geographic alignment of any horizontal line on a planar surface, and dip is the angle of slope at right angles to strike—thus the maximum slope of the surface. Note that strike is not unidirectional; that is, a strike line might be read either as N 30° E (azimuth of 30°) or S 30° W (azimuth of 210°). When using a quadrant compass, errors in transposition, recording, and plotting will generally be avoided if only the north half of the graduated circle is used.

Strike and dip can be measured in a number of ways, each having certain advantages and disadvantages, as noted in the following descriptions. Although the instructions specify beds, other planar structures are measured in the same way.

**1. Procedure for a single large outcrop.** Where beds trend across an outcrop for several meters or more, and locally jut out so as to give three-dimensional views of tops or bottoms, strike and dip can be taken so as to average out local irregularities and thus be accurate for that outcrop. The observer: (1) moves to a position several meters or more from the outcrop; (2) selects a planar bedding surface with an overall attitude representing the outcrop; and (3) moves gradually to just that position from which the two-dimensional surface appears as a line (Fig. 3-5A). The observer's eye is thus in the projected plane of the bedding surface to be measured. Staying in that exact position, the observer sets the clinometer as a hand level (Section 2-7) and takes a level sight to the trace of the planar surface. Because this sight is a horizontal line in the plane of the bedding surface, it is a strike line of that surface. The observer carefully notes the point determined by leveling and uses the compass to read the bearing (or antibearing) to that point (Section 2-5). *The compass can be held at waist height provided it is in the vertical plane that includes the level line of sight.*

The dip is measured from the same position by opening the lid and sighting arm of the compass and holding it at arm's length so that the edge of the lid and box appear to coincide with the trace of the bedding surface (Fig. 3-5B). The clinometer is then leveled and the dip read from it.

**2. Where beds dip more steeply than 60°** the bull's-eye level is precise enough to find the horizontal line needed for the strike. Thus after getting in a position from which a bedding surface appears as a line (method 1), the compass is held as in Fig. 2-4B, leveled carefully, and used to sight a bearing to the trace of the bedding surface.

**3. Leveling between two outcrops** gives a more accurate average measurement of strike and dip than method 1 and can be used wherever a bed is

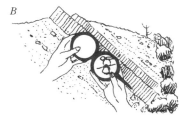

**Fig 3-5.** *A.* Making a level sight to a bedding surface that appears as a line (as in *B*). *B.* Measuring dip along the same line of sight.

continuous from one side of a valley to the other side. The observer stands at one outcrop such that his or her eye is level with the surface to be measured, and sights a level line (a line of strike) to the same surface across the valley (Fig. 3-6). The dip is measured by sighting from the same position.

**4. Beds truncated by a level surface,** as a stream bed or an unpaved road, permit a quick and accurate measurement of strike. The observer simply stands over the surface and aligns the compass with the linear traces of the beds, thus taking the strike direction. Dip can be read at a point nearby where the same beds crop out in three dimensions, or a pick can be used to dig out a planar surface against which the compass is placed directly (as in method 5).

**5. Procedures for small outcrops.** Where less than a meter or so of a planar surface is exposed, or where it is impossible to take sights as described in the foregoing methods, a measurement can be made by: (1) selecting a planar surface that is as representative as possible; (2) placing a notebook or map case against it to average out small irregularities; (3) placing the lower edge of the compass box directly against it, and (4) leveling the compass and reading the bearing or antibearing (Fig. 3-7). The dip is measured by placing the compass on the same surface and orienting it perpendicular to the line of strike. The map case and rocks should be checked to see if they affect the compass needle (Section 2-5). If the outcrop is small, the observer should also double-check to be sure it is not a block of float or part of a landslide.

**6. Around bodies of water,** strike can be measured directly where the edge of the body crosses a bedding surface, or along the mark of a former water-line on a bedding surface.

**7. At large exposures of bedding surfaces,** it may be convenient to use the compass as a clinometer and sight directly up or down the slope in order to measure the dip (Section 2-6). The strike is then taken as perpendicular to that line of sight.

**8. Gently dipping beds** are difficult to measure precisely by any method, for local irregularities cause major deflections of strike lines. If the lines of

**Fig. 3-6.** Sighting a strike line across a valley. The dip is taken from the same position and averaged over the entire outcrop on the far hillslope.

sight described in method 1 are a few degrees off horizontal, the resulting strike measurement may be in error by 5° to 15°. Used with care, methods 1 and 3 should give results accurate within a few degrees for planar structures dipping more than 5°. Where outcrops are small, it may be easiest to find the direction of dip first, as by pouring a small amount of water on the surface or placing the compass against the surface and moving it slowly while adjusting and observing the clinometer level until finding the direction of maximum slope (the true dip). The strike is then measured as a line perpendicular to this slope.

**9. The three-point method** can be used for accurate measurement of beds that dip gently and consistently over large areas. The method requires determining elevations and horizontal positions of three points on a single bedding surface, as the top of a distinctive bed or a contact between two conformable formations. The points must define a triangle with all internal angles greater than 20°, and if a topographic map is used to determine the positions and elevations of the points, the sides of the triangle should be greater than 1.5 cm (0.5 in.) as plotted on the map. If no suitable map is available, or if the outcrops are spaced too closely to use a map, the method requires: (1) reading the bearings from each point to the other points; (2) measuring the horizontal distances between any two pairs of points (Section 2-8); and (3) measuring the vertical distances by leveling (Section 2-7). These data are used to construct a triangle (Fig. 3-8), and a point (*D*), is located on one leg at the same elevation as *B*, as by using the relation

$$AD = AC \frac{\text{difference in elevation between } A \text{ and } B}{\text{difference in elevation between } A \text{ and } C}$$

The line *BD* is thus level and is a line of strike. Its bearing can be determined from the bearing of *BC* and the angle *DBC*, which can be read with a protractor. The dip can be determined by making a graphical construction as shown below the triangle of Fig. 3-8, or by constructing the line *AE*

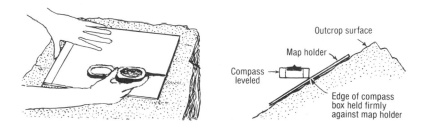

**Fig. 3-7.** Measuring strike of a planar surface by placing the compass against a map holder that is held against the surface. The map holder cannot have iron or steel parts.

perpendicular to $BD$ and solving the relation

$$tangent\ of\ the\ angle\ of\ dip = \frac{diff.\ in\ elev.\ between\ A\ and\ B}{AE}$$

The dip can then be read from a pocket calculator or table of tangents (Appendix 11).

If beds dip at angles less than 4° over large areas and accurate data are required, plane table methods are advised (Chapter 8).

### 3-6. Measuring Attitudes of Linear Features

Many kinds of linear grains and structures, often called *lineations*, are of value in interpreting rocks and geologic histories. Some indicate directions of currents that acted on sedimentary particles (Section 9-4). Others can be used to interpret flow in lavas (Sections 13-2 and 13-4), flow in intrusive magmas (Sections 14-2, 14-3, and 14-4), and flow in solid rocks (Section 12-2). Particularly useful linear structures may be formed during folding and faulting (Sections 12-3, 12-4, and 12-5). Because of the many ways in which they may form, each set of structures must be described exactly and not be given a specific genetic name unless there is little doubt about its origin.

Except where they are horizontal, linear features have a bearing sense (often called their *trend*) which is taken as the direction in which they point at an angle downward. The amount of this angle, measured from the horizontal, is called the *plunge* of the linear feature.

To measure the bearing (trend) of a lineation, the observer finds an outcrop surface that exposes the linear features in full length (as in Fig. 3-9A). For fault grooves this will be the fault surface, and for most igneous and metamorphic linear features it will be a principal surface of foliation. The observer stands on this surface (if possible) and aligns the compass by looking vertically down at the linear features (view Y in Fig. 3-9A). The sighting arm of the compass is pointed in the direction of plunge, and the north-seeking (white) end of the needle thus indicates the bearing (trend). The

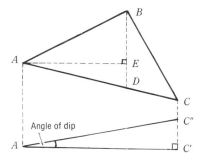

**Fig. 3-8.** Measuring strike and dip from three points of which $A$ is the lowest and $C$ the highest. In the elevation-view beneath, $AC'$ is a horizontal line perpendicular to $BD$, and $C'C''$ is equal to the difference in elevation between $A$ and $C$.

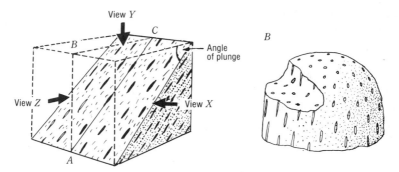

**Fig. 3-9.** *A.* Sloping surface exposing linear features in full length, with views used in measuring their bearing and plunge. *B.* Rounded outcrop showing the apparent shapes of identical spindel-shaped bodies that plunge nearly vertically.

angle of plunge is measured by moving to one side, looking perpendicular to the direction of plunge (view *X* in Fig. 3-9*A*), and using the clinometer of the compass as is done for measuring dip of beds (Fig. 3-5*B*).

If the linear features are exposed on the underside of a jutting surface, or are so steeply inclined that the observer cannot stand over them, their bearing may be measured at any position from which their trace in the surface appears vertical (view *Z* in Fig. 3-9*A*). The bearing is taken along this line of sight and transposed 180° if the surface slopes toward the observer.

Rocks that are lineated and not foliated commonly break in prisms with long axes parallel to the lineation, which makes them easy to measure. On the contrary, lineated rocks that do not cleave readily (as beds with linear imbrication) must be studied thoroughly in three-dimensional outcrops in order to identify the direction of lineation. Linear bodies will appear equi-dimensional, or most nearly so, on surfaces perpendicular to the lineation, and they will appear most elongate in sections parallel to it (Fig. 3-9*B*).

Appendix 7 illustrates map symbols used to represent bearings and plunges of lineations.

***Pitch*** is the acute angle between a linear feature and a horizontal line in the plane containing the feature (Fig. 3-10). Pitch can be measured by first marking a horizontal line on the surface by using a clinometer and a pencil,

**Fig 3-10.** Pitch is measured in the plane of the surface containing the linear feature.

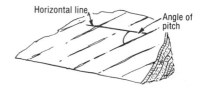

and then measuring the angle of pitch by laying a protractor on the surface. If this can be done easily at outcrops where it is difficult to measure the bearing and plunge of the lineation, pitch can be converted to plunge by using a stereo net as described in Fig. 3-11. Plunge can be converted to pitch by the inverse construction.

### 3-7. Finding and Collecting Fossils

Fossils may be used to determine environment of deposition as well as geologic age and should therefore be sought thoroughly in all sedimentary and pyroclastic rocks. Even metamorphic rocks may contain flattened or recrystallized forms that can be identified by a paleontologist (Section 15-1).

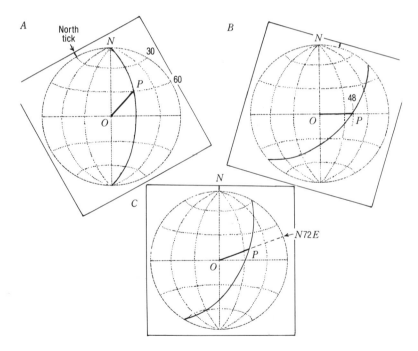

**Fig. 3-11.** Conversion of pitch to plunge with a stereo net (here dotted and simplified to 30°, 60°, and 90° arcs). The example used is for a surface striking N 30° E, dipping 60° E, and containing a lineation pitching 60° toward the NE quadrant. *A.* Mark a north tick on overlay and rotate overlay so that tick is 30° to the W of the N pole of the net; trace the meridian at 60°, mark the point 60° from the N pole, and draw a line from the origin *(O)* to the point *(P)*. This plot shows the planar surface and the lineation. *B.* Rotate overlay so that *OP* coincides with the equatorial line of the net; read the amount of plunge by referring to the meridian passing through *P* (here, 48°). *C.* Rotate overlay to north orientation and read the bearing of the lineation by extending line *OP* to the primitive circle of the net (here, at N 72° E).

In areas with unconformities, overturned folds, or low-angle faults, a dozen fossil finds may resolve more problems than anything else done in the study. The brief suggestions that follow may be supplemented from Kummel and Raup (1965), Raup and Stanley (1978), Brasier (1980), and other books on paleontology. Trace fossils are also important and are described briefly in Section 9-5.

**Macrofossils** give the advantage of being identifiable at the outcrop or in the field office. *Index fossils of North America* (Shimer and Shrock, 1944) illustrates a large number of important fossils and can be supplemented by copies of figures in specific papers or from the various volumes of *Treatise on Invertebrate Paleontology* (Moore, 1953-; Teichert, 1970-).

The chief tasks in the field are: (1) to find fossils and (2) to collect material that can be identified specifically. Fine-grained limestone, calcareous shale, and calcareous parts of sandstone are most likely to contain well-preserved forms. Well-sorted calcarenite and calcirudite may consist almost entirely of fossils, but typically of fragments that cannot be identified specifically. Dark, thinly laminated (unbioturbated) fine-grained rocks generally contain delicate fossils that would normally have been destroyed by bottom-dwelling scavengers. Concretions and nodules typically form around fossils at an early stage of diagenesis, thus protecting them from compaction and late-diagenetic solution. Whole fossils can be collected relatively easily from concretions that are softened by weathering, or by searching under ledges of resistant calcite-cemented siltstone and sandstone. Where permeable siltstone and sandstone have been leached of fossil material, molds of the fossils may be complete enough to be identifiable.

Collecting fossils can be guided further by knowing which kinds of rocks typically contain specific kinds, and which parts of the fossil organisms will be most useful for identifications. Pelecypods and brachiopods are most abundant in limestone and calcareous sandstone and mudstone, and are locally well preserved in chert. Some massive forms, as large oysters and rudists, may be preserved in conglomerate. Material needed for identification of pelecypods and brachiopods includes a complete view of the outer shell surfaces as well as fragments exposing the inside of the shell and the details of the hinge and beak, or pedicle. Gastropods occur similarly to pelecypods; collected material must show details of all structures around the aperture as well as the general form and ornamentation of the shell. Cephalopods are most abundant in limestone, calcareous shale, and calcite-cemented concretions in siltstone. They break easily and thus should be collected in adequate matrix, which may be cleaned later to reveal the sutures or the details of the aperture and siphuncle needed for identification. Corals are most abundant and best preserved in limestone and calcareous mudstone, and the most valuable samples are of whole colonies or completely preserved solitary forms. In addition to their overall shape and surface features, the internal structures are needed for identification, so that recrystallized or dolomitized material is generally inadequate.

Bryozoans are most abundant in thin-bedded limestone and calcareous shale and require study of internal structures for identification. Crinoids are locally abundant in limestone and sandstone but become disarticulated easily and thus provide few specimens suitable for assigning geologic age. Echinoids are commonly preserved intact in limestone and sandstone. Trilobites and graptolites occur chiefly in shale and thin-bedded limestone, where they are found by using a chisel to split the rock parallel to bedding. Graptolites are fairly common in low-grade slates and even in phyllite, and their uncompacted forms may be found in massive fine-grained limestone and chert.

Well-preserved fossil fish are most common in dark, unbioturbated shale, limestone, or diatomite of lakes, marshes, and dysaerobic marine environments. Rocks worth searching contain scattered fish scales, single bones, coprolites, or other phosphatic particles. Terrestrial vertebrate fossils occur mainly in mudstone, tuffs, and sandstone, and their teeth and case-hardened bones may be washed and sorted into rivulet-channels. The skull, teeth, and appendage bones are most valuable for identifications. Fossil land plants are most abundant in dark fissile shales associated with coal, and in thin-bedded lacustrine limestone, siltstone, diatomite, and tuff. Large collections are generally needed and should include all remains of flowers, seeds, catkins, or cones as well as whole leaves. Well preserved silicified wood may be valuable.

Each fossil locality should be examined to determine if the fossils: (1) were in living-position when they died; (2) accumulated at or near where they lived; (3) were transported a considerable distance before they accumulated; or (4) were eroded from older fossiliferous rocks. This study should be made before the outcrop is broken as well as during the process of collecting, and all evidence should be recorded in the notes. Fossils in living position are indicated by attached sessile habit, by abundant articulated specimens of bivalves, and by pelecypods oriented at right angles to bedding with beaks directed downward.

Evidence suggesting accumulation at or near a habitat includes: (1) many well-preserved specimens of one or a few species; (2) parts of complex individuals, such as crinoids, more or less in one place; (3) an assemblage of species thought to occupy a single ecologic niche; and (4) sediments suitable to the life-habits of the organisms. Exceptions to the first two criteria just given are plants and animals that tend to float after dying. Ammonites, for example, may be well preserved after long periods of transportation. Generally, however, extensive transportation is shown by: (1) rounded (abraded) fossils; (2) mixtures of ecologically diverse species; and (3) fossils restricted to beds that appear to be deposits of sediment gravity flows.

Reworked fossils are indicated by: (1) remnants of rock matrix (perhaps only inside the shell); (2) being weathered on one side; and (3) differences in preservation and appearance from other fossils in the same rocks. Fossils that seem stratigraphically out of place are not proof of reworking, for the age-ranges of many organisms are imperfectly known. Such finds must be collected carefully and studied thoroughly.

If fossils are abundant, as great a variety of species should be collected as possible. The relative abundances should be described in the notes, because this information will help to establish the age and environment of the assemblage. Generally, no attempt should be made to trim and clean specimens at the outcrop. Samples should be protected from wear during transport by packing them in soft paper, grass, or whatever is at hand. If no packing material is available, the fossils should be left at the outcrop and packed later.

Each specimen must be marked with the number of the field station, and the same number is written on the bag in which the sample is placed. Some geologists number fossils and rock samples with a compound number consisting of: (1) the number of the field sheet or aerial photograph; (2) the number of the field note; and (3) a letter to designate each of two or more samples collected at that site. In Fig. 3-3 the samples of note 2 would thus be numbered 3-2*A*, 3-2*B*, and 3-2*C*. Another system, one making it especially easy to find the field note for a given sample, uses a set of numbers for the month, day, and year, followed by the note number. A sample collected on March 15, 1954, and described in note 9, would thus be numbered 031554-9.

The location should be described in the field notes before leaving the outcrop, and this description must be complete enough so that the exact locality can be found again. A photograph of the outcrop with a hammer or scale placed at the exact point of collection will be helpful. The stratigraphic position of the collection site should be measured as precisely as possible (Chapter 11), and the outcrop must be described thoroughly with special attention to any features suggesting environment of deposition. Sooner or later, the latitudes and longitudes of localities should be determined exactly by reference to a topographic map, or the locations can be referred to a cadastral system such as that of Appendix 5. Paleontologists are more likely to help with identifications if all this information is provided.

**Microfossils** are exceptionally valuable because they are common in many fine-grained marine and lacustrine rocks that contain few macrofossils. They are virtually essential in the study of drill cuttings and drill cores. Biozones have been established at least locally on almost all kinds of microfossils, and several kinds of pelagic microfossils are the basis of worldwide correlations. Research on the least-known kinds is progressing rapidly, and separatory techniques are also being improved.

A drawback in the use of microfossils is that coccoliths, foraminifers, and ostracodes are generally leached from weathered permeable rocks. Augering or excavation may thus be necessary. Another drawback is that most pelagic microfossils cannot be seen with a hand lens when they are in rocks, so that sampling must be blind. The latter situation can sometimes be resolved by taking separatory materials and equipment to the field camp in order to test initial sets of samples. At the least, residues of limestone and dolomite that have been dissolved in weak acid can be examined with a binocular microscope for conodonts, radiolarians, diatoms, and silicoflagellates. If these fossils are reasonably abundant and the rocks are calcareous, then calcareous microfossils are likely to be present. Methods of separation are described by Brasier (1980), Kummel and Raup (1965), and in literature cited in those books.

Collecting may also be guided by the fact that certain rocks typically contain specific kinds of microfossils. Pelagic foraminifers, radiolarians, coccoliths, diatoms, and silicoflagellates are most abundant in claystone. They may impart a slippery (talc-like) feel to the rocks or, if the microfossils are the dominant constituent, a chalky appearance. Radiolarians may be visible with a hand lens in cherts, from which they can be separated by the methods described by Pessagno and Newport (1972) and Holdsworth and others (1978). Pelagic fossils are scarce in turbidite sequences except in thin pelagic layers of mudstone that accumulate between successive turbidity flows, thus lying at the top of each Bouma sequence (the Tep division in Fig. 9-6A). In the fine-grained facies of submarine-fan deposits (Section 9-9), the pelagic accumulations may be distinguished from thin turbidite layers by lack of grading, absence of current-generated lamination, and perhaps lighter color.

Diatom frustules are destroyed by diagenesis when buried to moderate depths (Section 4-2). Rock containing an abundance of diatoms is noticeably light to heft, and the larger of the diatoms are visible with a hand lens, appearing as silky disks. Ostracodes occur in a great variety of marine and lacustrine rocks. They are generally more visible in claystone and calcilutite, in which they look like small ooliths. Conodonts may be found in most kinds of marine rocks and are most abundant in limestone and mudstone. They are among the few fossils that are commonly preserved in dolomite, and many sandstones and siltstones contain them in moderate to small numbers. Although large enough to be seen with a hand lens, they are difficult to recognize in rocks.

Benthic foraminifers are commonly large enough to be seen with a hand lens and are most abundant in mudstone and fine-grained limestone. The largest foraminifers (fusulinids, orbitoids, and nummulitids) may be visible with the naked eye in a variety of marine rocks, especially in sandy limestone and calcareous sandstone. They can be identified specifically only if they have well-preserved internal structures.

Spores and pollen of land plants are most abundant in dark claystone and argillaceous limestone deposited in lakes, on deltas (as coal-bearing sequences), or in terrigenous marine deposits. They are too small to be seen with a hand lens and can be separated as described by Kummel and Raup (1965) and Brasier (1980). Dinoflagellates are typically in dark-gray marine beds, often together with spores and pollen, and are separated similarly.

Samples weighing 1 kg (2 lb) are ample for most studies, but samples of 2 kg should be collected for conodonts and other fossils that may be scarce. Beds that are clearly fossiliferous or are highly favorable for fossils are generally sampled in single chunks or sets of chips. In thick mudstone or limestone sequences where no microfossils are seen, channels are cut across a meter or so of section in the hope of sampling fossil-rich layers; these samples are thus a set of chips representing the full channel.

In sampling for small microfossils it is absolutely essential to avoid contaminating samples with dust from other sites. Sampling tools and hands must thus be scrubbed clean with soap and water, and clothing must be brushed and shaken of dust. Microfossils may also be washed down open fractures, root holes, and dissolved openings, which must thus be avoided in sampling. Samples should be placed in new plastic bags that are strong enough to remain unpunctured during transport. For bags of chips, a plastic identification tag should be placed in the bag and the same information written on the outside label. Describing and numbering samples in the notes are the same as for macrofossils.

## 3-8. Collecting Rock Samples

Rock samples can be used to: (1) make petrographic identifications; (2) measure small-scale planar and linear structures; (3) trace mineral reactions (Section 15-3); and (4) compare rocks from different parts of an area. If interest is in the principal (typical) rocks of an area (which is usual) they are the ones to collect. This will take effort, for typical rocks tend to become an unnoticed background to ones that are unusual or curious (Fig. 3-12).

In addition to being representative, samples must be as fresh as possible; however, additional weathered pieces may be needed in engineering studies or for fully characterizing rock units. The minimal size of a sample is pres-

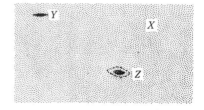

**Fig. 3-12.** Collect typical rocks (*always X*; and *Y* and *Z* only if you must). After C.M. Isaacs (oral presentation, 1981).

cribed by its purpose and grain size. In average situations, uniform rocks with grains less than 1 mm in diameter are well represented by pieces 3 X 5 X 5 cm or equivalent. If grains are 2 to 3 mm in diameter, samples should be about twice that large. Samples need not be trimmed unless they must be stored in shallow drawers. Important small features (e.g., crucial grains or small fossils) should be circled with a pen, because they may be difficult to find later. Rock samples are numbered, marked, and labeled in the same way as fossils (Section 3-7) and wrapped or placed in bags to protect them from abrasion.

**Oriented samples** are collected for later study of fabrics or small features that must be related to the geologic map or to major structures. Examples are sandstone with possible current oriented fabric; metamorphic rocks with folded lineations; igneous rocks cut by sets of thin veins; and orientations of crystallographic axes of quartz. Most samples can be collected by: (1) measuring the strike and dip of a planar surface on the rock and (2) drawing a strike-and-dip symbol on the surface before breaking the sample from the outcrop, or after fitting it back exactly to its original position. An arrowhead is added to the strike line and a geographic direction to the dip line to insure that the orientation will be unambiguous (Fig. 3-13A). The compass direction of the arrow is included in the notes, as is a notation as to whether the planar surface faces upward or downward (Fig. 3-3, Note 2). If the rock to be sampled has no planar surface, a clinometer can be used to draw horizontal and vertical planes across enough of the sample so that it can be reoriented later (Fig. 3-13B).

All features relating to samples should be described, perhaps with a drawing showing geometric relations. If the marked surface is not a primary one, the strike and dip of primary structures must be measured and recorded *for the sample site*. Linear features should also be measured, recorded, and described at the outcrop, because they are likely to be far less obvious when the sample is studied in the laboratory.

If the samples are intended for a systematic, thorough study, a sample or two might be analyzed before a large number are collected. The analyses may prove so time-consuming that only a moderate number can be com-

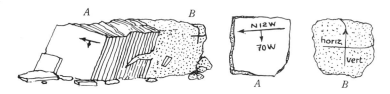

**Fig. 3-13.** Oriented samples of *(A)* a rock with a planar surface and *(B)* a rock that has none. The bearing of the vertical plane on *B* must be measured and recorded.

pleted, in which case sampling should be deferred until the field study has shown which sites will be most valuable.

## References Cited

Brasier, M. D., 1980, *Microfossils*: London, George Allen & Unwin, 193 p.

Gilbert, G. K., 1886, The inculcation of scientific method by example, with an illustration drawn from the Quaternary geology of Utah: *American Journal of Science*, 3rd Series, v. 31 (whole number 131), p. 284-299.

Holdsworth, B. K., Jones, D. L., and Allison, C., 1978, Upper Devonian radiolarians separated from chert of the Ford Lake Shale, Alaska: U.S. Geological Survey, *Journal of Research*, v. 6, p. 775-788.

Kummel, B., and Raup, D., editors, 1965, *Handbook of paleontological techniques*: San Francisco, W. H. Freeman and Co., 852 p.

Moore, R. C., director and editor, 1953-, *Treatise on invertebrate paleontology* (in many parts, each a separate volume): Lawrence, KA, The University of Kansas, and Boulder, CO, The Geological Society of America.

Pessagno, E. A., Jr., and Newport, R. L., 1972, A technique for extracting *Radiolaria* from radiolarian cherts: *Micropaleontology*, v. 18, p. 231-234.

Raup, D. M., and Stanley, S. M., 1978, *Principles of paleontology*, 2nd edition: San Francisco, W. H. Freeman and Co., 481 p.

Shimer, H. W., and Shrock, R. R., 1944, *Index fossils of North America*: New York, John Wiley & Sons, 837 p.

Teichert, C., editor, 1970-, *Treatise on invertebrate paleontology, revised* (in many parts, each a separate volume): Lawrence, KA, The University of Kansas Press, and Boulder, CO, The Geological Society of America.

# 4

# Identifying Rocks in the Field ■

## 4-1. General Rationale

Rocks must be identified at each outcrop in order to interpret other observed features (Sections 3-1 and 3-2). Identifying rocks is often easier in the field than in the laboratory, because many minerals and textures are more apparent on clean, weathered surfaces than on broken ones, and one can search through as much material as necessary to find diagnostic grains or other small features. In addition, rock associations and primary structures can help greatly in field identifications, as noted partly in this chapter and partly in Chapters 9, 10, 13, 14, and 15. Some features commonly thought to be microscopic are included in this chapter because they may be visible with a hand lens, providing it is used as effectively as possible (Section 2-2). Rock staining is another means of clarifying fine-textured relations as well as of identifying minerals (Section 2-3). Other equipment generally needed is described in Section 2-1, and Appendix 3 presents diagrams for estimating percentage composition.

As used in this book, the *texture* of a rock is the overall aspect imparted by the sizes, shapes, and arrangement of its grains, and *fabric* is the component of texture resulting from the relative sizes and shapes of grains, especially from the preferred orientation of platy or elongate grains. A *structure*, on the other hand, is a rock body or domain differing in texture or composition from other parts of the rock. Textures, fabrics, and small structures are emphasized in this chapter because they are often easier to recognize than minerals, and they may be more significant genetically.

## 4-2. Textures of Sedimentary Rocks

Three categories of texture are typically combined in sedimentary rocks: (1) clastic (fragmental) textures, imparted by detritus transported to the site of deposition; (2) crystalline textures resulting from mineral growth at the site of deposition; and (3) crystalline textures resulting from diagenesis. Many sedimentary rocks are classified on the basis of clastic texture. Textures of the other two categories are important, however, and should be described in the field notes.

*Clastic texture.* Grain size is the primary element of clastic texture because it suggests the level of kinetic energy during transportation and deposition. The size classification in Table 4-1 is used by most North American geologists, whereas other classifications are used by most engineers and by geologists in some other parts of the world (Pettijohn, 1975, p. 29).

A given sediment or rock is assigned to the size class that is the average (median) size for its grains, the amounts being measured either by weight or by volume. Generally, the average size is also the dominant size, with other size classes containing less and less material as they are increasingly coarser or finer than the average. Some sediments, however, have abundant grains in two distinctly different size classes. Such *bimodal* sediments are suggestive of the mixing of two separate batches of sediment, as by entrainment in a debris flow (Section 9-2) or by bioturbation (Section 9-5). The textural names of these sediments can be compounded, as *bimodal coarse and very fine sand*.

A simple device for determining grain size can be made by sieving sand and coarse silt into the size classes of Table 4-1 and gluing a small patch of each along one edge of a small card. The card is then held against a rock or sediment sample and the two are compared under a hand lens. Another method is to use a millimeter scale to locate several grains or spots in the sample that are exactly 0.5 mm and 1 mm in diameter, and then using a hand lens to estimate sizes of nearby grains. Grains coarser than sand can be measured directly with a scale.

**Degree of sorting** is a measure of how closely the clastic grains in a sediment approach being one size. Sorting provides a measure of the uniformity

**Table 4-1.** Classification of Clastic Sediments by Grain Size.

| Median diameter in mm | Phi scale | Sediment name | Group name |
|---|---|---|---|
| 256 | -8 | Boulder gravel | |
| 128 | -7 | Coarse cobble gravel | |
| 64 | -6 | Fine cobble gravel | Rudite |
| 32 | -5 | Very coarse pebble gravel | (psephite) |
| 16 | -4 | Coarse pebble gravel | |
| 8 | -3 | Medium pebble gravel | |
| 4 | -2 | Fine pebble gravel | |
| 2 | -1 | Very fine pebble gravel | |
| 1 | 0 | Very coarse sand | |
| .5 | 1 | Coarse sand | |
| .25 | 2 | Medium sand | Arenite |
| .125 | 3 | Fine sand | (psammite) |
| .06 | 4 | Very fine sand | |
| .03 | 5 | Coarse silt | |
| .015 | 6 | Medium silt | |
| .008 | 7 | Fine silt | Lutite |
| .004 | 8 | Very fine silt | (pelite) |
| | | Clay (any detritus this size) | |

and persistence of currents or waves and is thus invaluable in interpreting sedimentary processes. Degree of sorting can be estimated with a hand lens by determining the number of size classes included by the great majority (about 80%) of the grains. This determination can be made by noting the size of the largest grains that make up the finest 10% of the sediment, and the size of the smallest grains comprising the coarsest 10% of the sediment. The bulk (80%) must be distributed among all the classes between these two limits and the number of these classes gives a numerical measure of sorting. Figure 4-1 suggests names for the categories based on these numbers, and shows somewhat diagrammatic views of each sorting category.

The greatest difficulty in estimating sorting of sand arises where weak grains have been compacted between strong grains, intruding among them to form a *pseudomatrix* of fine material. The weak grains are typically fecal pellets or fragments of mudstone, slate, tuff, argillaceous calcilutite, or altered feldspar. Pseudomatrix can be recognized by its patchy nature or by noting grains that are only partly compacted among their neighbors (Fig. 4-2).

**Rounding of grains** by abrasion reflects duration of transport and is an aspect of grain shape that may be used as a textural adjective or described in the notes. Figure 4-3 suggests names for degrees of rounding. For material of a given kind, the rate of rounding by mechanical impacts generally

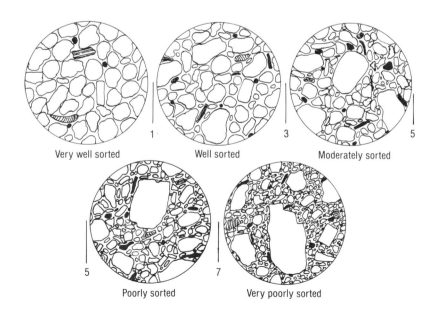

Very well sorted     Well sorted     Moderately sorted

Poorly sorted     Very poorly sorted

**Fig. 4-1.** Hand-lens view of one layer of detritus sorted to various degrees. The numerical limits are explained in the text.

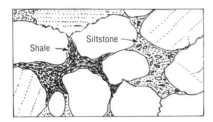

**Fig. 4-2.** Pseudomatrix of dark shale and siltstone pressed among tightly compacted quartz and feldspar grains of the same size as the distorted lithic grains.

increases with increasing grain size, and any departures from this norm should be noted. Distinctly different degrees of rounding of a given material within one size class indicate mixing of grains from two or more sources. Large angular grains should be noted because they imply either a nearby source, or transport in an unusually supportive medium, such as a mud-rich debris flow. At the other extreme, rounded or subrounded fine quartz sand implies an unusually long history of abrasion.

**Fabric** in rocks with only slightly platy or elongate grains will require careful study; however, even a small percentage of thin mica plates or strongly elongated fossil fragments may impart a recognizable fabric. Fabric is used in determining the orientation of bedding (Section 3-5) and of current direction (Section 9-4), and in interpreting sedimentary environments. Platy and linear grains that have settled in the absence of a lateral current tend to lie parallel to bedding, without a linear orientation. Grains that have accumulated from a strong to moderate current tend to impart a planar fabric dipping upstream, an arrangement called *imbrication* (Fig. 4-4). Gentle bottom currents may align elongate plant fragments and some other fossils parallel to the current and also to bedding.

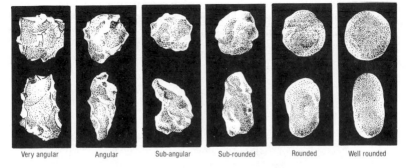

Very angular   Angular   Sub-angular   Sub-rounded   Rounded   Well rounded

**Fig. 4-3.** Degrees of rounding of sand-sized grains, the upper set equidimensional (spherical) and the lower somewhat elongated. After Powers, M.C. (1953). © Society of Economic Paleontologists and Mineralogists, copied with permission.

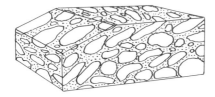

**Fig. 4-4.** Strongly developed imbricate fabric indicating a current that flowed from left to right. Elongate clasts may plunge upstream, as in the figure, or may have their longest axes horizontal and at right angles to the current direction.

*Textures formed by mineral growth at the site of deposition* are of great value in determining environments of deposition and may be classified into two genetic groups: inorganic and biogenic. In the inorganic group, *hypidiomorphic granular* texture is characterized by euhedral or subhedral grains among anhedral grains, or by anhedral grains with skeletal euhedral interiors or zones of pores (originally brine) lying along crystallographic planes. These various features indicate growth of crystalline grains from concentrated solution, as by evaporite sedimentation, rather than growth during diagenetic crystallization.

*Oolitic* and *pisolitic crystalline textures* result from inorganic precipitation of substances around separate cores, forming rounded, smooth-surfaced grains that commonly have internal radial fabrics or concentric shells (Fig. 4-5A). Oolites are sand-sized grains, and pisolites are larger. The precipitated minerals are aragonite, calcite, collophane (apatite), or chamosite. The grains may later be transported and mixed with other detritus, as indicated by abrasion and sorting.

*Crystalline pelletal texture* may result where certain substances, especially glauconite and apatite, accumulate by slow alterations of other grains on the substrate. These pellets differ from oolites and pisolites in being irregularly bulbous or in filling shells of protozoans or other small fossils.

**Biogenic crystalline textures and structures** result from biologic deposition of carbonates or collophane, the latter as bones. Especially abundant are deposits of calcareous algae, which encrust and otherwise hold sediment together. *Stromatolites* (Section 9-7) are included here for convenience, although most of them are formed by entrapment of fine carbonate detritus

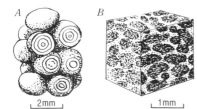

**Fig. 4-5.** *A*. Aggregate of oolitic grains, several cut partly away to show concentric internal structure. *B*. Pelletal limestone, each pellet a structureless aggregate of very fine-grained (micritic) calcite. The matrix around the pellets is also micrite.

between noncalcareous algal filaments. Corals in growth position, attached mollusks, and deposits secreted by worms are biogenic rock-forming structures. A common biogenic texture results from accumulation of skeletons of plants and animals that lived on the substrate or in the water directly above it. The evidence for this origin is lack of rounding and sorting (fragmentation alone is not necessarily evidence of transport because skeletal parts may be broken by predators). *Pelletal* biogenic texture results when animals excrete fine-grained sediment held together by mucous in rounded to elongate masses (Fig. 4-5*B*). Because these pellets would disintegrate if transported far across the substrate, they are a criterion for formation nearby or from the overlying water column.

**Diagenetic textures** form after a sediment is buried and while it is being transformed into rock. The specific processes are compaction, solution, cementation, grain growth or recrystallization, polymorphic transformation, and replacement.

Compaction of well sorted sand results in closer packing of grains, with extreme compaction causing grain deformation (Figs. 4-2 and 4-6*A*). These effects are not visible in fine-grained sediments, but are suggested by increase in bulk density (heft), toughness, and sound of a hammer blow (Appendix 4). The coalification series (Section 4-3) provides a means of comparing degree of compaction from one locality to another.

Solution under low to moderate loads produces holes having the shapes of dissolved grains, the most common being fossils. Partly dissolved sand grains may be recognized by corroded or honeycombed surfaces, and by pores too large to be primary. Partial dissolution of cement may produce pores with fluted or honeycombed surfaces, and pores formed by complete dissolution of carbonate may be coated with thin pigments of iron oxides. Under high loads, grains tend to be dissolved where they bear against one

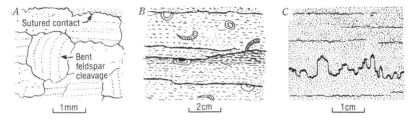

**Fig. 4-6.** Effects of directed pressure during diagenesis, shown here as though rocks were loaded vertically. *A.* Quartz and feldspar sand grains, some deformed and most showing moderately sutured contacts. *B.* Cross section of calcareous mudstone with pressure-solution surfaces marked by insoluble materials and partly dissolved fossils. *C.* Cross section of stylolite with residue of silt in silty limestone. The parallel thin lines are pressure-solution cleavage (Section 12-4).

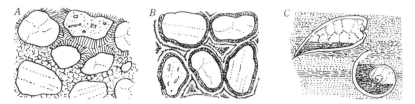

**Fig. 4-7.** *A.* Granular *(below)* and prismatic crystalline cements in sandstone. *B.* Banded colloidal cement in well-sorted sandstone. *C.* Large calcite crystals (dotted contacts) filling shells above deposits of fine sediment.

another, forming smooth to interlocking contacts (Fig. 4-6*A*). The contacts may join in sets of parallel surfaces that are a variety of spaced cleavage (Figs. 4-6*B* and *C*). *Stylolites* are highly irregular pressure-solution surfaces commonly developed in limestone (Fig. 4-6*C*).

Cementation textures are most apparent in coarse well-sorted sandstone and conglomerate. Crystalline cements are typically sugary or fibrous, and colloidal cements smoothly vitreous and often banded (Figs. 4-7*A* and *B*). Single-crystal overgrowths on quartz in partly cemented sandstone can be seen with a hand lens as bright geometric reflections from crystal faces. Large cavities in limestone commonly become filled with coarse (sparry) calcite or by fine sediment introduced by pore fluids during early diagenesis (Fig. 4-7*C*).

Besides forming cements, diagenetic crystallization may lead to increasing grain size of fine sediments and to devitrification of glass and colloidal substances. The diagenetic grains are generally anhedral because they grow against one another in the solid state, and they may become coarse enough in limestone and evaporites to show visibly granoblastic arrangements (Fig. 4-8*A*). Unaffected relics such as fossils, carbonaceous detritus, or well-sorted quartz sand are useful indicators of diagenetic crystallization (Figs. 4-8*B*

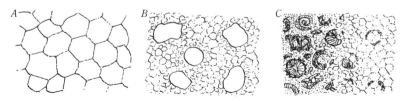

**Fig. 4-8.** *A.* Ideal granoblastic (mosaic) crystalline texture as seen under a hand lens. *B.* Well-sorted quartz sand "floating" in a crystalline limestone, indicating that the rock was originally a sandy calcarenite, and that the carbonate grains have recrystallized. *C.* Biomicrite (shelly micrite) partly recrystallized to a granular texture.

and $C$). In cherts, recrystallization of chalcedonic quartz to fine granular quartz is suggested by a change from waxy to dull stony luster.

Polymorphic transformations take place at certain temperatures and thus indirectly indicate depth of burial. Transformations of $SiO_2$ are especially valuable because they can be recognized in the field, as described by Isaacs (1981, 1982). In brief, the hydrated amorphous form of $SiO_2$ (opal-A) makes up the skeletons of diatoms, radiolarians, and some sponges, and it transforms to interlayered cristobalite and tridymite (opal-CT) at temperatures equivalent to about 750 m of burial. A sediment containing 80% or more of diatom skeletons is thereby changed from earthy, highly porous diatomite to vitreous, brittle opaline chert. At somewhat higher temperatures, diatomaceous sediment with 20% to 60% impurities is changed from friable mudstone to hard, moderately porous porcelanite (Section 4-3). With increased burial, opal-CT transforms to quartz at temperatures corresponding to depths of about 2000 m. The quartz is usually chalcedonic, giving transformed chert a waxy luster but leaving the appearance of impure (porcelanous) rocks unchanged.

Sedimentary minerals may be replaced chemically by other minerals during diagenesis, generally producing crystalline granular textures. Common replacive minerals are dolomite, calcite, ankerite, quartz, opal-CT, iron oxides, iron sulfides, and apatite. Some diagnostic features are: (1) fossils or oolites known to consist originally of aragonite or calcite that are now other minerals; (2) nodular or irregular bodies of replacive minerals crossing bedding or other primary structures; and (3) formations known by mapping to consist of certain primary minerals and now consisting locally of others. The approximate timing of diagenetic replacements may be judged from the degree of compaction of bedding around replaced bodies (Fig. 9-11$A$).

Many of these diagenetic changes can be mapped as zones, constituting an overprint on formations or other primary rock units. The zones will thus express systematic variations in depth of burial, diagenetic temperature, or action of fluids.

### 4-3. Naming Sedimentary Rocks

Sedimentary rocks are classified primarily on the basis of composition or grain size, as in the subtitled categories of this section. Genetic varieties of specific rocks are described in Chapters 9 and 10.

**Silicate-rich sandstone** may be classified texturally into *wacke*, which has silt and clay-sized detritus between its sand grains, and *arenite*, which is free of these matrix materials (Williams and others, 1982, p. 326). In *graywacke*, the fine matrix is so crystallized or indurated as to make the rock exceptionally tough and typically dark colored, regardless of the kinds of sand grains present.

Sandstone may be classified further on the basis of the kinds of sand grains comprising the rock, as in a scheme developed by W. R. Dickinson in 1978 for hand lens study (Fig. 4-9). To use it, one first estimates the proportions among: (1) quartz and quartzite grains, (2) feldspar grains, and (3) lithic (rock) grains, and then locates that composition in the diagram. An advantage of this classification is that each rock name indicates provenance (source materials) as well as relative weathering maturity, and thus suggests rate of uplift and erosion of the source area. A difference from classifications based on microscopic petrography is that chert grains, which typically look like other rock fragments, are included with the lithic grains rather than with quartz. In cases where chert grains can be recognized with a hand lens (and this typically will be when they are abundant), they are probably best added separately to the rock name, as *chert-grain subquartzose sandstone*. Sands derived largely from volcanic materials may be specified as *volcanic* or *andesitic*, and so on, and those with notable varietal minerals as *biotitic*, and so on.

Cementing substances have usually been referred to by adjectives such as *calcareous*, *dolomitic*, and *siliceous*; however, these terms might also imply accessory detrital materials, so that the unambiguous terms *calcite-*

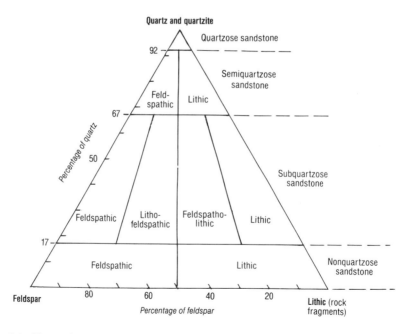

**Fig. 4-9.** Names for sandstones based on the proportions of three principal groups of detrital grains (see text for explanation). From W. R. Dickinson (personal communication, 1982).

*cemented, dolomite-cemented,* and *quartz-cemented* are recommended.

**Silicate-rich rudites** containing subangular to rounded clasts are *con-glomerate*, and those composed entirely of angular clasts are *breccia* or *sharpstone conglomerate.* The scheme proposed in Fig. 4-10 is based on degree of sorting. In a simpler twofold classification, *clast-supported* rudite is sorted well enough so that the large clasts touch, and *matrix-supported* is not. Clasts in the 2 to 6 mm range should be sought for and measured with care because they tend to look like sand. Pebble-sized fragments of fossil shells are also easy to overlook in coarse, sandy rocks.

Rudites are of particular value because their clasts are large enough to identify as rocks and thus to indicate provenance. If one kind of rock predominates, it is commonly used in the rock name, as *chert pebble conglomerate.* Thorough provenance studies require systematic pebble counts, as by marking out several square meters of an outcrop and inventorying all the gravel-sized clasts therein. Clast shapes may also indicate provenance; for example, somewhat abraded fragments of round clasts indicate conglomerate in the source area. Rudites are often termed *mature* if they contain little else besides tough, chemically stable fragments (chert, vein quartz, quartzite, quartzose hornfels, silicified or devitrified rhyolite, etc.). They are termed *immature,* or *polymictic* if they contain more than 10% of weak or chemically unstable clasts (mudstone, limestone, slate, dolomite, intermediate to mafic volcanic rocks, feldspathic plutonic rocks, schist, etc.).

**Silicate-rich lutite.** The grains in well-sorted *siltstone* can generally be seen with a hand lens, but clay-rich rocks require accessory tests. Well-sorted *claystone* has a smooth waxy aspect when cut or scraped with a knife, especially when moist, and is not gritty when rubbed between the teeth. Mix-

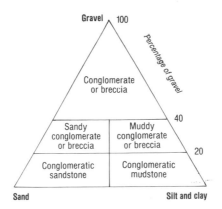

**Fig. 4-10.** Names for rocks containing gravel-sized clasts. Generally, rocks above the 40% line will be clast-supported and those below the 20% line matrix supported.

tures of silt and clay, called *mudstone*, may look like claystone because abundant clay-sized material screens the larger grains, but these rocks are gritty to the teeth or knife. Poorly sorted mudstone commonly contains sand that may be hidden on broken surfaces by the abundant clay but is visible on clean weathered surfaces. Sandy mudstone may grade to muddy fine sandstone (fine wacke) that looks like mudstone except on clean, weathered surfaces.

*Shale* is a lutite with a flaky cleavage (fissility) and, commonly, thin bedding laminations (but the name is also used for any silicate-rich lutite). Silty or sandy shales may have abundant visible mica flakes that lie parallel to the cleavage. The name *argillite* is sometimes used for tough lutite that breaks into angular fragments,and can be distinguished from calcite-cemented lutite because it does not effervesce and crumble in dilute HCl.

**Limestones** are composed mainly of calcite or aragonite, and those with clastic texture may be named according to grain size: *calcirudite* for sizes coarser than 2 mm; *calcarenite* for sizes between 2 and 1/16 mm; and *calcilutite* (or *micrite* or *lime mud*) for finer sizes. Dunham (1962) proposed an additional set of names based on sorting: *grainstone* is calcarenite or calcirudite with no micrite (calcilutite) matrix; *packstone* is calcarenite or calcirudite with a relatively sparse micrite matrix and with the larger grains touching (clast-supported); *wackestone* is a micrite-supported mixture containing more than 10% of sand-sized or coarser clasts; and lime *mudstone* is micrite

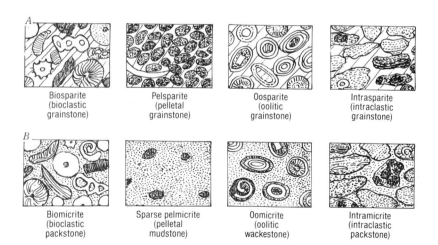

**Fig. 4-11.** Names for limestones with fragmental textures. *A*. Sparry (coarsely crystalline) calcite cement fills intergranular spaces (lined pattern represents calcite cleavage). *B*. Micrite (calcilutite) fills between larger grains. The upper names were proposed by Folk (1959, 1962) and the lower are based on the system proposed by Dunham (1962). In addition, Fig. 4-5*B* illustrates a *pelmicrite* or *pelletal packstone*.

with less than 10% of coarse clasts.

Limestone may also be classified according to five common kinds of particles (Folk, 1959, 1962): (1) *bioclasts* (skeletal grains); (2) *pellets* (also called *pelloids*); (3) *oolites* (or *ooids*) and *pisolites*; (4) *intraclasts* (fragments of limestone); and (5) *micrite* (all particles finer than sand size). The dominant kind of particle serves to name the rock (Fig. 4-11).

Two varieties of biogenic limestone are classified on a structural basis. *Reefal limestone, biohermite* (Folk, 1959), or *boundstone* (Dunham, 1962) forms thick lenses or steep-sided bodies composed mainly of skeletons of sessile organisms held together by encrusting or ramifying forms, typically algae. *Biostromal limestone* may be similarly composed but forms tabular sheets or thin lenses. Either kind of limestone can be specified further by adjectives such as *algal, stromatolitic,* and *coralline.*

Limestones with diagenetic textures are *crystalline* (or *crystallized*) *limestone.* Grain-size adjectives may be useful to indicate the degree of grain growth, as: *coarse,* larger than 1 mm; *medium,* 1 to ¼ mm; and *fine,* smaller than ¼ mm.

**Dolomite (dolostone)** consists predominantly of the mineral dolomite. Most varieties have crystallized diagenetic textures, so that the adjective *crystalline* need not be used before the rock name. Grain-size adjectives are generally useful, as just noted for crystallized limestones. Rocks with primary particles (see *limestone,* above) may be named accordingly (as *oolitic dolomite*). Rocks formed by replacement of a specific kind of limestone, as shown by gradations or by relict structures or residual materials, may be given compound names, as *dolomitized oolitic packstone.* Sections 9-7 and 9-9 describe primary structures that help in identifying varieties of dolomitic rocks.

**Phosphorite** consists largely of microcrystalline or cryptocrystalline apatite (collophane) in the form of bones, oolites, pellets, nodules, coprolites, and finely divided grains. When fresh, these materials are typically dark gray to black or brown in Mesozoic and older rocks and typically pale tan to medium brown or orange in Cenozoic rocks. The darker varieties can be recognized by a secondary coating ("bloom") that is white or pale gray and slightly bluish, and the pale varieties by local secondary green coatings. Pyrite and fish scales are common constituents. Cements are typically fine silica minerals, calcite, or dolomite. Bedded phosphorites are typically associated with organic-rich (kerogenic) laminated siliceous rocks, or with calcilutite, shale, and dolomite. Nodular deposits occur with glauconite in sandy sediments that accumulated very slowly and thus suggest unconformity.

**Siliceous lutites** that have not been strongly compacted or crystallized range from more or less pure *diatomite, radiolarite,* and *spiculite* (composed of sponge spicules) to ordinary mudstone or limestone with abundant siliceous skeletons (as *diatomaceous mudstone* and *radiolarian calcilutite*).

Rocks rich in siliceous skeletons are unusually light to heft, highly porous, friable, light in color, and insoluble in HCl.

The diagenetic origin of several derivative rocks is described in Section 4-2 and their field classification has been described by Isaacs (1981, 1982). *Opaline chert* is glassy, brittle, and light to heft. *Porcelanite* is tough, breaks like unglazed porcelain, sounds "clink" to hammer taps, and has so many minute pores as to have a dull luster and to stick to the tongue. *Chalcedonic chert* is heavier to heft (density close to quartz), has a waxy luster, and is commonly darker and more opaque in thin flakes than opaline chert. *Quartz chert*, in which grain growth has led to microgranular texture, has the stony appearance of fine-grained quartzite and has a hackly fracture.

These siliceous rocks commonly form distinct beds a few millimeters to 10 or 20 cm thick which are laminated and initially rich in organic (sapropelic) substances (Section 9-9). Distinctly different are the cherts that form nodular to irregular bodies in limestone, dolomite, and siliceous sediments. These concretionary and typically replacive bodies may show internal relics of carbonate fossils and bedding laminations, and may develop secondary banding subparallel to their nodular forms or arranged in bizarre oblique sets. Nodules formed at an early stage of diagenesis may consist of opal-CT, but most nodular cherts consist of chalcedony or microgranular quartz. Limestone and tuff may also be silicified through large masses to form chert with relict primary features.

**Volcaniclastic lutite.** Vitric tuff and fine detrital sediment rich in volcanic glass commonly devitrify diagenetically to hard, tough rocks easily mistaken for chert. Crystallized volcaniclastic lutite, however, has a duller, stonier luster than either opaline or chalcedonic chert, and can be recognized readily where the silicate minerals are strongly weathered. Patient hand lens study will almost always disclose minute feldspar subhedra or other igneous grains. Volcaniclastic lutite may also be associated with volcanic rocks or with sandstone containing igneous mineral grains or pumice lapilli.

**Evaporites,** where more or less pure, are named according to their mineral species: *anhydrite rock, gypsum rock, halite* (or *salt*) *rock*, and so on. The mineral names are used as adjectives in cases where other sedimentary materials predominate, as *gypsiferous shale, anhydrite-bearing dolomite*. Textures are crystalline in almost all cases, but grain sizes and textures formed by mineral growth at the site of deposition (Section 4-2) should be described in the notes. Any evidence of detrital origin, as in gypsum sands, should be described. Associations of evaporites are described in Section 9-7.

**Iron-rich rocks** can be recognized broadly by their high density, their surficial alteration to yellow-brown or reddish hydrated iron oxides, and, commonly, by their magnetism. Four principal varieties are: (1) bedded arenite composed of oolites or structureless pellets of hematite, magnetite, chamosite, greenalite, or glauconite, and named accordingly (e.g., *oolitic* *hematite arenite*); (2) lutite that is commonly laminated or nodular and consists of various mixtures of siderite, hematite, hydrated iron oxides, silica minerals, iron sulfide, iron-poor carbonates, and clay; (3) *laterite*, an earthy to nodular, red, residual deposit formed by extreme chemical weathering in humid, warm climates; and (4) sandstone composed of magnetite, hematite, and ilmenite concentrated as placers in beach or stream deposits, and named as other sandstones (e.g., *magnetite-rich nonquartzose sandstone*). Origin of the first two kinds of iron-rich deposits has been discussed by James (1966).

**Carbonaceous and kerogen-rich rocks.** Carbon-rich rocks are classified and named according to extent of compaction and heat-induced changes, resulting in a *coalification series*:

1. Peat—surficial deposits of decomposed and partly humified plant debris.

2. Lignite—friable brown coal that cracks markedly on drying and commonly includes recognizable woody or leafy plant remains.

3. Sub-bituminous coal—black to dark brown somewhat friable coal that gives a brown streak, contains no recognizable plant remains, and is weakly jointed perpendicular to bedding.

4. *Bituminous coal*—black to dark brown hard coal that is strongly jointed perpendicular to bedding and commonly laminated by dull and brightly reflective layers (except for *cannel coal*, which, being rich in sapropel, has a greasy luster and is massive).

5. *Anthracite*—black, hard, typically massive coal with semimetallic luster and conchoidal fracture.

The usable coals contain up to 50% silicate and carbonate "ash" (nonflammable impurities). Rock with 50% to 80% ash is called *bone coal*, and with more than 80% ash *carbonaceous shale* (or limestone, etc.).

Rocks containing abundant kerogen (brown to yellow microscopic bituminous residues) are commonly called *oil shale*. Most, however, are kerogen-rich dolomite, dolomitic limestone, and siliceous lutite, all with fine laminations and a brown color imparted by the kerogen.

## 4-4. Textures of Igneous Rocks

The diagnostic textural features of igneous rocks are those implying the former presence of molten silicate liquid: (1) glass or its alteration products; (2) well-formed crystals of minerals that are typically anhedral in metamorphic rocks (as feldspars or feldspathoids), implying growth in a melt; or (3) relative ages (suggested by shape, sizes, and reaction relations or overgrowth relations among grains) that can be matched to a known crystallization sequence in magma of the same bulk composition (Fig. 4-12). Igneous textures may be grouped into four categories: *phaneritic* (mineral grains are large enough to be visible without magnification), *aphanitic* (grains are too small to be seen without magnification), *glassy*, and *fragmental*.

**Phaneritic rocks** with *granular* texture are composed of grains that are not distinctly aligned, whereas rocks with *fluidal* texture contain distinctly aligned platy or elongate grains. Either textural type may also be *porphyritic* if it contains grains (*phenocrysts*) that are much larger than other (ground-mass) grains. *Seriate* rocks have inequigranular textures in which grains range more or less continuously from large to small.

The adjective *hypidiomorphic* (as in *hypidiomorphic granular* texture) is used where grains show a range of perfection of crystal forms, such that some are euhedral, some subhedral, and others anhedral, suggesting a crystallization sequence (Fig. 4-13A). In some mafic and ultramafic rocks, early-formed subhedra look like loosely packed sand grains surrounded by later-formed anhedra, a texture that may have formed by mechanical accumulation fol-

| Melt composition | Crystallization range per mineral |
|---|---|
| Water-poor basaltic | — Mg olivine —          quartz, K feldspar —⟶ ___ <br> ——— pyroxene ——— <br> ——— plagioclase ——— |
| Aluminum-rich basaltic | ——— plagioclase ——— <br> ——— Mg olivine ——— <br> ——— clinopyroxene ——— |
| Water-poor andesitic | ——— pyroxene ——— <br> ——— plagioclase ——— <br> quartz, hornblende, K feldspar —⟶ |
| Water-rich basaltic or andesitic | — hornblende — — — — mt., biotite —— <br> ——— plagioclase ——— <br> ——— quartz — |
| Dacitic | — pyroxene, hornblende, biotite — — — — mt. — <br> ——— plagioclase ——— <br> ——— quartz ——— <br> — K feldspar — |
| Water-poor rhyolitic, alkaline | —— Fe olivine ——      biotite, alk. amphibole⟍ <br> ——— Fe pyroxene ——— <br> ——— K-Na feldspar ——— <br> ——— quartz ——— |
| Water-rich rhyolitic, calcalkaline | ——— magnetite, biotite ——— <br> ——— plagioclase ——— <br> ——— K feldspar ——— <br> ——— quartz ——— |

**Fig. 4-12.** Typical crystallization sequences of common silicate melts in the upper crust. Solid lines represent dominant range of crystallization and dashed lines possible additional ranges. Magnetite *(mt)* typically forms early in water-rich melts and may continue to form at later stages. The plutonic equivalents of the seven volcanic melt names are gabbro, plagioclase-rich gabbro, pyroxene diorite, quartz diorite (or quartz gabbro, or mafic tonalite), granodiorite, alkaline granite, and calcalkaline granite.

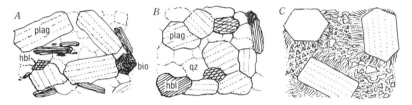

**Fig. 4-13.** *A.* Hypidiomorphic granular texture in granodiorite, indicating that plagioclase grew largely before quartz. Hornblende and biotite probably also grew before quartz; however, their shapes must be used with caution, because they may grow crystal faces against quartz and feldspar in the solid state (Table 4-3). *B.* Allotriomorphic granular tonalite. *C.* Granophyric groundmass with phenocrysts of alkali feldspar and quartz.

lowed by crystallization of interstitial melt (Fig. 4-14A). In *allotriomorphic granular* texture (Fig. 4-13*B*) all grains are anhedral, perhaps due to: (1) the minerals crystallizing simultaneously from a melt; (2) certain minerals continuing to grow until interstitial melt was used up or displaced (Fig. 4-14*B*); or (3) a hypidiomorphic rock recrystallizing when more or less solid.

The following grain-size terms may be useful in overall descriptions of phaneritic rocks: Smaller than 1 mm, *fine grained*; 1 to 5 mm, *medium grained*; 5 to 20 mm, *coarse grained*; and larger than 20 mm, *very coarse grained*.

**Aphanitic and glassy textures** are produced by rapid cooling or by rapid loss of dissolved vapor. These textures thus characterize most volcanic rocks and some shallow intrusions. The most rapidly solidified melts have *glassy* texture, and those with abundant microscopic crystals in glass have a waxy or resinous luster, as in *pitchstone*. Massive glass is generally called *obsidian*, whereas glass that separates into small pellets along a myriad of curving fractures is *perlite*. Glassy rocks with phenocrysts are *vitrophyric*.

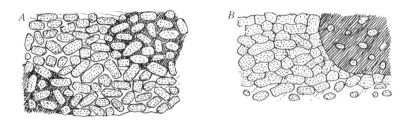

**Fig. 4-14.** *A.* Olivine subhedra (stippled) in large *poikilitic* crystals of pyroxene (dark) and plagioclase—possibly a cumulate texture. *B.* Olivine anhedra in pyroxene and plagioclase suggest partial reaction or solution of once-larger grains, whereas larger, closely packed olivine anhedra suggest continued growth of olivine grains like those in *A*.

*Aphanitic* rocks consist of crystals smaller than 0.25 mm in diameter and commonly accompanied by equally small patches or ramifying matrixes of glass. The abundant small crystals give the rock a stony, opaque appearance; however, the groundmass grains may be visible with a hand lens, especially in glass-poor basalt. The term *trachytic* may be used if the tiny plates and laths of the groundmass have a distinctly planar or linear fabric. Aphanitic rocks are *porphyroaphanitic* if they carry easily visible phenocrysts and *glomeroporphyritic* or *cumulophyric* if the phenocrysts tend to be joined in clots.

Vesicles may be so abundant as to constitute a textural aspect of glassy rocks, as in pumice, or of aphanitic rocks, as in scoria. Vesicles may also form ramifying interstices between small phaneritic crystals in basalts, a relation called *dikytaxitic* texture.

Aphanitic rocks may form from glassy rocks by *devitrification*, a change indicated by relict glass or by certain crystalline arrangements: (1) delicate, locally contorted *flow banding*; (2) *spherulites* (spherical stony bodies that may have internal radial structure and often lie along flow bands); (3) *lithophysae* (spherical stony bodies with concentric openings); (4) vapor-formed cavities lined by tridymite, cristobalite, magnetite, or hematite crystals; and (5) relict vitroclastic textures (see below). Fine-grained feathery or graphic intergrowths of quartz and alkali feldspar, called *granophyre*, form by devitrification or by late-stage crystallization of viscous melt or glass (Fig. 4-13C). Early devitrification of basalt, especially in the glassy skins of pillows, may form stony gray or green spherulites called *varioles*.

**Fragmental texture in phaneritic rocks** is important and widespread in some plutons (e.g., those of anorthosite), and is local or absent in others. Breccia, conglomerate, and finer fragmental rocks may be formed by explosions, by escape of vapor, by collapse, or by other displacements in nearly solid intrusions (Section 13-7). Nearly solidified margins of plutons may be pressed or deformed under simple shear so as to develop gneissose, schistose, or mylonitic textures. These rocks are sometimes called *protoclastic* when

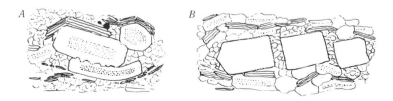

**Fig. 4-15.** Possible protoclastic textures. *A.* Bent feldspar (with dotted cores) and kinked or broken hornblende and biotite subhedra. *B.* Feldspar phenocryst that has been pulled apart and cemented by quartz and feldspar.

they result from deformation at a late stage of igneous crystallization (Fig. 4-15*A*). They may be distinguished from metamorphic rocks by late-stage igneous minerals that form small dikes or fill cracks in pulled-apart grains (Fig. 4-15*B*).

**Pyroclastic texture** results from any kind of volcanic process that leads to the eruption of fragments. The most common processes are volcanic explosions or continuous violent emission of volcanic gases. The textures are classified primarily on the basis of grain size (Fig. 4-16). The kinds of pyroclastic particles (all of which may be called *tephra*) also contribute to the texture, and they may be of one kind or a mixture of several. Because they are important genetic features, they should be listed or described whether or not they are used in the rock name (Fig. 4-17).

Additional pyroclastic textures result from compaction and welding of freshly deposited hot ash, and from vapor alteration of hot ash (Section 13-5).

## 4-5. Naming Igneous Rocks

Igneous rocks are named according to their texture and mineral composition or, if the rock is largely aphanitic, by an estimated mineral composition. Appendix 3 may be helpful in estimating amounts of minerals. Phaneritic rocks are classified by mineral composition in Fig. 4-18. Additional minerals that comprise more than 5% of the rock, or are genetically important, are used to modify the rock name, the most abundant being placed next to the name (as *biotite-hornblende granodiorite*).

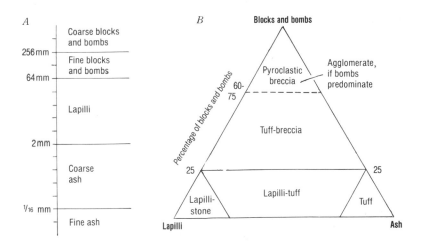

**Fig. 4-16.** Names for size categories of pyroclastic materials *(A)* and for pyroclastic rocks *(B)*. The dashed boundary in *B* is somewhat arbitrary (see Fisher and Schmincke, 1984, p. 92). *A* is after Fisher (1961) and *B* after Fisher (1966), with permission.

Several additional phaneritic rocks have special textures or compositions:

*Aplite*—typically equigranular and finer grained than 1 mm; contains less than 2% mafic grains; typically in thin dikes and other small bodies, often associated with pegmatite (Section 14-6).

*Pegmatite*—typically inequigranular, with many grains larger than 5 mm; textural and mineralogic variations patchy or in zones parallel to margins of body (Section 14-6).

*Granophyre*—groundmass consisting of feathery to graphic intergrowths of fine-grained feldspar and quartz; feldspar phenocrysts typical (Fig. 4-13C).

*Porphyry*—phenocrysts abundant and prominent in a phaneritic groundmass finer than 1 mm.

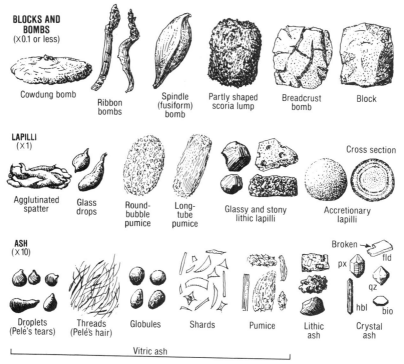

**Fig. 4-17.** Kinds of tephra (pyroclasts). In each row, viscosity increases from left to right. The cracked surface of breadcrust bombs is due to expansion of their interiors. Pumice and most shards result from vesiculation and disintegration of melt, and accretionary lapilli form by adhesion of fine ash in wet eruption clouds (Moore and Peck, 1962). The block and other lithic fragments are solid rocks derived from vent walls or beneath the volcano; some may be xenoliths brought up from great depths. Crystal ash may be of euhedra, as shown, but is more commonly of mineral fragments.

*Lamprophyre*—dark porphyry with abundant mafic minerals as phenocrysts and in the groundmass; biotite-rich *minette* can generally be recognized with a hand lens; other varieties may be named by their principal
mafic minerals, as *hornblende-pyroxene lamprophyre* (Section 14-6).

Igneous rocks that are largely aphanitic cannot be named directly from
Fig. 4-18 because some essential minerals may not have grown to visible size
before the magma solidified. However, if crystallization sequence is taken
into account (Fig. 4-12), phenocrysts can be used as a basis for field names
(Table 4-2). Additional suggestions of rock composition come from density
(heft), from primary structures, and from abundance of glass and other
features indicating viscosity of the original magma (Chapter 13).

Three kinds of altered volcanic rocks that retain many of their igneous
textural features are sometimes named specifically. *Spilite* is altered basalt
consisting of pale gray to greenish gray albite and abundant chlorite, often
with clinopyroxene unaltered and with calcite in amygdules. *Keratophyre* is
altered andesite or similarly plagioclase-rich rock consisting of the same

**Table 4-2.** Naming Volcanic Rocks on the Basis of Phenocrysts.

| | | |
|---|---|---|
| Rocks with quartz | Alkali feldspar : plagioclase ratio typically > 1:2; biotite or pyroxene generally < 5% | RHYOLITE |
| | Alkali feldspar : plagioclase ratio < 1:2 and alkali feldspar commonly absent; quartz may be scarce; hornblende, pyroxene, and biotite all likely | DACITE |
| Rocks without quartz, feldspathoids, melilite, or analcite | Alkali feldspar : plagioclase ratio > 4:1; biotite or pyroxene ± scarce olivine | TRACHYTE |
| | Alkali feldspar : plagioclase ratio < 4:1; hornblende, biotite, or pyroxene ± scarce olivine | LATITE |
| | Alkali feldspar absent; plagioclase abundant; pyroxene and (or) hornblende ± scarce olivine | ANDESITE |
| | Olivine and plagioclase abundant (high-alumina basalt), or pyroxene abundant and plagioclase and olivine abundant to scarce | BASALT |
| Rocks with feldspathoids, melilite, or analcite | Alkali feldspar abundant and > plagioclase; pyroxene, biotite, and amphiboles all possible | PHONOLITE |
| | Plagioclase abundant and > alkali feldspar; clinopyroxene abundant; no olivine | TEPHRITE |
| | Plagioclase abundant and > alkali feldspar; clinopyroxene abundant; with olivine | BASANITE |
| | Feldspathoids abundant; little or no feldspar; clinopyroxene abundant ± olivine | NEPHELINITE, etc. |

minerals as spilite but with mafic minerals distinctly subordinate. *Quartz keratophyre* is altered rhyolite or siliceous dacite, and is like keratophyre except for containing quartz phenocrysts or abundant quartz in the groundmass.

Pyroclastic rocks are named primarily according to grain size (Fig. 4-16), and secondarily by the predominant kind of fragmental material, which is used as an adjective: *vitric* (the sum of all glass fragments, including pumice); *crystal*; or *lithic* (the sum of all rock fragments other than pumice). A compositional name is added if possible (as *rhyolite vitric tuff*), and adjectives such as *welded* or *vapor-altered* as appropriate (Section 13-5).

### 4-6. Textures of Metamorphic Rocks

Metamorphic textures result from grain growth in solid rock, often during deformation, and from deformation of solid rock, which may be followed by recrystallization. Because the minerals grow simultaneously, they tend to include one another randomly rather than showing a sequential order like minerals in igneous rocks. Perfection of crystal form in metamorphic rocks

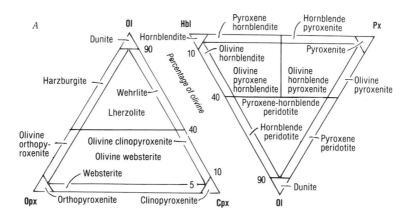

**Fig. 4-18.** Classification of phaneritic igneous rocks, based on proportions among the principal minerals. *A.* Ultramafic rocks (rocks with more than 90% mafic minerals). Ol, olivine; Px, pyroxene; Hbl, hornblende; Opx, orthopyroxene; Cpx, clinopyroxene. *B.* Rocks with less than 90% mafic minerals, composed mainly of quartz (Q), alkali feldspar (A), plagioclase (P), or feldspathoids (foids) (F). Modified slightly from Streckeisen (1973). To use the diagrams, (1) select the triangle with the appropriate minerals at its corners; (2) use the percentages of any two of these minerals to form a ratio (for example, if mineral A is 5% and mineral B is 20%, the ratio is 1:4); (3) use the ratio to locate a point on the appropriate side of the triangle (at a ratio of 1:4 it would be 1/5th of the distance from corner *B* to corner *A*; (4) do similarly for either of the other two mineral pairs; (5) draw or visualize lines passing from the two points to the opposite corners; and (7) use the intersection of the lines to plot or visualize a point, which will serve to name the rock.

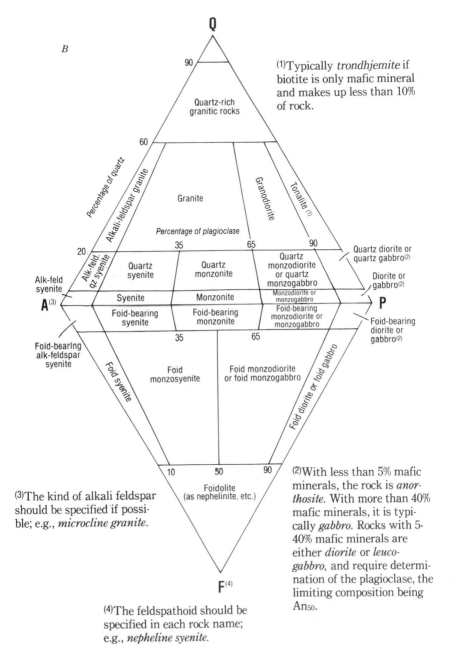

B

(1)Typically *trondhjemite* if biotite is only mafic mineral and makes up less than 10% of rock.

(3)The kind of alkali feldspar should be specified if possible; e.g., *microcline granite.*

(2)With less than 5% mafic minerals, the rock is *anorthosite.* With more than 40% mafic minerals, it is typically *gabbro.* Rocks with 5-40% mafic minerals are either *diorite* or *leucogabbro,* and require determination of the plagioclase, the limiting composition being An₅₀.

(4)The feldspathoid should be specified in each rock name; e.g., *nepheline syenite.*

typically depends on the growing strength of the mineral rather than the stage during which it crystallized. Because the degree of perfection is roughly predictable, it is a clue for recognizing metamorphic rocks (Table 4-3). Feldspars, for example, are typically euhedral or subhedral in igneous rocks and almost always anhedral in metamorphic rocks. Deformation gives metamorphic rocks another common textural characteristic: a linear and (or) planar fabric.

The principal metamorphic textures may be named as follows:

*Granoblastic texture*—phaneritic granular; grains approximately equidimensional and commonly of about the same size; if grain growth has been much more dominant than grain deformation, simple grain boundaries will predominate (Fig. 4-19*A*); grain growth during somewhat more rapid deformation will result in more irregular and platy or linear shapes Fig. 4-19*B*).

*Hornfelsic texture*—most metamorphic grains aphanitic or nearly so; large relict grains of the igneous or sedimentary protolith recognizable but recrystallized to fine aggregates; grains interlocked so tightly that rock is exceedingly difficult to break and rings like steel to hammer blows; broken surfaces are hackly at all angles to bedding or foliation and often have thin sugary coatings of rent grains that adhere so strongly they cannot be rubbed off.

*Schistose texture*—platy or elongate grains aligned and so abundant as to impart a fissility to entire rock, which splits in thin plates or elongate fragments and often forms tabular or elongate outcrops (Fig. 4-19*C* and *D*).

*Semischistose texture*—platy or linear grains less abundant or less perfectly aligned than in schistose rocks, so that rock breaks in unevenly platy or linear fragments; commonly developed in metamorphosed sandstone, conglomerate, and igneous rocks that show relict textural features (Fig. 4-19*E* and *F*).

**Table 4-3.** Crystalloblastic Series: Relative Degrees of Perfection of Crystal Forms in Metamorphic Rocks.

| | |
|---|---|
| Minerals typically euhedral against minerals lower in the series | staurolite, sillimanite, kyanite, rutile, chloritoid, ilmenite, tourmaline, pyrite, lawsonite |
| Minerals commonly euhedral against minerals lower in series | andalusite, garnet, sphene, epidote, zoisite, magnetite, other spinels, ankerite, idocrase |
| Minerals typically subhedral against minerals lower in series | micas and chlorites (platy forms), amphiboles and pyroxenes (prisms), wollastonite, dolomite, apatite |
| Minerals typically anhedral | quartz, feldspars, calcite, aragonite, olivine, cordierite, scapolite, humites |

*Cataclastic texture*—composed of brittle-fractured grains; unfoliated; typically the texture of a very poorly sorted breccia (Wise and others, 1984).

*Mylonitic texture*—groundmass aphanitic or nearly so and typically foliated but generally not fissile; typically with angular or rounded relicts (*porphyroclasts*) of protolith; surfaces broken parallel to foliation are commonly knobby due to porphyroclasts and commonly lineated by compositional streaks and minute folds.

*Polymetamorphic texture*—complex textures that may show age relations between two or more periods and perhaps two or more kinds of metamorphism (Sections 15-2 and 15-8).

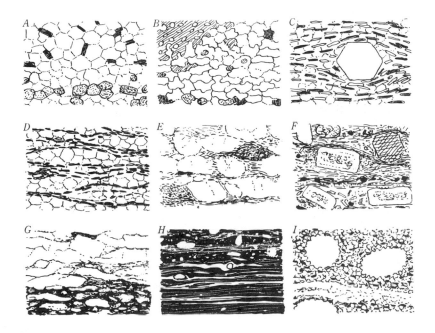

**Fig. 4-19.** Metamorphic textures, as seen through a hand lens. *A*. Granoblastic texture, in part a mosaic texture (with many triple junctions of grains with interboundary angles of around 120°). *B*. Granoblastic texture of irregular grains, with a large poikiloblast in the upper left. *C*. Schistose texture with a euhedral porphyroblast. *D*. Schistosity imparted by small flakes and lenticular granoblastic domains. *E*. Metasandstone with fine mica in the matrix, imparting a semischistose texture. *F.* Semischistose texture due to fine chlorite and actinolite in the groundmass of a blastoporphyritic metabasalt. *G*. Somewhat mylonitic granite passing downward into protomylonite. *H*. Orthomylonite passing downward into ultramylonite. *I*. Granoblastic texture in a blastomylonite, the latter indicated by the recrystallized margins of relict porphyroclasts (enlarged about 5 times relative to *H*).

**Grain sizes** have not been standardized for metamorphic rocks but these may be useful: less than ¼ mm, fine grained; ¼ to 1 mm, medium grained; 1 to 2 mm, coarse grained; and more than 2 mm, very coarse grained. Using actual sizes is often preferable because they give a firmer basis for judging specific cases, as degree of grain growth or completeness of mylonitization in a given area.

Inequigranular textures are of several kinds:

*Porphyroblastic*—where the larger grains (porphyroblasts) grew during metamorphism (Fig. 4-19*C*).

*Poikiloblastic*—where porphyroblasts include many small mineral grains (Fig. 4-19*B*).

*Porphyroclastic*—where the large grains (porphyroclasts) are relics in a fine mylonitic groundmass (Fig. 4-19*H*).

*Blastoporphyritic*—where the larger grains are relict phenocrysts of an igneous protolith (Fig. 4-19*F*).


### 4-7. Naming Metamorphic Rocks

Although certain rock names may connote specific metamorphic conditions or associations (Williams and others, 1982), field names should probably be based only on texture, thus:

#### Rocks with granoblastic texture

a. *Granofels*—a name that can be used for any granoblastic rock (Goldsmith, 1959). *Granulite* has been used in a similar way but is less desirable because it tends to imply certain metamorphic conditions.

b. *Skarn*—granoblastic rock, often of uneven grain size, consisting of calcsilicate minerals, especially garnet, clinopyroxene, and epidote (Section 15-7).

c. *Marble, quartzite, amphibolite*—rocks of specific mineral composition that are taken to be granoblastic unless otherwise modified, as *schistose marble*, and so on.

#### Rocks with hornfelsic texture

a. *Hornfels*—the basic rock name for all hornfelsic rocks; where relict features are well-preserved, however, the adjective *hornfelsic* might be used with the name of the protolith, as *hornfelsic metarhyolite*.

#### Rocks with schistose texture

a. *Schist*—grains are phaneritic.

b. *Phyllite*—microscopically schistose, as indicated by fissility and sheen caused by alignment of platy or acicular grains.

c. *Slate*—microscopically schistose; splits readily in sheets (or elongate fragments if lineated); cleavage surfaces have a dull luster.

### Rocks with semischistose texture

*a. Semischist*—basic name for the group; however, rocks with relict features can be named as modified protoliths, as *semischistose metarhyolite*.

*b. Gneiss*—most parts of the rock are granoblastic, but separate folia or elongate groups of minerals impart a crude cleavage and a semischistose aspect (Section 15-5).

### Rocks with cataclastic texture

*a. Fault gouge*—grains chiefly finer than sand; coherent but may be friable or plastic when wet; may become silicified or altered otherwise.

*b. Fault breccia*—coherent but may be somewhat friable (unless recrystallized or cemented); fragments have a large range of sizes.

### Rocks with mylonitic texture (after Wise and others, 1984)

*a. Mylonite*—the general name for the group.

*b. Protomylonite* —fine, tough matrix that is at least locally foliated and includes at least 50% of porphyroclastic mineral grains or lenses and chunks of the protolith (Fig. 4-19*G*).

*c. Orthomylonite*—fine, tough matrix that tends to be foliated strongly and includes 10% to 50% porphyroclasts (Fig. 4-19*H*, *top*).

*d. Ultramylonite*—same but with less than 10% porphyroclasts (Fig. 19*H*, *bottom*).

*e.* Varieties of mylonite with visibly schistose or granoblastic matrix have also been called *blastomylonite* (i.e., a mylonite with large amounts of late-stage grain growth) (Fig. 4-19*I*).

### Rocks dominated by relict textures

Nonschistose low-grade rocks that are not hornfelsic and are dominated by relict textures generally are given names such as *metarhyolite, massive metasiltstone*, and so on.

***Mineral modifiers.*** Metamorphic rock names are completed by adding the names of the principal minerals before the rock name. They are placed in order of abundance with the most abundant next to the name. A *diopside-plagioclase-hornblende granofels* is thus a granoblastic rock in which hornblende is most abundant, plagioclase less so, and diopside least abundant.

The mineral names are important because they may indicate metamorphic grade or facies. Some, as the *index minerals* of metamorphic mineral zones (Section 15-3), should be included in rock names even if they are scarce. The minerals in the name should be those that formed together at the most advanced stage of metamorphism. Relict and retrograde minerals may be included by brief modifying statements, as *partly chloritized garnet-biotite-quartz schist.*

***Rocks affected strongly by metasomatism*** commonly have variable textures and include abundant veins, coarse clots, and zoned bodies (Section 15-7). *Skarn* is described above and in Section 15-7, and a few other metasomatic rocks have been named specifically: *serpentinite* (consisting mainly of serpentine minerals and formed typically from peridotite or pyroxenite); *greisen* (quartz-muscovite rock, often with topaz; see Section 15-7); *adinole* (pale, fine-grained albite-rich rock typically formed near diabase bodies); and *rodingite* (white to pale green, heavy rock consisting mainly of hydrogrossularite, tremolite, and chlorite, and occurring in serpentinized peridotite near contacts). Other metasomatic rocks may be named descriptively, as *coarse quartz-schorl rock, silicified tuff*, and so on.

## References Cited

Dunham, R. J., 1962, Classification of carbonate rocks according to depositional texture, p. 108-121 *in* Ham, W. E., editor, *Classification of carbonate rocks*: American Association of Petroleum Geologists Memoir 1.

Fisher, R. V., 1961, Proposed classification of volcaniclastic sediments and rocks:- *Geological Society of America Bulletin*, v. 72, p. 1409-1414.

Fisher, R. V., 1966, Rocks composed of volcanic fragments and their classification: *Earth-Science Reviews*, v. 1, p. 287-298.

Fisher, R. V., and Schmincke, H.U., 1984, *Pyroclastic rocks*: New York, Springer-Verlag, 472 p.

Folk, R. L., 1959, Practical petrographic classification of limestone: *American Association of Petroleum Geologists Bulletin*, v. 43, p. 1-38.

Folk, R. L., 1962, Spectral subdivision of limestone types, p. 62-84 *in* Ham, W. E., editor, *Classification of carbonate rocks*: American Association of Petroleum Geologists Memoir 1.

Goldsmith, R., 1959, *Granofels*, a new metamorphic rock name: *Journal of Geology*, v. 67, p. 109-110.

Isaacs, C. M., 1981, *Guide to the Monterey Formation in the California coastal area, Ventura to San Luis Obispo*: Pacific Section American Association of Petroleum Geologists, v. 52, 91 p.

Isaacs, C. M., 1982, Influence of rock composition on kinetics of silica phase changes in the Monterey Formation, Santa Barbara area, California: *Geology*, v. 10, p. 304-308.

James, H. L., 1966, *Chemistry of the iron-rich sedimentary rocks*: U.S. Geological Survey Professional Paper 440-W, 61 p.

Moore, J. G., and Peck, D. L., 1962, Accretionary lapilli in volcanic rocks of the western continental United States: *Journal of Geology*, v. 70, p. 182-193.

Pettijohn, F. J., 1975, *Sedimentary rocks*, 3rd edition: New York, Harper & Row, 628 p.

Powers, M. C., 1953, A new roundness scale for sedimentary particles: *Journal of Sedimentary Petrology*, v. 23, p. 117-119.

Streckeisen, A. L., chairman, 1973, Plutonic rocks: classification and nomenclature recommended by the IUGS Subcommission on the Systematics of Igneous Rocks: *Geotimes*, v. 18, no. 10, p. 26-30.

Williams, H., Turner, F. J., and Gilbert, C. M., 1982, *Petrography: an introduction to the study of rocks in thin section*, 2nd edition: San Francisco, W. H. Freeman and Co., 626 p.

Wise, D. U., and seven others, 1984, Fault-related rocks: suggestions for terminology: *Geology*, v. 12, p. 391-394.

# ■ Mapping Rock Units and Structures

## 5-1. A Geologic Pace and Compass Traverse

This chapter describes procedures used in finding, walking out, and defining rock units and structures, regardless of the base on which they are mapped. Methods of locating and plotting these features on topographic maps and aerial photographs are the chief topics of Chapters 6 and 7, respectively, and this section describes mapping by a compass traverse at scales larger than topographic maps and aerial photographs typically permit. A compass traverse is a suitable introduction to geologic mapping in general, because it is often used to explore, measure, and describe rock units that may then be mapped more widely and at smaller scales. The map resulting from the traverse may be only a narrow strip (Fig. 5-1); however, additional traverses could be used to expand the map. Mapping by closed traverse loops is described in Section 6-3, as is the use of triangulation to control large-scale mapping with a compass. Another method to consider for unusually detailed maps of large outcrops is that of laying out a grid directly on the ground (Section 3-4). Plane table mapping is another alternative for detailed studies; however, it would require more than one person, expensive equipment, and additional time (Chapter 8).

The purpose of the traverse described here is to investigate and measure a sequence of layered rocks, although it could also be used to explore any rock body or structure in detail. If the rocks to be studied are formations or other formal (named) units, Section 5-3 should be read at the outset. The first step in the field is to find a place where the rocks are well exposed along a course that is roughly perpendicular to the strike of bedding or other rock layers. The survey itself consists of measuring the bearing and distance from a

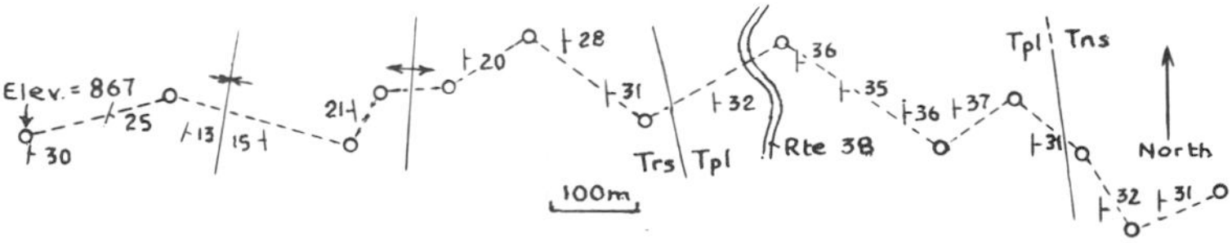

**Fig. 5-1.** Map of traverse across several formations (indicated by letter symbols), with small circles marking traverse stations and dashed lines traverse legs. The scale of the figure is about one-third that of a typical field sheet.

station at one end of the course to some station ahead, and from that station to another, and similarly to the far end of the course (Fig. 5-1). As the figure suggests, geologic features are examined and plotted sequentially along the traverse, thus forming a skeletal geologic map.

A reconnaissance will usually be needed to select the best course for the traverse and to examine the rocks in a preliminary way. Because measurements will be made by pacing, the course should be as unobstructed and smooth as possible, as along a road, trail, beach, or open ridge. Stream courses may provide abundant exposures but are commonly too overgrown for pacing. Compass traverses along railroads and near power lines are likely to be inaccurate because of magnetic anomalies.

The rocks are reconnoitered in order to subdivide them into more or less uniform units and to determine if any of their contacts is a fault. A decision should also be made as to the smallest features that are to be plotted to scale on the traverse map and sections, because this size will prescribe the minimum scale of the traverse map. Units and features less than 2.5 mm (0.1 in.) across on a map are difficult to plot accurately; thus if the thinnest unit to be mapped is 10 m across on the ground, the map scale can be no smaller than 1:4000 (1 cm = 40 m or 1 in. = 333 ft).

The equipment needed for the survey will vary somewhat with the kinds of rocks or deposits being studied and can be selected from the descriptions in Chapter 2. Good quality 8.5 × 11 in. quadrille paper is suitable for plotting a traverse map, and line-ruled paper for taking notes. Basic operations with a compass and clinometer are described in Chapter 2 and methods of studying outcrops and taking notes in Sections 3-1 to 3-3.

The traverse can be started at either end of the course, but differences in elevation can be measured more easily when traversing upslope. The ends of the traverse should be marked with stakes or firmly set stones so that they can be located easily. Traversing may then proceed by these steps:

1. Standing at an end station (station 0) sight ahead to select the clearest course for the first traverse leg, then walk ahead to the farthest point from which station 0 is visible.

2. Mark this point (station 1) with a stake or stone and read a bearing back to station 0, using the first method described in Section 2-5.

3. Record the bearing, and pace and record the distance back to station 0.

4. At station 0, read the bearing to station 1, record it, and pace back to station 1; if the bearings agree within about 1° and the paced distances within 1 part in 100, use their averages to plot the first traverse leg. Use the rulings on the quadrille paper as north-south and east-west lines, and plot the first leg so that the entire traverse will fall on the sheet.

The survey can then be continued by similar steps, or the geology along the first leg can be studied and mapped at once. The choice between the two procedures is determined by the uniformity of the rock units and the amount of detail to be plotted. If units are uniform and thick, the traverse legs should be surveyed at least as far as the first contact before describing the rocks. If the sequence consists of a variety of thin units, or if many details in thicker units are to be mapped, the rocks should probably be examined and described as each traverse leg is completed.

Geologic features next to the traverse course are located by the paced distance to that point. Outcrops more than several meters from the traverse course generally require offsetting by a compass bearing and a paced distance.

Section 3-1 describes steps in examining the rocks themselves. Contacts between rock units are plotted as lines (Fig. 5-1). Strike and dip of bedding should be measured and plotted wherever attitudes change significantly, such as by 10° in strike or 5° in dip. Folds will be indicated by bedding symbols and should also be mapped by the bearing and plunge of their hinge lines (Sections 5-4 and 12-3). Sections 5-4 and 12-5 describe ways of recognizing faults, which are plotted as lines with dip arrows and other symbols indicating the sense of displacement along them (Appendix 7).

Notes taken on traverses are traditionally numbered by the distance from the last traverse station. Thus, data observed 21 m along the first traverse leg would be numbered 0 + 21, and an outcrop 8 m along the second leg, 1 + 8. Section 3-3 describes note-taking and systematic descriptions of rock units.

A ground profile will be needed for the cross section, and it is measured when the map is complete enough so that a section line can be selected and drawn on it. The location of this line is a compromise between two needs: (1) it should be about perpendicular to the strike of bedding or other planar structures, and (2) it should lie as close as possible to the measured data. If the traverse is along a straight road or ridge that crosses the beds at about right angles, the profile can be surveyed by measuring differences of eleva-

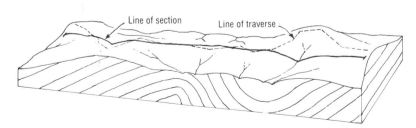

**Fig. 5-2.** Traverse (dashed) and cross section line (solid) used to measure a sequence of folded strata exposed along a high ridge.

tion along the traverse, as by hand leveling (Section 2-7). In the more usual case, the traverse is sinuous, so that the profile must either be measured by a separate survey of elevations along the section line or be estimated by rough measurements of slope angles made from the traverse course (Fig. 5-2). The second procedure is less time-consuming and is typically accurate enough for rock sequences that are structurally simple. In every case, an elevation above sea level should be carried to the traverse by leveling from a nearby bench mark or by estimating from a topographic map.

The geologic information obtained by the traverse can be presented as a cross section and columnar section arranged with the traverse map on one plate (Fig. 5-3). The constructions should be made first in pencil. The equipment needed includes a drawing board, a large drafting triangle, a T-square (or a second triangle), a 3H pencil and eraser, a ruling pen (or any pen that will give an even line when guided along the edge of a triangle), black waterproof ink, an accurate scale, and a protractor. Lettering guides and a contour pen (for drawing evenly weighted freehand lines) may be helpful.

The following steps are suggested:

1. Tape the traverse map to the drawing board and upon it tape a second piece of paper for the cross section. On the latter, draw a pencil line exactly parallel to the section line on the map; this will be the base line of the cross section. Make short ticks (cross lines) at the ends of the section line on the map, and project the two ends to the cross section (Fig. 5-4A). Use the elevation data from the survey to plot the ground profile about 3 cm above the base line.

2. Check the notes and map to be sure that all data have been plotted, and

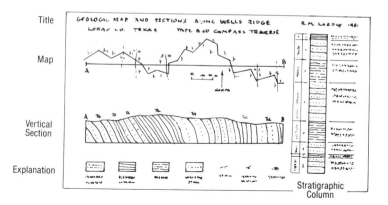

**Fig. 5-3.** Layout of a traverse map and constructed sections.

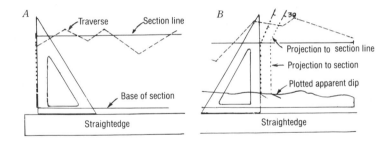

**Fig. 5-4.** *A.* Using a triangle and straightedge to project end of cross section line to end of section. *B.* Projecting structures from the section line to the profile of the section.

then project each structure to the section line on the map, curving the projection lines as necessary (Fig. 5-5*A*).

3. From each of the points thus marked on the section line, project the structures to the ground profile of the cross section, as shown in Fig. 5-4*B*.

4. Use a protractor to plot the dip of each structure, making lines about 1 cm long at each of the points just transferred (Fig. 5-4*B*). Apparent dips must be used for structures that strike oblique to the section, and they are defined and tabulated in Appendix 13. Fig. 5-5*B* shows how to obtain dips along curving strike lines.

5. If contacts or faults strike approximately parallel to the section line and thus do not cross it, they may project into the cross section below the surface. Section 6-5, step 9, describes how to add them to the cross section.

6. Complete the geology of the cross section by extending bedding and other structures to the base of the section. Lines that cannot be extended with reasonable confidence may be dashed, questioned, or omitted.

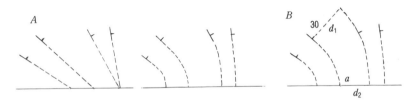

**Fig. 5-5.** *A.* Projected surfaces that converge unnaturally *(left)* are probably parts of a fold and should be curved *(right)* so as to maintain unit thickness as nearly as possible. *B.* To determine the dip at the section line, calculate a thickness from $d_1$ by the relation $t = d_1 \times \sin \angle$ dip; then use this thickness and the distance $d_2$ to calculate the dip at the section line *(sin $\angle$ dip $= t/d_2$).* For the case shown, the dip at *a* is approximately 40°.

7. Figure 6-8 shows other additions that may be appropriate.

8. If strike of bedding is about perpendicular to the section line on the map, scale thicknesses of units directly from the cross section. Otherwise, measure the outcrop widths of the units perpendicular to strike, and calculate thicknesses by the appropriate equation in Fig. 11-12.

9. Using the unit thicknesses, construct a stratigraphic column in pencil by starting at a ruled base and adding the units in order of decreasing age. Use as large a scale as the final plate will permit so that the rocks can be shown in as much detail as possible. Graphic symbols for rocks are shown in Appendix 8, and Fig. 11-15 illustrates the general style and specific parts of a stratigraphic column. Important items to include are a vertical scale; a brief description of each unit, which is lettered to the right of the column; and the rock-unit names and ages on the left. An explanation will not be necessary if the unit descriptions cover all rocks plotted in the column.

10. Assemble the pencil copies of the three parts of the plate so that the base of the cross section is parallel to the section line on the map, and the stratigraphic column is perpendicular to it (Fig. 5-3). Tape a piece of tracing film over them with its base parallel to the section line and trace the illustrations in ink. Add at least these items: (1) a title at the top of the plate (as *Cross Section and Stratigraphic Column of Middle Miocene Rocks Along Shady Ridge*); (2) your name; (3) date of the project; (4) the method used (as *By pace and compass traverse*); (5) a bar scale for the map and cross section; (6) a north arrow next to the map; (7) elevations at the ends of the cross section; (8) brief labels or symbols for all features plotted on the map; and (9) an explanation that identifies all symbols (see Appendixes 7, 8, and 9).

## 5-2. Finding and Tracing Contacts Between Rock Units

Any rock that forms bodies large enough to plot on the base being used may serve as a mapping unit. Rock units are mapped by walking out their contacts with adjoining units and by locating and drawing the contact traces as lines on the map or photograph. This procedure should not be delayed until all outcrops have been visited, nor until contacts are "understood" and ideal rock units have been selected (Section 5-3). Rock units are bound to be refined and perhaps redefined drastically during the course of mapping. The most productive procedure at the outset is to start mapping obvious contacts, and thus let the map take form with as little bias as possible.

A sharp contact between two distinctly different rocks is an ideal one to start mapping. In the usual situation the contact will be exposed only locally, so that it must somehow be followed across areas where it is concealed. If outcrops are fairly closely spaced, the contact can be followed by walking a zig-zag course in the general direction predicted by the strike and dip of the contact where it is exposed. The strategy is to locate all outcrops of the two limiting rock units. If the limits so defined are less than 10 m apart, and the base map has a scale of 1:24,000, the contact can be plotted as precisely as if it were totally exposed.

If outcrops are a good deal farther apart, the position of the contact must be worked out by using accessory evidence. The commonest indicators are float fragments of one or both rock units, which can be used even where downslope creep has displaced them (Fig. 5-6A). Weakly resistant rocks that rarely form float may be brought up at animal burrows or may leave residual concretions or fossils in the soil. The soil itself commonly varies from one rock unit to another, often by containing specific refractory mineral grains, such as quartz grains that might mark the position of a granite or a sandstone next to a gabbro or a limestone. Other soil properties, such as color, texture, and compactness may also be useful in certain kinds of soils (Section 10-9). Vegetation may differ across an unexposed contact, and this relation is generally more visible from a distance or by examining aerial photographs. Concealed contacts may also be indicated by a change in the angle of slope or by aligned springs or patches of wet ground.

A sharp, well-exposed contact is drawn on the map by a solid thin line. Irregularities that cannot be drawn to scale must be generalized, but otherwise the line shows the contact's exact position. Strike and dip of the contact surface should be measured where possible and plotted by the symbols shown in Appendix 7. If outcrops and other firm evidence are so widely spaced that the contact line may be in error by 1 mm (1/25 in.) on the map (24 m or 80 ft on the ground at 1:24,000 scale), the line can be dashed to indicate the uncertainty. Dashes approximately 3 mm (1/8 in.) long are commonly used where the uncertainty is under 2 mm (1/10 in.) on the map, and dashes 1 mm (1/25 in.) long are used for less certain contacts (Appendix 7). Dashed

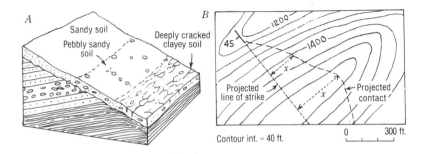

**Fig. 5-6.** *A.* Using upper limits of pebbles and *(below)* of clay-rich soil to map two contacts. *B.* Strike and dip measured near a contact *(upper left)* are used to project the contact ahead by the relation $x = diff.$ *in elevation* $\div$ *tan* $L$ *dip.*

contact lines are not always used on published maps because of drafting
costs or the difficulty of scribing dashes, but they are valuable as a field
record.

**Use of strike and dip to locate contacts** is helpful where one or both units
are bedded or foliated parallel to their contact. The method is based on the
assumption that the strike and dip remain the same for some distance
viewed ahead. In open country the procedure is as follows: (1) stand on the
trace of the contact, facing in the direction of a line of strike measured
nearby; (2) set the clinometer at the measured dip and hold the compass as
though measuring dip, sighting ahead along the projected trace of the contact; (3) visualize the planar surface in the mind's eye and thus estimate
where it would cross the topographic forms that lie ahead; and (4) walk
ahead on and near this line, searching in places where the units might be
exposed. In areas where one cannot get a view ahead along the projected
contact, a predicted line can be constructed on a topographic map by using
strike and dip (Fig. 5-6*B*). Neither of these methods should be used to map a
contact, for the contact might turn or be displaced along a fault at any point.
They help, however, in the search for crucial outcrops or other indications
of the contact's actual position.

**Units too small to map to scale** include many dikes, sills, intrusive pipes,
veins, tuff beds, and thin sedimentary units, such as coal seams, fossiliferous beds, and conglomerate beds. Where tabular, these units are plotted as
single lines that represent their mapped traces. Color pencils may be used to
distinguish different kinds of units. Where dikes or other thin units are too
closely spaced to be plotted separately, lines may be used to represent
swarms of more or less parallel units. Intrusive pipes and other nontabular
rock bodies can be plotted as small circles, ellipses, or lenses that can be
identified by a letter symbol or a spot of color. Because the size of all these
units is exaggerated, their actual dimensions must be entered in the notes.

**Gradations between rock units** are as important as sharply defined contacts. Sharp contacts imply distinct changes in conditions when the rocks
were formed or abrupt changes in the materials being emplaced at a given
site. In sedimentary rocks they suggest a hiatus in deposition and should
always be examined for evidence of unconformity (Section 9-6). Gradations,
on the other hand, imply continuity of accumulation, or materials already in
place becoming mixed with newly deposited materials. Gradations resulting
from gradual changes during accumulation are especially common in sedimentary rocks but apply to other kinds of sequences, such as lavas that
become successively more silicic due to evolution of magma at depth.

Gradations between rock units must be mapped by criteria that are considered carefully and used consistently. Where a single contact line is drawn,
it is generally located in the middle of the gradation. In some cases, a simple
physical criterion can be used as the basis for the line. For example, if sand-

stone grades upward into shale through a sequence of interbedded sandstone and shale, the line might be placed at the top of the highest sandstone bed exceeding a certain thickness. Units that grade to one another normal to bedding, however, are also likely to grade laterally, and thus a single bed such as that just mentioned is not likely to retain the same thickness for a great distance. Figure 5-7 illustrates a typical case.

Some broadly gradational relations may be mapped by plotting two contacts, one where each unit begins to show changes from its normal characteristics. The two lines thus encompass the total zone of gradation and if desirable a midline can be drawn as the final contact line (Fig. 14-1). Spots of color may be effective in marking the limits of each rock because they are readily distinguishable from other penciled data in the same area.

Compton (1962, p. 64) suggested using a thin band of hachures for gradational contacts. Although generally not used in published maps (perhaps because it is not a crisp line), this symbol may make field maps more informative. Appendix 7 shows two other line symbols that have been used for gradational contacts.

### 5-3. Refining and Correlating Geologic Units

Mapping is best started with units that seem obvious, natural, and suit the scale of the mapping, rather than forcing contacts to fit formations mapped in other areas. As a study progresses, however, these initial units are typically modified for several possible reasons. Thin units may have to be combined to complete mapping on schedule. A thick unit that seemed uniform may prove to be separable along an unconformity proven by fossils of greatly different ages. Some units may be impossible to map throughout the area because of lateral changes or inconsistent contact relations. Still others may prove to be entirely secondary, as a dolomite body superimposed across one or more limestone units.

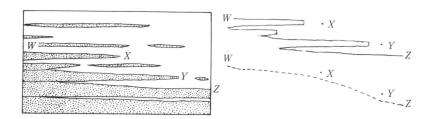

**Fig. 5-7.** A vertical and lateral gradation between sandstone and shale units *(left)* can be mapped at the top of sandstone beds with a certain minimal thickness *(upper right)* or generalized by a dashed line.

An important question to consider is whether or not mapped units correspond to formations or other units established elsewhere in the region. Established (*formal*)units may or may not be ideal (truly useful), but one should lean toward using them in order to keep regional geologic relations as simple and clear as possible. References to the names, original descriptions, and current usage of formal units in the United States have been computerized and are available by phoning, writing, or visiting the Geologic Names Committee representative in any regional office of the U.S. Geological Survey. Information on units in other countries is generally available through that nation's geological survey and is usually also available through provincial geological surveys.

Rock units that prove useful but have not been formalized by the rules of the stratigraphic code are generally termed *informal*. Formal or informal, the only units suitable for the sort of mapping described in this book must have lithic characteristics or distinctive contacts that can be recognized in the field. The latest stratigraphic code (North American Commission on Stratigraphic Nomenclature, 1983) proposed four categories of units that meet these requirements, and they are defined briefly below:

1. *Lithostratigraphic units* are composed of sedimentary rocks, extrusive volcanic rocks, or metamorphic variations of unmistakable sedimentary or volcanic protoliths. The basic formal unit (the one generally used in mapping) is the *formation*, and it may be divided into *members* and may be part of a unit called a *group*, which is composed of two or more formations..

2. *Lithodemic units* are composed of intrusive rocks, of highly deformed rocks, or of metamorphic rocks too metamorphosed to be mapped as sedimentary or volcanic protoliths. The basic formal unit is the *lithodeme*, which may be divided into informal subunits and may be part of larger lithodemic units called *suites*.

3. *Allostratigraphic units* are composed of sedimentary or extrusive volcanic rocks that are lithologically indistinguishable from adjoining units but are mappable because of distinctive contacts that are typically unconformities. The basic unit is the *alloformation*, which may be divided into *allomembers* and may be part of an *allogroup*.

4. *Pedostratigraphic units* are buried soil profiles or parts of profiles, and are too thin to map to scale in most studies but are important stratigraphically (Section 10-9).

*Biostratigraphic units* and *magnetostratigraphic units*, also defined in the 1983 code, are of great value in determining geologic age but require intensive laboratory study of samples and are thus impractical for typical mapping in the field.

The rules for naming and defining these various kinds of units are too comprehensive to include here; moreover, they are likely to be modified from time to time in succeeding stratigraphic codes. If the current code cannot be read before or during the field season, yet certain units being mapped seem likely to be proposed as new formal units, the following should be done in the field:

1. A type section should be selected and measured (Chapter 11) if the unit is lithostratigraphic or allostratigraphic, and a type locality or area should be selected if it is lithodemic or pedostratigraphic. This selection is based on two needs: (1) the rocks must be truly representative of the unit, and (2) they should be as well exposed as possible.

2. The unit should be described fully, as suggested in Section 3-3 and parts of Chapters 9, 10, 13, 14, and 15.

3. The contacts should be mapped as extensively as possible and studied and defined exactly.

4. Fossils and rocks that can be used for assigning geologic or numerical ages should be collected.

All rules in the code can then be met after the field season.

**Correlation** consists of working out the equivalence of rock units from one area to another, based on one or more of these criteria: age, stratigraphic position, lithologic characteristics, and fossil content. Correlations are generally made when a field study has reached the stage that local rock units have become well enough known to be compared with units established elsewhere.

Fossil collections and isotopic data are typically the surest means of correlating units by age; they can be in error, however, so that equivalence should be checked by as many of the following as possible: (1) specific thin units that can be identified with certainty, such as a tuff bed, a limestone bed, a phosphate-rich bed, a claystone, or a layer rich in unusual detrital minerals; (2) geomagnetic polarity; (3) climatic markers such as glacial deposits (Section 10-5), specific kinds of paleosols (Section 10-9), or fossils indicative of changes in ocean temperature; (4) durations of time expressed by varves or other annual layers in sediments or fossils; (5) intrusive or extrusive igneous rocks recording unique igneous events; (6) evidence of rapid marine transgression or regression, such as sequences of primary structures like those described in Section 9-8; (7) major unconformities; (8) unique sequences of lithologic units; (9) tectonic events, as indicated by deformed rocks or by rapidly accumulated coarse sediments; and (10) age relations among deposition, deformation, metamorphism, and igneous activity. Although all of these means of correlation can be fallible, certain ones may work well in a given area. Additional age-relations that may be useful are described in Chapters 9, 10, and 12 through 15.

*Geochronology* consists of assigning specific geologic ages (geochronologic units) to rocks, generally on the basis of fossils or numerical ages. Appendix

10 can be used to assign geologic ages on the basis of numerical ages; however, some of the numbers may be modified from time to time. Lacking suitable materials, ages can be assigned tentatively by other means (see *correlation*, above).

### 5-4. Mapping Geologic Structures

Folds, faults, unconformities, and intrusions are indicated by the mapped shapes of rock contacts and by patterns of strike-and-dip symbols or plunge symbols of lineations. Adjacent symbols that are not parallel indicate a structure of some kind and imply a need for additional information. Structural attitudes should thus be recorded in the notes and plotted on the map or aerial photograph as soon as they are measured. Strike and plunge lines can be plotted quickly by orienting the protractor as shown in Fig. 6-1*A*. As soon as it is plotted, each symbol must be checked by holding the map or photograph so that it is oriented parallel to the terrain, and by comparing the symbol with the outcrop. Does the bed dip in the direction plotted? Is the strike line in the right quadrant?

In structurally disturbed areas each symbol must be located so that the place occupied is exactly at the point where the strike and dip lines join, or is at the back end of a plunge arrow.

Nonparallel structure symbols commonly correlate with tectonic structures (see below) but may be due entirely to: (1) soft-sediment or other intraformational folding; (2) landslides; and (3) creep. Closely spaced structure symbols may also differ because of variously inclined initial dips, such as those of channel-facies sandstone in a small, steep alluvial fan, or of tephra on the flanks of a cinder cone. In areas where structural attitudes are moderately variable, the overall attitude of a contact or layer-surface can be determined by the three-point method (Section 3-5).

**Folds** of large size will generally be shown by the curving or zig-zag traces of mapped contacts. Individual distinctive beds (*key beds*) or intraformational contacts may have to be mapped additionally to work out the folds' full shapes and positions. Folds within single rock units are suggested by opposed dips of structure symbols, and these suggestions can be verified by finding small-scale fold hinges or cleavage-bedding relations, which are generally coaxial with large folds (Fig. 5-8). These relations and other suggestions for working with folded rocks are described in Chapter 12. Section 3-6 describes measurement of lineations, and Appendix 7 shows line symbols used for folds on maps.

**Faults** are rarely seen in natural exposures because the broken or weakened rocks along them are preferentially weathered and eroded, especially along faults with large displacements. In fairly well exposed areas, faults may be suggested at the outcrop by: (1) rock units or parts of units that are clearly out of place; (2) sudden and unlikely changes in lithology; (3) abrupt changes in orientation of bedding, foliation, or lineations; (4) brecciation; (5) sudden appearance of altered rocks; (6) abrupt increase in jointing; and (7) fracture surfaces with local slickensides or other indications of displacement. As the mapping progresses, faults with large displacements will become obvious on the map, which will show one or another of: (1) abrupt offsets of rock units; (2) repetition of units or parts of units; (3) missing units or parts of units; or (4) angular discordance of bedding or other kinds of layering. Faults must be walked out completely, for they may turn abruptly, may terminate in another fault, may die out in flexures or splays of many minor faults, or may pass into bedding-plane faults that cannot be located by mapping rock-unit contacts.

A variety of features may suggest faults in poorly exposed areas or where faults lie entirely within one rock formation: fragments of rocks that are stratigraphically out of place; float of brecciated, crudely foliated, or slickensided rock; silicified rock fragments or rock altered to clays or other secondary minerals; lines of springs or water-seeking plants; aligned breaks in slope; aligned saddles in ridges; valleys that are oblique to the main structural grain; and lineaments on aerial photographs or other imagery. Because some of these features form along fractures that are not faults, each possible fault must be tested by walking single rock-layers up to it, and seeing whether or not they cross it without offset.

Once a fault has been discovered, exposures should be sought in stream banks, road cuts, and excavations in order to measure its strike and dip. Exposed fault surfaces may have slickened grooves, striations, or major corrugations that indicate the direction of latest fault movement, and the plunge or pitch of these linear structures should be measured and recorded

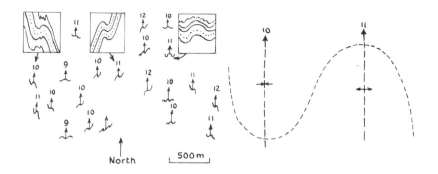

**Fig. 5-8.** Map of outcrop-size folds *(left)* and approximate forms, plunges, and axial surfaces of larger folds interpreted from these data *(right)*. Each symbol in the map on the left is drawn as the small fold would appear on a horizontal surface (the three boxes show horizontal outcrops from which the respective symbols were derived).

(Section 3-6). Rock fragments foreign to the wall rocks should be noted and collected. Cataclastic rocks and mylonitic rocks can be classified as suggested in Section 4-7. Section 12-5 describes structures and relations suggesting sense of movement as well as features indicating recent movement on faults.

Symbols for plotting faults are given in Appendix 7 and should be no more specific than justified by mapped and observed relations (Section 12-5). Known faults should be plotted even though they do not offset contact lines at the scale of the map. Such offsets may be exaggerated slightly on the map to show readers the sense of displacement.

**Foliation, joints, and tabular intrusions** are measured in the same way as bedding, and each is plotted with a distinctive symbol (Appendix 7). The structures must always be plotted as mapping proceeds, in order to determine the overall shapes of rock bodies, the locations of faults, and the principal trends of fracture systems. Bearings and plunges of a variety of linear structures are measured and plotted for the same reasons. During mapping, these various structures must be identified no more specifically than evidence permits. General groupings and specific (genetic) varieties are described in Chapters 9, 10, and 12 through 15. Symbols for many varieties have not been standardized but may be keyed on the field map by small letters placed next to structure symbols or by using colored structure symbols.

## 5-5. Rapid (Reconnaissance or Regional) Geologic Mapping

The thorough mapping described in the preceding sections may take too much time for some projects, or may be impossible because of limited access. More rapid methods may be used, and they will result in accordingly approximate or incomplete maps. The term *reconnaissance* is often applied to such mapping, especially when it precedes more thorough studies. Rapid mapping methods may also be used to develop the regional geology of a detailed study or to supply a regional framework for other measurements. In addition, projects consisting of detailed mapping may close with brief mapping excursions into surrounding areas to check the continuity of rock units or major structures.

Rapid geologic mapping should be preceded by study of aerial photographs (Section 7-3). Data from the photographs and from existing geologic maps and reports should be penciled tentatively on a base map. The field mapping can then be planned on the basis of key areas described in the literature, as well as of major exposures and routes of access.

The mapping can often be started by traversing along an open ridge (or perhaps a road or trail) that crosses the structural grain at a large angle. As each unit-contact and structure is plotted along the traverse course, its mapped trace is extended as far as possible on the basis of topographic forms and vegetation. Field glasses will help greatly in identifying outcrops and soils along each extended line. As the traverse is continued, views back over the terrain from different vantage points may improve the mapped contacts and structure lines.

A second traverse is then made along an adjoining ridge or road. Contacts and structures are connected to the lines drawn during the first traverse, some of which can be corrected by views back to features not visible from the first traverse line. When the second traverse is completed, units generally become easier to recognize from a distance and the spacing of traverses may be increased.

Rock units and structures which are finally mapped throughout an area are typically larger and simpler than those that would be mapped in a thorough survey. Generally, however, the data collected along the first traverse should be quite thorough, so that contacts can be recognized confidently on the next few traverses. Later, when stratigraphic details have become familiar, small units may be grouped into larger and simpler units. These larger units should be as genetically coherent as possible. For example, a variety of silicic lavas, pyroclastic rocks, and minor intrusions might be grouped into a unit of related silicic volcanic rocks. Continuity and distinctiveness are as important as size or thickness in selecting map units. An ideal map unit, for example, might be a persistent thin limestone that forms a distinctive ledge on slopes.

### 5-6. Outcrop Maps, Maps of Surficial Deposits, and Bedrock Maps

*Outcrop maps* show individual exposures plotted to scale and thus are valuable in showing exactly what is visible in the field (Fig. 5-9). Persons using these maps can find important outcrops easily and can make their own interpretations of concealed contacts, faults, and folds (Kupfer, 1966). Outcrop maps are commonly made for studies of surficial deposits, for engineering geologic studies, and in cases where new kinds or new sets of data may be added from time to time, as in mapping a mineral deposit that is being prospected and developed into a mine.

In order to show geologic relations at individual outcrops, outcrop mapping is typically done at scales larger than 1:24,000. Contacts between map units are drawn as solid lines in outcrop areas and as dotted lines in covered areas (Fig. 5-9). Letter symbols or colors are used to identify units within outcrop areas, and pale color tints can be added to show interpreted continuity of concealed units. On field sheets, some geologists draw colored lines around outcrops to distinguish these contacts from those within the exposed area. Others draw a black line around the outcrop and add a color line on the outside to clarify the positions of rock and cover in large irregular exposures.

The continuity and significance of bedrock structures can be interpreted and plotted most effectively if outcrops are mapped systematically across areas. The method of predicting contact traces by using a measured strike and dip (Section 5-2) may be helpful in following concealed structures from one outcrop to the next.

**Surficial deposits** encountered during general geologic mapping are mapped as specific units if time permits. Small patches are usually generalized and sometimes omitted, because they may be difficult to map accurately or may interfere with structural patterns in the underlying units.

On the contrary, accurate, complete maps showing all kinds of surficial deposits as well as their stratigraphic subdivisions are essential for some geologic studies. Examples of such studies are interpretations of Quaternary history, measurements of recent deformation, selection of sites for construction, exploration for surficial economic materials, and land-use planning. Mapping methods are basically similar to those described elsewhere in this chapter; however, they require patience and thoroughness, because surficial deposits are commonly unbedded and so weakly consolidated as to form few outcrops and to spread widely as float. A working knowledge of a variety of genetic units and features is desirable (Chapter 10).

Mapping is best started at quarries, road cuts, or other large excavations where a sequence of deposits is exceptionally exposed. Tracing contacts across country typically requires augering or trenching and study of soils and soil maps (Section 10-9). Well-drillers, farmers, engineers, and local

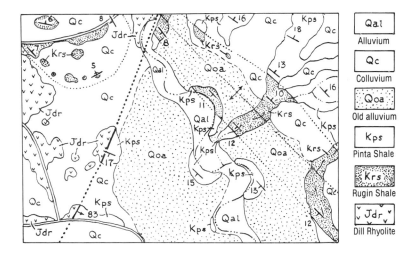

**Fig. 5-9.** Fragment of an outcrop map with contours omitted. The double lines are roads and the structure symbols are explained in Appendix 7.

government surveys may be able to supply geologic or geophysical data from well logs, test holes, or former excavations, such as for pipelines, tunnels, and building foundations.

Surficial deposits commonly comprise two or more age-units (as a sequence of loess sheets) and quite commonly more than one kind of unit (as alluvium overlain by till). Contacts in multiple units are locally emphasized by cut-and-fill relations, by lags of coarse particles, and by paleosols, as described more fully in Chapter 10. These stratigraphic sequences are of great value and may be plotted on field sheets next to their location (Fig. 5-10).

A system of color spots or letter symbols for keying outcrops on field sheets should be devised with extra care because of the variety of deposits that may be encountered. The $Q$ may be omitted from symbols for Quaternary deposits because they will be the dominant age class. Letter symbols should be as brief and informative as possible, as in the system described in

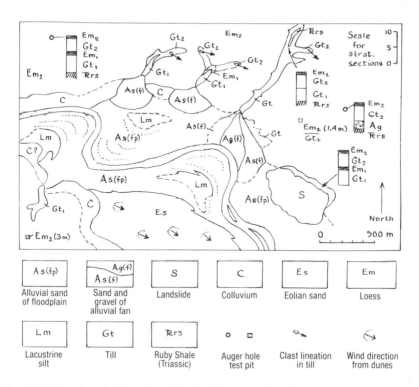

**Fig. 5-10.** Map of surficial units, with simplified stratigraphic sections at exceptional exposures. Contours omitted. Surficial unit symbols based on the system in Table 5-1.

Table 5-1, which was devised for engineering geologic mapping. Multiple sequences can be coded by numbers (Fig. 5-10).

**Bedrock maps** show the distribution of more or less solid (indurated) mapped units, whether the units crop out at the earth's surface or are covered by loose or moderately indurated surficial deposits. Bedrock units are typically much older than the surficial deposits; however, an age distinction may be arbitrary in areas where solid rocks are being formed more or less continuously, such as in long-active volcanic fields.

The purpose of bedrock maps is to show the continuity and thereby the structure of the bedrock units. They are usually constructed for areas covered largely by surficial deposits. Figure 5-9 shows such an area, and this map could be converted to a bedrock map by making the dotted (covered) lines solid and coloring or patterning the bedrock units over the entire map. It is useful to retain the contact lines of surficial deposits, and to show the positions of small bedrock outcrops in areas of surficial deposits by marking each location with a small x. The bedrock units may be shown in full color where they crop out and in a pale tint of the same color where they lie under surficial deposits.

### 5-7. Mapping Engineering Geologic Units

To be as useful as possible for non-geologists and particularly for engineers, geologic maps must show properties of earth materials as well as all of the usual geologic features. This information is needed for site selection, construction, and land-use planning. Ordinary geologic maps are always pertinent to such projects, and complete maps of surficial deposits may be essential. However, a single rock unit on a geologic map commonly includes materials with widely different physical properties, such as separate bodies of sandstone and mudstone. In addition, map units that are uniform to a geologist may be critically nonuniform to an engineer. An example would be a specific granite that is solid and widely jointed in some areas, highly fractured in other areas, and deeply decomposed in still others.

Thus, engineering geologic mapping typically involves two categories of units: (1) rock and surficial units of ordinary geologic maps and (2) units based on features indicative of one or more of these basic properties of materials: (1) strength (resistance to deformation under loading); (2) ductility (ease of more or less plastic flow); (3) hardness (resistance to abrasion, cutting, and drilling); (4) durability (resistance to decay and disintegration); and (5) permeability. Attewell and Farmer (1976) have described ways of measuring these basic properties in the laboratory and in the field, but these measurements usually are not made during initial mapping. Instead, the basic properties are estimated from specific characteristics, simple field tests, and features such as those listed below:

**Table 5-1.** Engineering Geologic Symbols for Surficial Deposits*

A. SYMBOLS FOR GENETIC CLASSES

| | | |
|---|---|---|
| A – Alluvial | G – Glacial | S – Slide |
| C – Colluvial | L – Lacustrine | V – Volcanic |
| E – Eolian | M – Marine | |
| F – Fill (man-made) | R – Residual | |

B. LITHOLOGIC SYMBOLS (always used and placed after genetic symbol; symbols for mixed materials can be composed in three ways: (1) for simple combinations of sizes, the symbol of the most abundant material is placed first, as *ms* for sandy silt; (2) where sizes show a large range, a dash is used, as *c-g* for all sizes from clay to gravel; and (3) where different materials are mixed as structural domains, a slash is used, as *m/p* for interbedded silt and peat)

| | | |
|---|---|---|
| c - clay | k - cobbles | e - erratic blocks |
| m - silt | b - boulders | p - peat |
| s - sand | r - rock rubble | o - organic material |
| g - gravel | t - trash or debris | d - diatomaceous material |

C. QUALIFYING GENETIC SYMBOLS (may be used for specific kind of units; placed in parentheses after lithologic symbol, as *Ag(f)* for gravel of an alluvial fan)

*Alluvial* - (f), fan; (te), terrace; (fp), floodplain; (p), pediment; (df), debris flow; (de), delta

*Colluvial* - (sw), slope wash; (ra), rock avalanche; (ta), talus; (cr), creep deposits

*Eolian* - (d), dune morphology; (1), loess

*Fill* - (u), uncompacted; (e) engineered

*Glacial* - (t), till, undifferentiated; (lt), lodgment till; (at), ablation till; (es), esker; (m), morainal ridge; (k), Kame; (o), outwash; (i), ice contact

*Lacustrine and marine* - (b), beach; (et), estuarine, undifferentiated; (sp), swamp; (de), delta; (ma), marsh; (tc), tidal channel; (o), offshore, undifferentiated

*Residual* - (sp), full soil profile; (bh), B horizon; (ch), C horizon; (sa), saprolite; (wp), weathering products, undifferentiated

*Slide* - (ro), rotational; (tr), translational; (fl), earthflow; (fa), fall; (sl), slump or soil slip

*Volcanic* - (af), airfall; (pf), pyroclastic flow; (s) surge; (py), pyroclastic, undifferentiated; (1), lahar; (pw), water-deposited pyroclastic; (pc), pyroclastic cone

D. THICKNESS is indicated by placing the thickness in meters or feet in parenteses at the end of the symbol, as *Gs(o)(14m)* for 14 meters of glacial outwash sand; symbols may be stacked, with intervening horizontal lines, to show stratigraphic sequence at any one place

E. PHYSICAL MODIFIER SYMBOLS (placed before the genetic symbol, as *cAg(f)* for cemented alluvial fan gravel)

| | | |
|---|---|---|
| c - cemented | e – expansive | h – hydrocompactible |

*Modified after Galster (1977) and Keaton (1984)

*Strength:* (1) response to tests of firmness, as those of Appendix 4; (2) spacing of joints; (3) degree of fissility, which is likely to be high in shale, slate, schist, and foliated igneous rocks; (4) porosity; and (5) steepness of topographic slopes.

*Ductility:* (1) clay content of rocks, deposits, and soils; (2) creep deformation on slopes; (3) tendency to form earthflows rather than blocky slide masses (Section 10-7); (4) swelling, shrinking, cracking, and slaking; (5) soil consistence (Section 10-9); and (6) presence of specific ductile rocks, such as shale, claystone, gypsum, salt rock, ice, altered vitric tuff, altered slate and mica-rich schist, and mélange deposits.

*Hardness:* (1) resistance to scratching with a steel point; (2) in firm rocks, hardness of the principal minerals (Appendix 4).

*Durability:* (1) resistance to erosion, as expressed by topography; (2) thickness of disintegrated materials on moderate slopes; (3) grain size of disintegrated materials; (4) fracture spacing; (5) proportion of cemented pores and kind of cement; (6) susceptibility to gullying, slumping, and sliding; (7) all other indications of high ductility, softness, and low strength.

*Permeability:* (1) size and continuity of pores; (2) spacing and openness of fractures; (3) grain size and sorting of aggregates; (4) rates of drying after thorough wetting; (5) rates of infiltration of water; (6) presence of vegetation requiring rapid, deep drainage (generally large plants) as opposed to prolonged shallow moisture (generally small plants).

The procedures used in making an engineering geologic map depend on the uniformity of geologic units and on whether or not a geologic map of suitable scale is already available. If an accurate geologic map is available, and field checking shows that each of the mapped units is uniform with respect to engineering properties, the geologic map can be converted to an engineering geologic map by measuring or estimating and describing the engineering properties of each unit and by plotting specific data at sites of borings or other tests, as in a map prepared by Radbruch (1969).

In cases where many geologic units have been mapped, and field study shows that each is uniform and physically similar to others of the units, the geologic map can be converted to an engineering geologic map by grouping similar units and describing the engineering properties of each group, as in a map by Briggs (1971).

In the third and perhaps commonest case, geologic units are nonuniform and must thus be divided into units based on engineering geologic properties, using one or more criteria such as those described above. The mapping should be done on a clean topographic base or aerial photograph. If a standard geologic map is available, it should be checked and improved during the mapping. If a geologic map is not available, one should be made, because it will be needed for a full interpretation of structural relations. Field studies may include unusually large-scale mapping (1:50 to 1:200), as described briefly in Section 3-4 and for trenches and other excavations by Hatheway (1982).

In some cases the engineering geologic data plotted on the map will be based on measurements or estimates of actual quantities; in other cases they will consist of broader qualitative judgments. Examples of measurements or estimates of quantities are the numbers of Appendix 4, the thickness of surficial ductile deposits, and the spacing of fractures. Qualitative judgments are usually rated in five categories (very low, low, moderate, high, and very high) and may involve one kind of information (as porosity) or all information relating to a basic property (as ductility). Both kinds of information can be recorded at field sites by letter symbols, color spots, or colored numbers (Tables 5-1 and 5-2, Fig. 5-11). Contact lines are mapped in the field where properties change from one category to the next. Varnes (1974) and the International Association of Engineering Geology (1976) have described additional categories of information plotted on engineering geologic maps, and several specific mapping systems have been presented in the August 1984 issue of *The Bulletin of the Association of Engineering Geologists* (vol. 21, no. 3).

The mapping should also indicate places where on-going processes may affect structures or prohibit certain land uses. Examples are sites of rapid

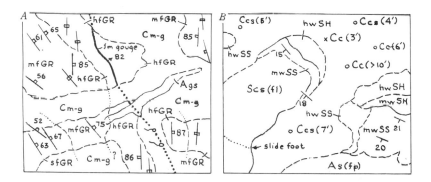

**Fig. 5-11.** *A.* Part of a field sheet made to determine distribution of large-scale strength of bedrock in an area underlain by granite (*sfGR,* slightly fractured; *mfGR,* moderately fractured; and *hfGR,* highly fractured granite; *Cm-g,* colluvium ranging in grain size from silt to gravel; *Ags,* alluvial sandy gravel; other symbols explained in Appendix 7). *B.* Part of a map showing distribution and thicknesses of surficial ductile materials (hwSS, highly weathered sandstone; mwSS, moderately weathered sandstone; hwSH, highly weathered shale; mwSH, moderately weathered shale; Cc, colluvial clay; Ccs, colluvial sandy clay; Scs (f1), sandy clay of earthflow; and As(fp), sand of floodplain). Small x's are outcrops and small circles are auger holes.

erosion or deposition; faults and landslides that may become active; areas susceptible to flooding by rivers or tides; potentially active sinkholes; significant deposits of uncompacted fill; soils undergoing rapid creep; slopes prone to gullying, loess units susceptible to hydrocompaction and thus to collapse; and ground that would develop unusually strong motion during earthquakes. Some critical areas can be recognized by comparing the present terrain with features shown on old aerial photographs, ground photographs, or maps, or described in historical records.

All of the information described above will be of immediate use for a specific project and may be enhanced for other uses. Data on ductility and per-

**Table 5-2.** Engineering Geologic Symbols for Rocks*

SEDIMENTARY ROCKS

| | | |
|---|---|---|
| SS – sandstone | DT – diatomite | DO – dolomite |
| ST – siltstone | SH – shale | CK – chalk |
| CG – conglomerate | LS – limestone | |
| CH – chert | CS – claystone | |

IGNEOUS ROCKS

| | | |
|---|---|---|
| GR – granite (granitic rock) | AN – andesite | RH – rhyolite |
| GA – gabbro | BA – basalt | VO – volcanic |
| FE – felsite | DI – diorite | IG – undifferentiated |
| TU – tuff | SY – syenite | |

METAMORPHIC ROCKS

| | | |
|---|---|---|
| QT – quartzite | SL – slate | MA – marble |
| SC – schist | AR – argillite | SE – serpentinite |
| GS – greenstone | GN – gneiss | ME – undifferentiated |
| PH – phyllite | HO – hornfels | |

MAN-MADE ROCKS

| | | |
|---|---|---|
| CC – Portland cement concrete | AC – asphaltic concrete | PA – undifferentiated pavement |

SYMBOLS FOR MIXTURES

Mixed domains too small to be mapped to scale can be indicated by a slash, as SS/ST for interbedded sandstone and siltstone

THICKNESS

Thicknesses of units that cannot be mapped to scale are indicated by numbers in parentheses placed after the rock symbols. Symbols may be stacked, with intervening horizontal lines, to indicate stratigraphic or structural sequence at any one place.

MODIFIER SYMBOLS, if used, are placed in front of the rock symbol, as in Fig. 5-11. Symbols based on quantified systems, as that of Williamson (1984), are preferred.

*After Galster (1977) and Keaton (1984)

meability, for example, can be combined with slope angles to produce a map showing probable slope stabilities. Slopes should be measured if possible, because contour maps may show moderately steep to steep slopes to be gentler than they actually are. Varnes (1974) described how to analyze several sets of data in order to locate areas suitable for specific uses.

*Environmental assessments* predict how and to what degree specific engineer-projects will affect the natural and cultural environment. Basically, the assessments describe: (1) the existing environment; (2) the probable impacts of a project on it; and (3) ways (if any) of mitigating against adverse effects. Typical nongeologic topics include the local cultural history, specifics of the flora and fauna, and interactions of local people with their surroundings—a topic ranging from air pollution to aesthetics. Although brief checklists for preparing an assessment are available (Jorgenson, 1982), the nongeologic topics vary so greatly with the kind of project, the specific needs of people, and local politics, that state and local codes should be consulted in all cases.

The geologic topics of environmental assessments have their basis in engineering geologic studies such as those already described. In addition to describing rocks, structures, and surficial deposits, the assessments should include thorough descriptions of topographic features, hydrologic systems, surficial processes, and the effects of adding or removing materials. Topics relating to topography include: (1) details of genetic forms (floodplains, terminal moraines, karst, beach ridges, periglacial ground, etc.); (2) lateral stability of slopes underlain by certain rock or surficial units; (3) distribution of all landslides and soil slips; (4) locations of cracked, hummocky, or patterned ground; (5) slopes susceptible to gullying or unusually rapid creep; and (6) stability of excavated surfaces and of natural slopes placed under heavy structural loads.

Topics to consider in describing hydrologic systems include: (1) kinds of water bodies (streams, ponds, swamps, springs, etc.); (2) their sources; (3) their distribution relative to other geologic features and units; (4) rates of flow in streams, springs, and groundwater systems; (5) diurnal, seasonal, and long-term variations in amounts of flow or sizes of water bodies, citing specific evidence of long-term variations (historical records, mineral deposits, topographic forms); (6) interactions with flora and fauna; (7) effects of water on properties of rocks and surficial deposits; (8) effects on landsliding and creep; (9) water chemistry; (10) freeze-thaw regimes; (11) variations in height of the water table; (12) production of groundwater relative to long-term variations in precipitation; and (13) confined systems and their piezometric surfaces.

Some topics relating to on-going processes are mentioned in the foregoing section on engineering geologic maps and the references cited there. Additional topics in areas subject to earthquakes are: (1) direct effects of fault movements (ground rupture, subsidence, uplift, etc.); (2) effects of shaking on the various rocks and deposits (liquefaction, sliding, cracking, settling due to compaction); (3) effects related to groundwater (fountaining, liquefaction, sliding); (4) and indirect hazards (seiching, tsunamis, flooding). Additional suggestions for overall geologic-seismic studies are available in Note Number 37 of the California Division of Mines and Geology (reproduced in Slosson, 1984).

**References Cited**

Attewell, P. B., and Farmer, I. W., 1976, *Principles of engineering geology*: New York, John Wiley & Sons, 1045 p.

Briggs, R. P., 1971, *Geologic map of the Orocovis quadrangle, Puerto Rico*: U.S. Geological Survey Miscellaneous Geologic Investigations Series Map I-615.

Compton, R. R., 1962, *Manual of field geology*: New York, John Wiley & Sons, 378 p.

Galster, R. W., 1977, A system of engineering geology mapping symbols: *Association of Engineering Geologists Bulletin*, v. 14, no. 1, p. 39-47.

Hatheway, A. W., 1982, Trench, shaft, and tunnel mapping: *Association of Engineering Geologists Bulletin*, v. 19, p. 173-180.

International Association of Engineering Geology, 1976, *Engineering geological maps; a guide to their preparation*: Paris, UNESCO Press, 79 p.

Jorgenson, D. B., 1982, Outline for environmental impact statements, Data sheet 27.1 *in* Dietrich, R. V., Dutro, J. T., Jr., and Foose, R. M., *AGI data sheets*: Falls Church, VA, American Geological Institute.

Keaton, J. R., 1984, Genesis-Lithology-Qualifier (GLQ) system of engineering geology mapping symbols: *Association of Engineering Geologists Bulletin*, v. 21, no. 3, p. 355-364.

Kupfer, D. H., 1966, Accuracy in geologic maps: *Geotimes*, v. 10, no. 7, p. 11-14.

North American Commission on Stratigraphic Nomenclature, 1983, North American stratigraphic code: *American Association of Petroleum Geologists Bulletin*, v. 67, p. 841-875.

Radbruch, D. H., 1969, *Areal and engineering geology of the Oakland East quadrangle, California*: U.S. Geological Survey Quadrangle Map GQ-769.

Slosson, J. E., 1984, Genesis and evolution of guidelines for geologic reports: *Association of Engineering Geologists Bulletin*, v. 21, p. 295-316.

Varnes, D. J., 1974, *The logic of geological maps, with reference to their interpretation and use for engineering purposes*: U.S. Geological Survey Professional Paper 837, 48 p.

Williamson, D. A., 1984, Unified rock classification system: *Association of Engineering Geologists Bulletin*, v. 21, no. 3, p. 345-354.

# 6

# ■ Geologic Mapping on a Topographic Base

### 6-1. Topographic Maps

Topographic maps delineate landforms and approximate elevations above sea level by means of contours, which depict level lines spaced at a useful vertical interval called the *contour interval*. Each line lies at a fixed elevation above sea level, and this number generally is used to label every fifth contour. In addition, specifically determined, and typically more exact, elevations are shown for some hilltops, road crossings, or survey stations. Most of these points are marked with a small x and a number for the elevation, which is shown in black if the elevation has been checked and in brown if it has not. On U.S. Geological Survey quadrangle maps, contours are printed brown; drainage lines and all water bodies are in blue; geographic boundaries, bench marks, triangulation stations, roads, buildings, mines, and all other human culture are in black; areas covered by trees, brush, and crops are in patterns of green; and highways, certain fencelines, and all survey lines and corners of the township-section cadastral system (Appendix 5) are in red. Lavender or pink may be used for areas of closely spaced buildings in urban areas, and overprints of purple are applied to show up-dated cultural features, such as new highways and buildings. The patterns used for specific kinds of features are defined in a *Topographic Maps Symbol Sheet,* which is available free on request from any of the NCIC addresses given below.

Short lines in the margins of quadrangle maps show exact positions of lines of latitude and longitude, lines of the Universal Transverse Mercator Grid, and lines of state coordinate systems. Besides the map title, the margins also carry labels for townships and ranges (Appendix 5) and, in the lower margin, a scale ratio and bar scale, the contour interval, arrows indicating the directions of true and magnetic north, a location map, notation of mapping dates and methods, and perhaps other notes referring to geographic features. Ellis (1978) has given descriptions of the grid and coordinate systems, the construction of map projections, and the methods of making the maps and charts of the United States. This reference also shows full-scale examples in color of topographic maps, shaded contour maps, coastal charts, orthophoto maps, and other kinds of modern maps.

Some advantages of using topographic maps in geological mapping are given in Section 1-5. Mapping should ordinarily be done on maps with as large a scale as available, and maps with scales of 1:24,000 or 1:25,000 are generally the most recent and accurate. The most widespread coverage in the United States is by 1:24,000 7.5 X 7.5-minute quadrangles; the new metric maps have a scale of 1:25,000 and most are 7.5 X 7.5-minute and the others 7.5 X 15-minute quadrangles. Maps with scales of 1:50,000 (the basic map series in Canada), 1:62,500 (available for many parts of the United States), and 1:63,360 (covering much of Alaska) are adequate for detailed geologic mapping if enlarged 2 or 3 times, and are excellent for less detailed mapping. Maps with scales of 1:100,000, 1:125,000, or 1:250,000 have been made for most parts of the world and are geologically useful for planning, for reconnaissance, for regional geologic mapping, and for compiling geologic data from other sources. Topographic maps with scales larger than 1:24,000 are made occasionally for local government agencies or private companies and may be available from them. Unusually large-scale maps without contours (*planimetric maps*) are commonly made for urban and suburban areas, and their detailed views of buildings, roads, and waterways make them excellent geologic base maps.

**Accuracy of topographic maps** of the United States is prescribed by national standards adopted in 1941. Maps are tested for compliance and bear a statement to this effect in the lower margin. The horizontal accuracy must be such that not more than 10% of the well-defined map points tested are more than 0.5 mm (1/50 in.) out of correct position. This tolerance corresponds to 12 m (40 ft) on the ground for 1:24,000-scale maps and 31 m (104 ft) for 1:62,500-scale maps. Standards for vertical accuracy require that no more than 10% of the elevations of test points interpolated from the contours shall be in error more than half the contour interval. Thus, even the most precise methods of locating geologic features from map features may occasionally lead to errors somewhat greater than these tolerances.

**Sources of topographic maps** of the United States are the headquarters or regional offices of the National Cartographic Information Center (NCIC) of the U.S. Geological Survey. The NCIC can also supply information about all maps and charts, aerial and satellite photographs and imagery, map data in digital form, and geodetic control data obtained by other federal agencies. The addresses are: NCIC–Headquarters, U.S. Geological Survey, 507 National Center, Reston, VA 22092; NCIC–East, U.S. Geological Survey, 536 National Center, Reston, VA 22092; NCIC–mid-Continent, U.S. Geological Survey, 1400 Independence Rd., Rolla, MO 65401; NCIC–Rocky Mountain, U.S. Geological Survey, Box 25046, Federal Center, Stop 504, Denver, CO 80225; and NCIC–West, U.S. Geological Survey, 345 Middlefield Rd., Menlo Park, CA 94025. Inquiries should include a small-scale map or a description giving the exact latitudes and longitudes bounding the area of interest. Quadrangle names and scales are shown on indexes published periodically for each state, and are available on request. The indexes also show locations of maps being planned or made, and preliminary editions of the latter are sometimes available.

In addition to topographic maps printed on paper, maps printed on transparent materials and enlarged to the user's specifications (up to a limiting size) can be ordered from the U.S. Geological Survey. Of particular value for geologic projects are *chronoflex* prints, which are monochromatic prints of maps, made on a stable plastic base that has a burnished upper surface suitable for ink and pencil work. Green chronoflex prints have proven especially useful. They should be ordered with the map printed on the bottomside of the sheet in order that corrections of geologic mapping on the upper surface will not erase the base map.

Canadian topographic maps of all scales are shown on three indexes: No. 1, Canada east of 96° longitude and south of 62° latitude; No. 2, Canada west of 88° longitude and south of 68° latitude; and No. 3, the remaining northern parts of Canada. The indexes give prices and instructions for ordering and are available on request from The Canada Map Office, 615 Booth St., Ottawa, Ontario, K1A OE9.

### 6-2. Preparations for a Mapping Project

Chapter 1 presents the general philosophy and organization of a geologic mapping project and Sections 1-3, 1-4, and 1-5 describe important preparations. In addition to the basic equipment described in Chapter 2, some special equipment might be considered. If the terrain is wooded and slopes are typically gentle, traversing by pacing may be a principal procedure and a small tally counter will be helpful. If the area is one of mafic and ultramafic igneous rocks, a sun compass may be essential (Section 2-4). A Jacob staff is recommended for projects requiring detailed stratigraphic measurements (Chapter 11). The equipment list of Appendix 1 should be examined in light of other special needs and of expected weather conditions.

The base map may have to be prepared in various ways. Paper maps can be waterproofed inexpensively with a spray of clear lacquer, or a print of the map can be obtained on a stable waterproof sheet (Section 6-1). Enlarged map sheets may be needed for unusually detailed studies. Maps can be cut into sheets that will fit a map case or notebook, and if each sheet is cemented to a sheet of quadrille paper that shows on all margins, the grid will be helpful in plotting bearings and strike lines. Several extra copies of the map should be taken to the field, preferably in a waterproof container. Topographic maps of surrounding quadrangles may be needed for reconnaissance, and a set of aerial photographs or other remote-sensed imagery should be obtained for prefield photogeologic reconnaissance and for use in the field (Sections 7-1, 7-2, and 7-3).

### 6-3. Locating Points in the Field

From the first outcrop studied, information is plotted at points on the map that correspond with locations on the ground. Regardless of how detailed and successful the study of each outcrop may be, the accuracy of the map — even its meaning — depends on locating these points exactly. If done patiently at the outset, the methods of location become familiar in a week and second nature in a month. The following steps are recommended for making these locations:

1. Use a compass to find the geographic direction of north.

2. Face in that direction and hold the map flat with its north (top) edge pointed northward.

3. Compare the map with the surrounding terrain, carefully identifying nearby features on all sides.

4. Continue to compare the map with surrounding features until the location can be estimated and marked lightly on the map with a pencil.

5. Use two or more of the following methods to refine the location:

   *a.* If the place occupied proves to be at a unique feature, such as a distinctive bend in a stream, road, or narrow ridge, a junction of streams or roads, a hilltop, and so on, a final point can be marked on the map with confidence.

   *b.* If the place occupied is along a linear feature, such as a stream, a road, or a narrow ridge, its position on the map can be fixed by a bearing line from another visible feature, as a hilltop, that can be located accurately on the map. The location is made by drawing an antibearing line from the observed feature to intersect the linear feature occupied (Fig. 6-1*A*). For an accurate location, the linear feature and line of sight should cross at angles near 90° and never less than 30°.

   *c.* If the location is along a linear feature that slopes steeply, such as a mountain road, it can be fixed on the map by determining its elevation and referring to the contours that cross the linear feature on the map. The elevation of the location can be determined by: (1) sighting to nearby hilltops and other distinctive features with a hand level (Section 2-7); (2) finding one that is at the same elevation as the place occupied; and (3) reading its elevation from the topographic map. The elevation can also be read from an altimeter.

   *d.* If the place occupied is near a feature that can be identified with certainty on the map, it can be located by reading a bearing to the feature, pacing the distance to it, then checking both results by taking a bearing back to the place to be located and pacing the distance back. After the bearing line is plotted, a scale is used to lay off the paced distance.

   *e.* If the place occupied gives a reasonably full view of the surroundings, it can be located on the map by reading bearings to three or more features that can be identified with certainty on the map. The lines plotted are antibearings from the three map points (Fig. 6-1*B*), and the accuracy of the location will depend on their intersecting at large angles.

   *f.* Under conditions like those for method *e,* a location can also be made by reading and plotting one bearing line to a distant feature and determining the elevation of the place occupied as described in method *c.* The point is located where the bearing line crosses the appropriate contour line or a line interpolated between two contours.

**Locating outcrops by compass traverses** may be necessary in wooded areas where long sights are generally not possible and where relief is so subdued that contours give inaccurate locations. The details of a pace and compass traverse are described in Section 5-1, and a tape can be used if greater precision is needed. If possible, the traverse should be brought around to its starting point or to another point that can be located independently on the map. When this is done, the two points commonly will not coincide. Such errors of closure can be ignored if they are small or within the limits of precision of the method. If they are large enough to displace geologic data seriously, the traverse should be corrected as much as possible. Plotting errors may be eliminated by checking the traverse against the recorded data, and if this does not resolve the error, the traverse must be resurveyed. On long traverses, errors in pacing or mistakes in reading and recording bearings sometimes can be located quickly by making connecting traverses that will localize the error (Fig. 6-2). If the traverse cannot be corrected, it should not be replotted by distributing the error among the traverse legs as described in most sur-

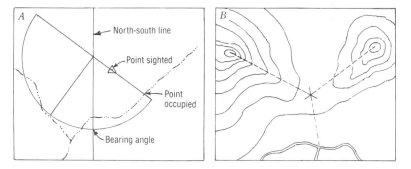

**Fig. 6-1.** *A.* Using a protractor to draw an antibearing from a point sighted to the point occupied. If the point occupied lies beyond the protractor, a plastic scale can be placed along the base of the protractor to extend its reach. Strike lines may be plotted in the same way. *B.* Locating a point by three antibearing lines. The dashed parts of the lines need not be drawn.

veying texts. Later traverses or other locations may correct the locations in question.

**Locating outcrops with an altimeter** is especially suited to areas where relief is at least moderate and forest hides much of the terrain. The method entails first establishing a line along which the outcrop is known to lie, then using an elevation read from the altimeter to locate the point by referring to the corresponding contour or interpolated contour on the map. The line along which the point lies can be established: (1) by inspection, if it occurs along a ridge, creek, road, or trail shown on the map; (2) by a bearing line to any feature that can be located accurately on the map; or (3) by making a compass traverse from any point that has been located nearby (see the preceding subsection).

The altimeter must be adjusted fairly often because it will respond to pressure variations due to changes in temperature, local winds, and broad weather conditions. Adjustment is made by going to a point that can be located confidently on the map and setting the altimeter scale so that the indicated elevation corresponds to the local elevation read from the map. The precision of the adjusted altimeter should be to the nearest few meters, and because contours may be in error by as much as one-half a contour interval (Section 6-1), the checkpoints should, if possible, be places where an elevation is recorded on the map, typically at bench marks, hilltops, or road junctions. The checks should be made every half hour or so, depending on how rapidly atmospheric conditions are changing. If checkpoints cannot be visited easily and wind and weather conditions seem constant, the effects of temperature can be corrected by reading temperatures from a thermometer. If elevations are in feet, the change in elevation since the last adjustment is multiplied by 0.002 for each °F, and if elevations are in meters, it is multiplied by 0.0006 for each °F or by 0.0011 for each °C. The resulting number is added when the temperature has risen and subtracted when it has fallen.

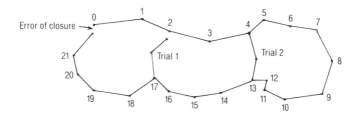

**Fig. 6-2.** Traverse map with closing error and two connecting traverses which suggest (trial 1) that the error is prior to station 17, and (trial 2) that it is somewhere between stations 13 and 17.

An altimeter used for mapping at scales of 1:24,000 or 1:25,000 should have a reading precision of about 2 m and must have an elevation range suitable to the area being studied. It should be tested before the field season to make sure it will respond with the necessary precision during normal field handling. Details of maintenance and tables for temperature corrections should be obtained from the manufacturer.

***Locating outcrops by use of triangulation stations*** may be necessary where topographic relief is subdued and the base map has only a few features that can be located exactly in the field. The method first requires that a number of easily visible markers not shown on the map are surveyed onto it by triangulation (intersection) (Fig. 6-3*A*). The markers may be any features, natural or manufactured, that can be seen and identified over large areas. Examples are distinctive lone trees, dead spars, prominent outcrops, rock cairns, lone buildings, transmission poles, and crossings of roads or trails. For average mapping conditions, they should be spaced about a mile apart and located so that lines of sight among them intersect at angles between 30° and 150°.

The survey must be started at two adjacent points (as *A* and *B* in Fig. 6-3*A*) that can be located accurately on the map by inspection or other means. Each of these points is occupied in turn and compass bearings are read to all adjacent station-markers, which are located by intersections of bearing lines plotted lightly in pencil. Each intersected station is then occupied and bearings are read to all visible stations, including the two used as starting points. These readings serve to check the initial bearing lines and to provide three-line intersections of stations such as *C* and *D* in Fig. 6-3*A*. If intersections result in triangles that are too large for the precision required, the bearings must be reread and corrected. When the intersections are acceptable, the survey can locate additional stations (as *F*, *G*, and *H*). The located stations can then be used to locate outcrops by methods *d* or *e*, above, and thus to map the geology.

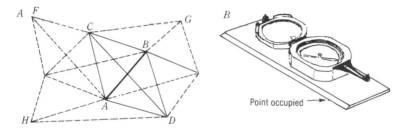

**Fig. 6-3.** *A*. Triangulation network based on two points of known location (*A* and *B*). *B*. Brunton compass mounted on a ruler so that the line of sight is parallel to the ruler's edge, and with lid and sighting arm in position for direct sighting.

Triangulation errors that cannot be corrected are generally caused by magnetic anomalies. This source of error can be avoided by triangulating graphically with a plane table and telescopic alidade (Section 8-7) or with a peep-sight alidade and traverse board (a small plane table). Lacking a peep-sight alidade, a Brunton compass can be mounted on a heavy plastic ruler (Fig. 6-3*B*) or a Brunton alidade protractor (see a supplier's catalog, Section 2-1) which has a straightedge parallel to the Brunton sighting axis. The sighting arm and lid are then opened so that direct sights can be made. The method is basically like the one described in Section 8-7.

### 6-4. Interpretation of Geologic Lines on a Topographic Base

Following contacts between rock units is described in Section 5-2, and finding and mapping various structures is presented in Section 5-4. Most of the other information in Chapter 5 will also be pertinent to geologic mapping on a topographic base. Geologic relations will be easier to recognize on field sheets if the rock units are colored lightly and if a color (customarily red or green) is used for fault lines and perhaps for other important kinds of structural symbols, particularly if they occur among numerous symbols for other kinds of structures.

An advantage of plotting contacts and faults on a topographic base is that the map clearly and precisely depicts geometric relations between mapped surfaces and topographic forms. Direction and approximate amount of dip can be seen at a glance by noting the relation between a contact or fault trace and the contours (Fig. 6-4). Studying the diagrams, one can derive some simple rules for attitudes of dipping surfaces, which will be simplified here as contacts: (1) contacts that cross topographic forms as straight lines must be vertical; (2) contacts that remain parallel to curving contour lines must

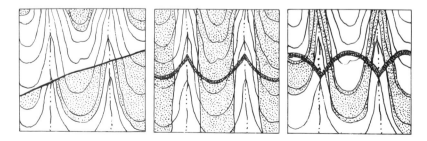

**Fig. 6-4.** Structural attitudes expressed by V patterns on contour maps. *Left,* horizontal tabular units and vertical fault; *middle,* tabular units dipping parallel to stream gradients, and dike dipping upstream; and *right,* tabular units dipping downstream less steeply than stream gradient, and dike dipping steeply downstream.

be horizontal; (3) contacts parallel to lines depicting streams must dip the same as the stream gradient; (4) contacts that form V's pointing downstream must dip downstream more steeply than the stream gradient; (5) contacts that form V's pointing upstream must either (a) dip upstream, (b) be horizontal, or (c) dip downstream less steeply than the stream gradient.

The strike and dip of any planar feature on a contour map can be determined exactly as described in Fig. 6-5. This procedure is crucial in areas where contacts can be mapped on the basis of float and soil but outcrops are so crumbled that reliable structural attitudes cannot be measured. If rock units and contacts are truly hidden in some parts of an area but are well exposed in adjacent parts, the method described in Figure 5-6*B* may be used to project a contact across the unexposed part.

A topographic map can also be used to determine the thickness of a mapped rock layer by the steps that follow:

1. Construct a strike line for the layer by drawing a line between any two points where one of the contacts crosses a given contour.

2. Draw a line perpendicular to the strike line, extending this "dip line" at least from one contact past the other.

3. Using the dip line as a line of section, construct a ground profile and a vertical cross section of the layer (Section 6-5); then scale the layer's thickness from this section.

4. If more convenient: (a) determine the vertical distance between the top and bottom of the layer as shown by the contours intersected along the dip line drawn in step 2; (b) scale the horizontal distance between the contacts along the dip line; (c) in the three equations that follow, let the vertical distance be $x$, the horizontal distance $y$, and the amount of the dip $d$; if the slope and dip are in opposite directions, *thickness* = $cosd(x + y \tan d)$; if the slope and dip are in the same direction and the dip is the steeper, *thickness* = $cosd(y \tan d - x)$; and if the slope and dip are in the same direction and the dip is the steeper, *thickness* = $cosd(x - y \tan d)$.

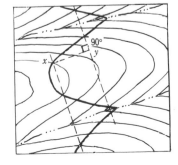

**Fig. 6-5.** Contour map with mapped trace of a planar surface. Strike can be determined by drawing a line joining two or more points where the surface crosses a given contour (such as the dashed lines). Dip can be determined by drawing a line (*xy*) perpendicular to the strike lines and calculating from the relation: $tan\angle dip = vertical\ distance\ between$ $x\ and\ y \div horizontal\ distance\ between\ x\ and\ y.$

## 6-5. Office Routines; Constructing Vertical Cross Sections

Routine tasks in the field camp or office and specific steps in completing a field project are described in Sections 1-6 and 1-7. Preparing accurate, detailed cross sections is one of the more important procedures late in the field season. Cross sections not only develop and clarify geologic relations for the investigator, they are necessary for others who will read the report or use the map. In many cases, traverses may have to be made along the lines of section in order to check important relations and add attitudes of structures. In all cases, the sections must be prepared and studied thoroughly when it is still possible to go back to the field and look. The sections may be constructed from the topographic-geologic map as follows:

1. Select a section line that will develop as much of the mapped geology as possible and pass through or near to areas where data are especially reliable and abundant. In order to show more or less true dips and true thicknesses, orient the section line within 20° of perpendicular to the strike of structures. Consider all possibilities thoroughly, because selection of a line of section is by far the most important step in the procedure. Note that some features not crossed by the line will dip or plunge into the section below the surface.

2. Draw the section line on the map using a sharp pencil and an accurate straightedge. Hold the pencil consistently vertical so that the line will be straight, and check its straightness by laying the straightedge along the opposite side of the line. Draw short crosslines (ticks) perpendicular to the section line at its two ends.

3. Prepare a strip of drafting film or tracing paper wide enough to include the highest and lowest points along the section line plus space for geologic features under the ground profile. Use quadrille-ruled material if the divisions can be calibrated to contour intervals, at map scale, or rule a set of horizontal lines on blank material, spacing them at some contour interval. Do not use an exaggerated vertical scale except in the rare cases where it is needed; for example, to show sequences of surficial deposits in detail (Fig. 16-6).

4. Lay the strip over the map and parallel with the section line. Orient the strip so that the right-hand end is either the more easterly end or is oriented due north. However, if two or more subparallel sections are oriented roughly north-south, all their northerly ends should be on the right. Trace the ends of the line on the strip exactly and extend them as end lines perpendicular to the horizontal lines of the strip (Fig. 5-4A). Label the horizontal lines with elevation numbers.

5. Plot points for the ground profile by either (a) raising or lowering the strip until the elevation line equivalent to a contour lies over the section line (Fig. 6-6A), or (b) taping the strip in place and using a triangle and straight-

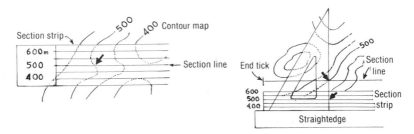

**Fig. 6-6.** *A.* Cross section strip in position for marking elevation points for the 500 m contour (*as at dark arrow*). *B.* Projecting the elevation point for the 500 m contour, using a triangle placed against a straightedge parallel to the section line.

edge to project each point where the section line is crossed by a contour to the corresponding elevation line on the strip (Fig. 6-6*B*). If relief is at least moderate, plot elevation points at about 40 meter intervals, and draw the profile between these points by visual inspection from the other contours. Draw the profile line in pencil, using a straightedge or drafting curve for reasonably smooth slopes. Check the line carefully against the map; then ink it so that it will not be smudged when geologic data are added.

6. Lay the strip over the map again, exactly matched at its ends, and mark the profile at each point where a contact, fault, hinge line, or strike line

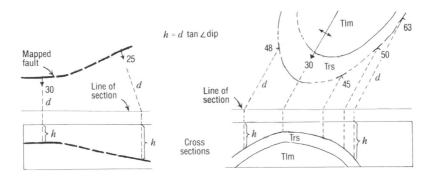

**Fig. 6-7.** Projecting structures that cut across sections beneath the surface, which in the cases shown is a level plain. Depths (*h*) are calculated by the equation shown. In calculating *h* for the fold, the angle of plunge is used rather than the angle of dip. For a contact or fault (*left*), the distance *d* is measured parallel to the local dip, and for a fold (*right*) it is measured parallel to plunge. The apparent dip to be plotted at any point can be read from Appendix 13 or by projecting a number of points and connecting them by a smooth line. If the ground surface is not a level plain, the difference in elevation between the original map point and the point projected to the line of section must be added or subtracted to *h*.

crosses the section line, using either method shown in Fig. 6-6. Use a protractor to construct a line about 6 mm (¼ in.) long, showing the dip of the feature below the profile. If the section line is oblique to the direction of strike, plot the apparent dip, which can be determined from Appendix 13.

7. Remove the strip and project geologic data to the section line on the map, including only the data that are close enough to the section line, or persistent enough, to be projected reliably. Where structures have approximately parallel strikes or plunges, projection lines will be straight. Where structures appear to converge, follow the instructions of Fig. 5-5 to project the structures and to correct their dips.

8. Lay the strip over the map and use the methods of step 6 to add the features projected to the section line in step 7.

9. Examine the map thoroughly for contacts or other structures that do not project along the surface to the section line but dip or plunge into the section below the surface. The more common will be contacts, faults, and beds or other layers that strike about parallel to the section line and dip toward it. The depths at which features meet the section may be determined trigonometrically or by accessory cross sections (Fig. 6-7).

10. Project plotted features to as great a depth as seems reliable. Typically, some parts of the section may be drawn to greater depths than others. Layers in parallel folds (Fig. 12-12) may be projected as circular arcs, as by the *Busk*

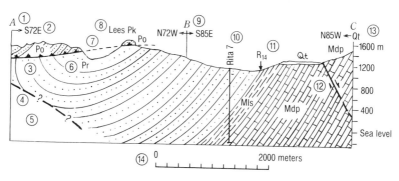

**Fig. 6-8.** Geologic cross section illustrating general form and certain specifics: (1) letter symbol designating end of section line on map; (2) bearing of section line, if section does not accompany map; (3) bedding lines and lithologic patterns; (4) questioned structure where control is weak; (5) blank areas where units and structures are not well known; (6) unit symbols; (7) projected structures where helpful in clarifying relations; (8) geographic names; (9) hinge where section line changes direction; (10) wells (dashed where projected into section); (11) position of important samples; (12) arrows showing relative movement on faults: (13) elevations (include elevations in feet at one end of section if map contours are in feet); (14) bar scale.

*method*, described in most structural geology texts. The less reliable lines may be dashed or queried (e.g., faults without observed dips). Large areas under surficial deposits may have to be left blank or perhaps completed with queried lines.

11. When the section is completed in pencil, trace it in ink and add some or all of the items shown in Fig. 6-8. If the section will not be on a plate with the map and will not be bound in a report, also include with it: a title, giving the geographic location; the name(s) of the geologist(s); the date of the survey; and a full explanatory key to the rock units, structures, and any special symbols used.

### References Cited

Ellis, M. Y., editor, 1978, *Coastal mapping handbook*: U.S. Geological Survey and National Ocean Survey, Washington, Government Printing Office, 200 p.

# 7

# Use of Aerial Photographs and Other Remote Imagery ■

### 7-1. Conventional Aerial Photographs

Geologists use remotely sensed images of the earth's surface to identify geologic features before or during the field season, to evaluate terrains for mapping routines, and as a base for geologic mapping in the field. In addition to these uses, which are described in later sections of this chapter, modern topographic maps are constructed from remote imagery. The imagery used most commonly by geologists and topographers are black and white photographs taken from an airplane with an optical camera. Most of these are *vertical photographs*, taken with a camera pointing precisely downward, thus giving a map view of the earth's surface. Vertical photographs of almost all of North America and much of the rest of the world have been made at one time or another, and many areas have been photographed several times and at different scales. *Aerial obliques*, taken with a camera pointing obliquely downward, give more nearly perspective views of topography but are not generally available except for parts of Alaska and northern Canada.

Most available photographs can be purchased from government agencies, either as contact prints or enlargements. In the United States, information on photograph coverage and scales can be obtained from the National Cartographic Information Center (see Section 6-1 for addresses). The inquiry should specify acceptable scales as well as the exact location of the area. The instructions sent by NCIC will include information on costs and availability of photograph indexes and enlargements. The photo index of an area is needed to select the photographs, and therefore the initial inquiry should be made at least eight weeks before the prints will be needed. Additional coverage, some of it of unusually large scales, may be available from local governments, air survey companies, or mineral and fuels exploration companies. In Canada, information on aerial photograph coverage and cost of prints and enlargements is available from The National Air Photo Library, Room 180, 615 Booth St., Ottawa, Ontario, K1A 0E9.

Some parts of the United States and Canada are covered by *orthophotomaps*, also called *orthophotoquads*, made by rectifying distortions due to tilt of the camera axis and to relief within the landscape. The rectified imagery is printed on a map sheet that also shows some of the features normally shown on topographic maps (see Ellis, 1978, for examples). The photographic imagery is commonly colored to enhance differences in vegetation or soil color. The maps make excellent bases for geologic mapping in areas of low relief where vegetative pattern is variable, as in marshland and complexly cultivated areas. They do not have the sharp detail and subtle tonal images of individual contact prints, but geologic mapping can be transferred from prints to an orthophotomap more easily than to a topographic map.

*Photomosaics* are made from contact prints by cutting away all but the central part of each print and gluing the central parts onto a controlled base, thus producing a nearly map-true image of a large area. Mosaics show all the details of contact prints and provide a view of large structures and topographic features that would be difficult to see and impossible to measure by superimposing a set of contact prints. A disadvantage of mosaics is in not showing a stereoscopic view of the terrain. They are used mainly for compiling data from contact prints and for interpreting large features or extensive image patterns. Photomosaics are typically prepared for local government agencies and private companies, and may be available from them.

**Stereoscopic viewing.** Vertical photographs are taken in parallel strips (*flight lines*) spaced so that the photographs of adjoining lines overlap sideways, and at intervals that result in an overlap of about 60% between adjoining photographs of one line (Fig. 7-1). Any two consecutive photographs of a flight line are called *stereo pairs* because they can be used to see stereoscopic views of images in the overlap area. This is generally done with a *stereoscope*, a simple optical instrument that directs one eye toward one photograph and the other eye toward the other photograph. Because the two photographs were taken from different positions, their images are melded in the viewer's brain as a three-dimensional view, as in normal vision of nearby objects.

The folding *pocket stereoscope* (Fig. 7-2) is used in field studies because it is compact, inexpensive, and gives a moderately magnified view of the terrain.

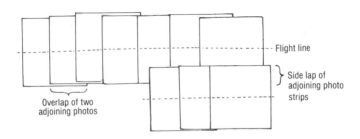

**Fig. 7-1.** Parts of two adjoining flight lines of aerial photographs, showing typical overlaps.

In order to avoid eye strain, it should be used as follows:

1. Adjust the stereoscope so that the centers of the lenses are the same distance apart as the pupils.

2. Place a stereo pair on a smooth surface and in the same order as in their flight line (Fig. 7-1).

3. Select a distinctive feature lying near the center of the overlap area.

4. Place one photograph over the other so that images are superimposed; then shift one photograph in the direction of the flight line until the features selected are as far apart as the distance between the centers of the lenses.

5. Place the stereoscope over the photographs so that the two images lie under the centers of the two lenses.

6. By looking into the stereoscope, the viewer should be able to see a three-dimensional image; if not, the two two-dimensional images can be brought together by shifting one photograph in the direction of the flight line or by rotating one of the photographs slightly. The latter adjustment is typically required for images with great relief, only small parts of which can be viewed stereoscopically at one time.

7. To view the concealed part of the overlap area, bend back the upper photograph, but not so sharply as to crease it.

Relief in the stereo view is exaggerated because of the geometric relation between the altitude of the camera and the focal length of the lens. Slopes and dips up to 60° appear twice as steep as they are, and those between about 70° and 75° 1½ times as steep.

Two additional kinds of stereoscopes used by geologists are the *mirror stereoscope* and the *single-prism stereoscope*. The mirror stereoscope (Fig. 7-2) is a table-top instrument that permits viewing an entire stereo-overlap area at one time. The image, however, is transmitted by two sets of mirrors and by optical eyepieces, and thus minor amounts of moisture or dirt on these

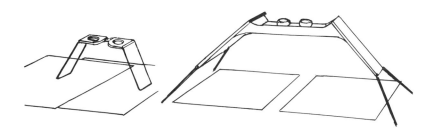

**Fig. 7-2.** Folding pocket stereoscope (*left*) and mirror stereoscope, with photographs in position for stereo viewing.

surfaces reduce the crispness of the image. The single-prism stereoscope consists of two backboards that carry the photographs. The prism directs one eye to one image, whereas the other eye views the other image directly. The instrument can be carried in the field, shows the entire overlap area at once, and does not require finding a flat surface on which to place the photographs. A drawback is its rather high price. It is currently available from Mechanical Technology Co., Inc., 5530 Port Royal Rd., Springfield, VA 22141.

A three-dimensional image can be seen without a stereoscope, although this normally requires experimentation and practice. If one looks fixedly for a minute at the point midway between the two dots in Fig. 7-3 and then relaxes the eyes (as in day-dreaming), the eyes will turn gradually to the position of rest, and as they do so the images of the two dots will move toward one another and merge. The single dot will be blurred at first but will shortly come into sharp focus, and the set of figures beneath the dot will appear as a three-dimensional form. To get this effect with aerial photographs, it is necessary at first to select distinct images, as a road-crossing that contrasts with its surroundings. The photographs must be placed so that the two images are no farther apart than the two pupils and are at a comfortable reading distance. With practice, the entire overlap area of two photographs can be examined quickly and easily — a skill worth developing because it saves a large amount of time and eliminates the need to carry a stereoscope in the field. Nonetheless, the magnifying power of the pocket stereoscope may be necessary to locate points based on details, as single trees or shrubs.

**Data printed on aerial photographs** include the date of photography and a unique number, as *DGW-10-23*, which is composed of: (1) a designation for the project (here, *DGW*); (2) the number of the film roll (*10*); and (3) the number of the photograph on that roll (*23*) (Fig. 7-4*A*). The average scale, the altitude of the aircraft, the focal length of the lens, and the full project designation are in some cases stamped on the back of prints and in other cases given on the photo index of the project. If only the focal length of the lens and the average altitude above the ground are given, the average scale can be calculated from the relation shown in Figure 7-4*B*. Because $La = Lb$, the ratio between the focal length and the altitude is the same as the ratio

**Fig. 7-3.** Dots and a pair of simple images that can be viewed stereoscopically.

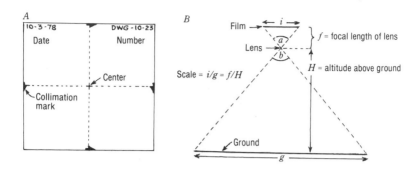

**Fig. 7-4.** *A*. Aerial photograph with typical data and construction lines for locating its center. *B*. Relations between camera and photographed terrain, with formula for calculating scale.

between any image dimension, *i*, and its ground dimension, *g*. The latter ratio is equal to the photograph scale. For example, if the focal length of the lens is 6 in. (0.5 ft) and the altitude is 10,000 ft, the scale is 0.5:10,000, or 1:20,000.

Each print also includes four black figures called *collimation marks* at the center of each margin. These marks can be used to locate the center of a photograph by aligning a straightedge between each opposite pair of marks and drawing two short lines to form a cross (Fig. 7-4*A*). This center point, often called the *principal point* (abbreviated P.P.), is where the optic axis of the camera intersected the earth's surface. If the camera was not tilted, the

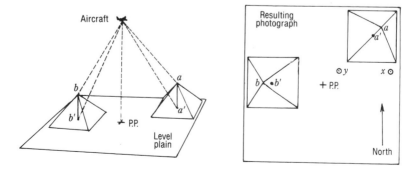

**Fig. 7-5.** Displacements of the tops of two pyramids (*a* and *b*) from their map-true positions (*a'* and *b'*).

point represents a ray of light arriving vertically from the ground. All other light rays that produced the photo image were inclined obliquely upward from the ground, increasingly so as their image points are distant from the principal point (Fig. 7-5).

*Image displacements* are caused partly by the obliquity just described and partly by differences in elevation of points on the ground surface. In Fig. 7-5, for example, the tops of the pyramids are displaced radially away from the center of the photograph compared with their actual map positions. The amounts of the displacements are proportional to (1) horizontal distance from the principal point and (2) vertical distance between the image point and the general level of the area photographed.

The displacements result in untrue bearings between image points at different elevations. The bearing from point $x$ in Fig. 7-5 to the top of the pyramid next to it is thus distorted from N 30° W to N 15° W. The example is not exaggerated and is typical of the outlying parts of large-scale photographs taken where relief is large compared to the altitude of the aircraft.

Such displacements can be reduced by using the central parts of photographs in measuring and transferring bearings. They can be eliminated entirely by taking bearings between points at the same elevation, or by sighting lines that are radial with respect to the principal point. The bearing from point $y$ to the top of the upper right pyramid is close to radial and thus nearly the same as a true bearing. The bearings between the tops of the pyramids remain true because the two points are at the same elevation.

Another result of image displacement is a noticeable difference in shape or width of features in adjoining photographs (Fig. 7-6). These differences are the basis of the stereoscopic image, and a stereo view will help the eye to compensate for the differences. Contacts and other geologic data will be easiest to compile or transfer accurately, however, if plotted on the image closest to the center of a photograph.

*Scale variations.* The camera lens is designed so that all features at one

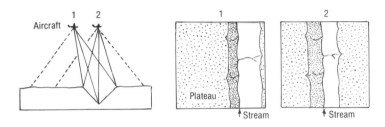

**Fig. 7-6.** Geometric relations (*left*) and two consecutive photographs (*right*) of a deep V-shaped canyon.

elevation will be shown at the same scale in all parts of a photograph. The scale varies directly, however, with the distance from the ground to the camera (Fig. 7-7). The actual variations in scale can be calculated by substituting the difference in elevation for $H$ in the relation: $scale = f/H$ (Fig. 7-4$B$). If, for example, the scale in one part of a photograph taken with a 0.5 ft lens has been determined to be 1:17,500 (Section 7-5), another area lying 200 ft lower would have a scale of 1:17,500 + (0.5/200) or 1:17,900, and an area 100 ft higher would have a scale of 1:17,500 − (0.5/100) or 1:17,300.

*Tilt* of the camera from vertical at the instant a photograph is taken results in systematic distortion of scale and of most directions. If the camera is tilted to the left of vertical, the scale will increase systematically from the left side of the photograph to the right side, and lines that are parallel on the ground will converge from right to left in the image. Camera mountings are designed to compensate for aircraft motion, and therefore photographs are rarely tilted more than a degree from vertical. Tilt distortions are removed in rectified images, as orthophotos. The photographs generally used for geologic mapping may have tilt distortions that would shift bearings by 1° and introduce small changes in scale. These distortions must generally be ignored during mapping. They will be removed when the mapped data are transferred to a topographic map or compiled by controlled photogrammetric methods.

### 7-2. Other Kinds of Remote-sensed Imagery

Remote-sensed imagery other than aerial photographs may be available in photographic black and white prints, in color prints, and as negative transparencies. The brief descriptions given here can be supplemented by reference to Sabins (1978), Lillesand and Kiefer (1979), or Colwell (1983). Current literature should be checked in any case, because additional kinds of imagery

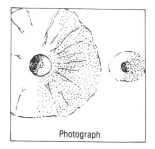

**Fig. 7-7.** Scale variation due to elevation: The craters of the two volcanoes have the same diameter but their images in one photograph differ greatly in size.

or products will certainly be developed. Information on obtaining federal imagery is available from the EROS Data Center (U.S. Geological Survey, Sioux Falls, SD 57198). Inquiries should include an exact description of the area and the kind of imagery needed. Aerial survey companies may be able to supply additional coverage.

**Daytime photographic infrared** imagery is available in color and black and white, and is best suited for delineation of vegetation. Tones and colors are thus suggestive of certain soils, surficial deposits, rock formations, and faults or fracture systems defined by vegetation. The most useful coverage is generally taken in the spring or fall, the seasons when major vegetative changes take place. A variety of scales may be available.

**Landsat multispectral scanning** (MSS) is imagery transmitted from orbiting satellites. Green, red, and two infrared wavelength bands are composed on 70 mm film (scale of 1:3,369,000) from which enlargements are made on paper or film in black and white or color. Each frame covers 185 x 185 km, with resolution that makes enlargements with scales of 1:1,000,000 to 1:100,000 most usable. Many frames are spoiled by clouds or mechanical defects, but each frame has been repeated enough times so that good to excellent images are available of almost all areas. The better frames can be viewed at U.S. Geological Survey regional offices. The images are used for study of major topographic units, geologic structures, waterways (especially on infrared bands), and vegetation differences. Frames recorded over periods of several years can be used to interpret on-going processes. The special value of the infrared imagery is similar to that noted above.

**Thermal infrared scanning** (IR) senses day or night radiation in the thermal range, and the data may be processed to black and white prints. Scales range from very large to very small, depending on the project. The imagery is generally flown for specific purposes, as studies of forest cover or wet ground, and much of it is not generally available. Because thermal emissions from certain surfaces and substances are distinctive, the imagery may show the distribution of certain kinds of rocks, vegetation, and surficial deposits. It is invaluable for locating areas of hot rocks and hot water, as in regions of potential volcanic activity.

**Radar sensing** by side-looking airborne radar (SLAR) records continuous strips that cover large tracts of land at intermediate to small scales (original image-scales are 1:100,000 and less). Because radar penetrates clouds, images are complete and may be flown at any time. Topographic forms are more enhanced than in ordinary photographic images, and topographically expressed fault lines and other lineaments are unusually distinct. The surface textures of the actual terrain correlate with brightness of the images, so that images may be used in reconnaissance mapping after field studies have identified specific kinds of surfaces. The imagery is available for much of the United States.

## 7-3. Photogeologic Studies

Aerial photographs and other remote imagery provide the only direct view of an area short of being there, and thus they are an invaluable introduction to its geography and geology. In addition, photogeologic study in the field is of great value because the scale and orientation of aerial views permit seeing features that are invisible on the ground. Once field study has established the causes of certain photograph tones, patterns, or lines, they can be used to interpret other parts of an area. A compelling concern is failure to notice these features on photographs until *after* the field season. The purpose of this section is to give a brief description of photogeologic interpretation based on a set of photographs, or other imagery, and a stereoscope. More thorough accounts and additional methods have been given by Ray (1960), von Bandat (1962), Miller and Miller (1961), and Smith and Anson (1968). Terrain properties of interest to engineering geologists were included by Way (1978).

Photogeologic studies should be based on stereo pairs and also on single photographs. Stereo views are used to determine the relief, shape, and orientation of landforms, which are commonly controlled by the kinds of rocks, the shapes of rock bodies, and the foliations and fractures within them, as will be described shortly. Stream densities and patterns may be especially expressive of certain rocks or deposits, and streams are typically oriented by the structural grain of a given area. Uncontrolled but useful drainage maps can be made by tracing streams on a transparent overlay after assembling photographs by carefully superimposing details along their flight lines.

Viewing single photographs is essential because tonal differences and patterns may go unnoticed in a stereo image due to its dramatic three-dimensional quality. It is important to examine tonal areas closely, even though their significance is totally unknown at the outset. Some are smooth-toned darker or lighter areas, and others are textured by dots, lines, or perhaps a fuzzy appearance. The separate areas may have elongate or other distinctive shapes, and they may be joined along straight or evenly curving lines and may be repeated in patterns. Straight alignments are clearest when the photograph is tilted away from the eye, so that the view is along the linear feature. Subdued patterns sometimes can be seen by first looking closely at parts of the photograph and then looking toward the photo center while relaxing the eyes to see the entire photograph at once (using one's peripheral vision, as when driving in heavy traffic).

The causes of the tonal and textured areas may relate directly to: (1) rocks, deposits, or soils developed on them; (2) natural vegetation, which is commonly controlled to some extent by the geology; or (3) burned areas, agriculture, or other human activities. Tonal and textured areas in arid regions and areas at altitudes above the tree line generally translate directly to certain rocks or deposits, and stereo viewing may reveal structural orientations of contacts or strike and dip of bedding or foliation. In semiarid regions, photo images are generally caused by some combination of plants and underlying materials. Areas covered by grass and brush tend to develop geologically related patterns or contrasts early in the growing season. Trees in these areas occur only locally and may be controlled by geologic features or by topography, in the latter case often being on the cooler or moister sides of ridges. In contrast, geological features under temperate forests or tropical jungles are likely to be expressed topographically, but not by distinct tonal or textural patterns.

*Identifying rocks from photographic images* requires careful study of texturing and of topographic expression. Subparallel elongate patterns, which may be clear only locally, suggest bedded or strongly foliated rocks. Distinct vegetative bands suggest sedimentary rather than metamorphic rocks, because the latter rarely have large contrasts in permeability. Bands underlain by sedimentary rocks with low permeability, such as clay-rich rocks in general, are typified by small forms of vegetation and by closely spaced small-scale drainage networks. These bands may be marked by landslides in recently uplifted regions and by subdued, smooth topographic forms in stable regions. Except for areas of pale-toned tuff, soil tones tend to be dark because of retained moisture.

Bands underlain by permeable rocks, especially sandstone and conglomerate, tend to be lighter in tone, to have coarser drainage networks, and to support large trees where precipitation is adequate. Their high permeability leads to rapid drying in the summer, thus causing lighter tones than in adjoining impermeable (moist) materials. Thick-bedded, widely jointed sandstone units are likely to form ridges in relatively stable areas and to crop out boldly in recently uplifted areas. Thick-bedded horizontal sandstone may also show parallel lines of vegetation along master joints.

Limestone commonly supports abundant vegetation and may be recognized by solution sinkholes, discontinuous drainage lines, and steep-sided hills (karst) in regions with ample rainfall. Limestone is likely to underlie valleys in humid regions, whereas in arid and semiarid regions it commonly forms ridges, steep slopes, and light-toned areas. Gypsum and rock salt also lead to sinkhole topography but not typically to the other features just mentioned.

Recent volcanic deposits are usually obvious from their association with eruptive landforms and flow forms (Chapter 13). Basaltic lava accumulations of Tertiary age may be characterized by extensive parallel layer-patterns, by occasional steep bluffs of dark-toned rock with talus aprons, by large upland areas with sparse drainage lines, by major springs issuing from deeply cut valley slopes, and by long lines of vegetation marking vertical feeder dikes and associated vertical fractures. Eroded terrains of intermediate to silicic lavas are suggested by pale tones; by scattered hills and linear ridges based on volcanic necks, pipes, and dikes; by curving patterns of flow structures (Section 13-4); and by closely spaced variations in vegetation and landforms. Volcanic rocks older than Cretaceous are likely to be altered, relatively impermeable, and closely jointed, thus appearing like any unit of layered hard rock.

Intrusive igneous bodies are suggested by uniformity of tone, texture, and drainage networks. Silicic and intermediate plutons are noticeably light in tone and often associated with light-toned dikes. Curving trellis drainage systems may parallel foliation in marginal facies, and crisscross drainage lines and tree lines suggest prominent joints within plutons (Section 14-7). Gabbro and ultramafic rocks show dark tones, and serpentinized peridotite is likely to support only sparse vegetation.

Aureoles of hornfels around intrusions commonly stand higher than either the intrusions or the metamorphic rocks outside the aureoles. Metamorphic rocks otherwise may be recognized by widespread foliation and layering that impart a parallel linear grain to the topography and drainage. Pronounced, closely spaced joints approximately perpendicular to ridges and linear patterns suggest folded metamorphic rocks. Terrains of high-grade schist and gneiss are suggested by overall dark tones interrupted here and there by pale bands of quartzite and pale-toned small plutons and dikes. In arid and semiarid regions, marble may also stand out as pale-toned ridges. Complexly curving landforms based on folded units are common in many metamorphic areas.

**Surficial deposits** of late Pleistocene age generally can be recognized by landforms and other features associated with them. Some examples are end moraines, drumlins, eskers, and pitted outwash glacial deposits; patterned ground and thermokarst in a present or past periglacial environment; alluvial terraces along river valleys; sets of beach ridges and associated dune fields; coastal terraces with former stacks and subdued sea-cliff faces at their landward margins; and extensive flat plains once covered by shallow Pleistocene lakes. Many of these deposits and features are described briefly in Chapter 10 and more fully in the references given there. Their photogeology is illustrated by superb stereo pairs collected and described by Mollard (1973?).

**Structures** of large size must be viewed on small-scale imagery, and SLAR imagery is typically ideal, as described in Section 7-2. Topographic forms, drainage patterns, and tonal patterns are used in conjunction to recognize large plutons, homoclinal sequences of strata, large folds, foliated metamorphic terrains, block-faulted areas, and major faults.

On intermediate to large-scale imagery, actual linear outcrops and detailed tonal or textural bands are indicators of stratiform sequences and therefore of folds, foliations, and faults. Direction of dip is indicated by the directions in which lines or bands cross topographic forms, exactly as with units mapped on topographic maps (Fig. 6-4). Where such lines are not visible, sets of asymmetric ridges suggest homoclines dipping in the direction opposite the steeper set of slopes; however, asymmetric ridges may be caused solely by moisture or temperature regimes. Amounts of dip are exaggerated by the stereo image but can be estimated approximately as noted in Section 7-1. Dips can be determined more accurately by measuring vertical distances on topographic maps of the same area or by photogrammetric instruments (von Bandat, 1962).

Faults are expressed by offset tonal or textured bands, by isolated lines of springs or trees, by abrupt breaks in slope, by abrupt changes in drainage patterns, and by streams with anomalously straight or smoothly curving courses. Features typical of recently active faults may be visible (Section 12-5).

Major joints are commonly widened by weathering and erosion and are thus visible as vegetative or other tonal lines. They do not offset other linear units, and they may form parallel systems or sets crossing at characteristic angles. Joint spacings commonly depend on the thickness of rock layers and on lithology, and are thus of great value in determining the extent of certain rock units. Their patterns in plutons and folded metamorphic rocks have been mentioned above. Joint sets should help to distinguish between surficial deposits and solid rocks; however, individual lineaments in bedrock may show through thin surficial deposits.

### 7-4. Equipment and Preparations for a Field Project

Chapter 1 describes the organization of a field project and some typical preparations before the field season. Kinds of aerial photographs and suggestions for obtaining them are described in Section 7-1. Even if enlargements will be used for plotting geology, a full set of contact prints will be needed for photogeologic reconnaissance (Section 7-3) as well as for stereo viewing in the field. Viewing will be easier if the two margins perpendicular to the flight line are trimmed from the prints. If photo indexes are not available, the geographic coverage of prints should be outlined and numbered on a map, as a 1:62,500-scale topographic map.

In addition to the usual field equipment (Chapter 2), a pocket or prism stereoscope will be needed for viewing in the field, and a mirror stereoscope may be useful in the office (Section 7-1). The map case that will be used in the field must be somewhat larger than the photographs and should close tightly enough to keep them clean and dry. For prolonged wet weather, a sheet of matte-surfaced drafting plastic (as Mylar) can be fitted to the base of the carrying case and taped on three margins to form a transparent envelope and drawing surface for each print. Fine-tip ink-flow pens (Section 2-1) can be used to mark all photograph surfaces and are required for plastic-base photographs. Paper-base photographs can be marked by pens or soft pencils, which should be tested before the field season. The photograph emulsion is softened by moisture and must then be marked gently.

If possible, mapped geology should be transferred from aerial photographs to a topographic base map as the mapping progresses. If the area is not covered by a map with a scale of approximately 1:24,000, an enlargement of a 1:62,500-scale map or a remotely sensed image that covers the entire area can be used.

## 7-5. Determining Photograph Scales and Orientations

Several photograph scales may be needed during field projects: (1) an approximate average scale for groups of photographs; (2) an average or typical scale for one photograph; and (3) scales for specific parts of one photograph. The first of these is given on photograph indexes and descriptions of photography, and can be calculated from the elevation of the aircraft and the focal length of the camera lens (see the subsection *Data printed on aerial photographs*, Section 7-1). In areas of low to moderate relief, the average scale of an aerial photograph will be nearly the same as the average scale determined from a flight line including that photograph, which is determined as follows:

1. Find two points on a map that can be identified exactly on two photographs within one flight line (Fig. 7-1).

2. Fix one of these photographs to a smooth surface, using paper drafting tape.

3. Lay the adjoining photograph over it by carefully superimposing images that are at about the same elevation and near the flight line; tape the photograph in place.

4. Continue similarly with successive photographs until the one with the second point is in place.

5. Measure the distances between the two points on the photographs and on the map, and calculate the scale by the relation

$$photo\ scale = map\ scale \ \text{X} \ \frac{distance\ scaled\ on\ the\ photos}{distance\ scaled\ on\ the\ map}$$

Scales in specific parts of single photographs can be determined in several ways: (1) by measuring the distance between two photo points that can be located precisely on a map of the same area, scaling the map distance, and calculating the scale by the relation given above; (2) by comparing any scaled distance or dimension on the photograph with the actual distance measured in the field; (3) by scaling photo-image distances along roads or crop boundaries marking the north-south and east-west lines of the township-section cadastral system (Appendix 5); (4) by determining the difference in elevation between a part of the photograph where the scale is unknown with a part where it has been determined (see the subsection *Scale Variations*, Section 7-1); and (5) if the scale in one photograph is known and that in an adjoining one is not, by scaling distances between two points in the overlap area and using the ratio of the distances to compute the scale from one photograph to the next.

**North arrows** are drawn on photographs to serve as a basis for plotting bearings and structure symbols. Roads and field boundaries of the Bureau of Land Management survey system (Appendix 5) can commonly be used as north-south lines; however some are out of alignment. Any straight stretch of a more or less level road or waterway can be used to transfer an orientation from a large-scale map to a single photograph by reading the bearing of the line on the map with a protractor. Lacking such a feature, the bearing measured between any two points can be used similarly, provided the points are at the same elevation or the line between them is about radial to the center of the photograph (Section 7-1).

Once a north arrow has been drawn on one photograph, it can be transferred to an adjoining photograph by referring to the bearing between any two points on or near the flight line (which is a radial line) or any two points that have the same elevation.

If a north arrow has not been drawn on a photograph before mapping is started, it can be plotted in the field by: (1) selecting two points at about the same elevation; (2) measuring the compass bearing between them; (3) aligning a protractor between the two points on the photograph; and (4) using the measured bearing to lay off a north-south line. Data can usually be plotted anywhere on the photograph if the line is approximately 6 cm long and is inked near the center.

### 7-6. Locating Photo Points in the Field

The great detail of imagery on aerial photographs with scales of 1:20,000 or larger makes it possible to pinpoint most field locations by inspection alone and often without stereo viewing. Individual trees and large shrubs, minor roads and trails, gullies, clearings, and buildings are examples of small distinctive features that can be recognized directly. The date of photography must be noted in order to allow for seasonal changes in vegetation, growth of vegetation, and changes due to fires or construction. The general procedure for making locations is to start by orienting the photograph parallel with the actual terrain, as by pointing the north arrow in the direction of north determined with the compass. Large and distinct features in the surroundings are then identified on the photograph so that the point occupied can be approximated. Finally, details in the immediate surroundings are used to pinpoint the location.

If the point is on a ridge or any other distinct break in slope, a stereoscope should be used at once to help in making the location. The magnifying lenses of the pocket stereoscope will disclose features too small to see otherwise, and the vertical exaggeration of the image will make large trees and subdued landforms obvious.

**Compass bearings** can be used to make locations on aerial photographs by the methods described in Section 6-3 if the limitations explained in Section 7-1 are taken into account. In brief, a compass bearing read between two points on an aerial photograph will correspond exactly to one read with a compass or from a map under any of three conditions:

1. The area being mapped is a level plain.

2. The elevations of the two points are the same.

3. The two points fall on a line radiating from the center (the principal point) of the photograph (Fig. 7-5).

Situations similar to one of these conditions will give accurate locations in the central part of a photograph.

**Measured distances** can be used to make locations on aerial photographs exactly as on maps if the photograph scale is known along each line. In general, differences in elevation greater than 30 m will result in measurable differences in scale. Section 7-1 describes the calculation of photograph scales at different elevations, and Section 7-5 describes how local photograph scales can be determined in the office or in the field.

### 7-7. Geologic Mapping in the Field

Geologic aspects of mapping on aerial photographs are described in Chapters 1, 3, and 5. Photogeologic methods (Section 7-3) provide a powerful means of planning mapping as well as of predicting the position of rocks, contacts, and faults during the course of field work. Some contacts show far more clearly on aerial photographs than on the ground, and photographs are thus an ideal base for rapid (reconnaissance) mapping (Section 5-5).

Contacts and faults are drawn as thin pen or pencil lines that are true to the local photograph imagery. For example, a contact that lies along a tonal boundary on a photograph is drawn exactly along that boundary, even though this line might slope so steeply that its image is oriented 5° or more off its actual bearing (Section 7-1). Such lines are corrected later when mapping is transferred to a topographic base or to a planimetric map compiled from the photographs.

Strike lines, on the other hand, are plotted map-true and must therefore be based on a true-north arrow (see Section 7-5). Strike lines can be plotted over a large part of a photograph by using one north line, a transparent protractor, and, if necessary, a transparent scale to extend the line (Fig.

6-1*A*). Structural data must also be recorded in the notes, or perhaps on the back of the photograph if large areas are being mapped rapidly and points are spaced widely. The recorded data are used later to plot all structures accurately on a topographic map or a compiled planimetric map.

Field locations are generally marked by pricking a small hole through the photograph with a needle, which can be mounted as in Fig. 2-1*B*. The note number is written on the back of the photograph. The needle holes on the image surface can be located by holding the photograph up to the light or by inserting a needle in the hole from the back of the photograph. Another method is that of marking and numbering points on the photograph; however, the marks obstruct the photo image and plotted geology and may damage the surface more than a needle hole.

Where a large number of photographs are being used, it may be helpful to record on the back of each photograph each date when it was used and the location numbers added for each day. These data can be used at any time to locate the field note applying to any location on any photograph. Geologic lines and symbols should be plotted as mapping progresses in order to point up errors and discover structural relations immediately. If a soft pencil is being used, the lines should be inked every day or so because they rub off easily. Ink work should be as fine as possible. Faults may be inked in a color rather than as broad black lines, and red and green lines are generally more legible than blue ones. Each part of the area should be mapped and inked on only one photograph. This system preserves as much of the photo imagery as possible and makes the geologic lines appear clearer in stereo images.

The central part of a photograph provides the most nearly map-true image and therefore is generally the preferred part to use in mapping. If relief is low to moderate, however, no part of a photograph will be seriously distorted, and there are several advantages in using entire photographs for mapping: (1) larger areas can be examined synoptically; (2) photographs can be laid out in overlapping sequences without covering most of the mapped relations; and (3) a set of alternate photographs can be left completely unmarked, which is desirable for photogeologic studies.

### 7-8. Compiling Data from Aerial Photographs

Geologic data should be transferred to a base map or smaller-scale photo image from time to time during the mapping season. Important relations might otherwise be missed and some parts of the area might well go unmapped. Compilations are most valuable when made on a topographic map because geometric relations are shown clearly and cross sections can be constructed easily. If only a small-scale topographic map is available, enlarging it to approximately the same scale as the photographs will make transferring much easier.

Photograph imagery is so nearly map-true in areas of low to moderate relief that transferring can be based on any geographic features shown on both the map and the photographs. Many of the transfers can be made by inspection; for example, where a structure symbol is located at a distinct bend in a creek or road. For points that cannot be transferred by inspection, a pair of proportional dividers can be set so as to convert the local photograph-scale to the map scale. The dividers are then used to obtain distances between recognizable points along streams or other geographic features and the points where contacts or faults cross such features. Lacking proportional dividers, triangles can be drawn among points that can be located on both a photograph and the map, and distances scaled along triangle legs can be used to control contacts and other data (Fig. 7-8).

In areas of high relief, transferring may be done largely by comparing the stereo image of the photographs with the contoured topography of the map. Points where contacts and faults cross ridges and irregular hill forms can usually be estimated closely. Data along streams or other areas that are more or less at one elevation can be transferred with proportional dividers as just described. Distortions due to differences in elevation are corrected by drawing from the stereo image. If slopes are so steep that only parts of the stereo image are clear at one time, the transferring must be done patiently, completing one small area at a time.

A planimetric map may also be used for a base if it shows drainage and cultural features in detail. If no suitable map is available, small-scale aerial photographs or other remote-sensed imagery can be used (Section 7-2). Small-scale images are far less distorted by relief than those of large-scale

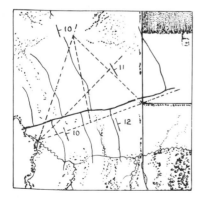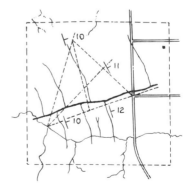

**Fig. 7-8.** Transferring geologic data from an aerial photograph (*left*) to a map by drawing a triangle (*dashed*) among three distinct points and using proportions along these lines, and along bisectrix lines, to make locations.

photographs, so that data can generally be transferred by inspection, as described above. Once geologic features are transferred from a set of large-scale photographs to a few small-scale photographs, the latter can be taped down so that their images are superimposed and the geology traced on a transparent overlay. This planimetric compilation can be completed by adding streams and cultural features.

### 7-9. Cross Sections from Aerial Photographs

Structural relations on aerial photographs can be viewed only one stereo pair at a time, so that broader relations must be checked by constructing cross sections across groups of photographs. If the geology can be transferred to a topographic base map, cross sections are constructed as described in Section 6-5. Otherwise, partially controlled cross sections can be constructed from the photographs and a planimetric base map by the following steps: (1) select a line of section and prepare a cross-section strip as described in Section 6-5; (2) transfer from map to strip the points where the section line crosses streams and ridge lines; (3) use elevations given on the planimetric map to plot positions of those points on the cross section; (4) transfer the line of section from the map to the photographs by comparing geographic features in the central parts of the photographs; (5) add topographic detail and geology from the aerial photographs as described in steps (4) through (8) in the following paragraph.

If no base map is available, approximate cross sections can be prepared from photographs as follows: (1) select the photographs through which the section passes and tape them down so that photo images near the flight line are superimposed and the north arrows are parallel; (2) draw a line of section across the assembly and prepare a strip of tracing paper or plastic drafting film for the section; (3) lay the strip along the line of section and mark on the strip the points where the line crosses streams or other distinct geographic features; mark at least two points per photograph; (4) disassemble the photographs and examine them stereoscopically in order to locate the topographically lower points (typically the main stream valleys) and the higher ridges in the terrain; estimate their elevations and plot their positions on the section strip; (5) using these points as a general control, estimate the vertical positions of intermediate streams and ridges and plot them on the section strip; (6) complete the profile by noting the shapes of slopes between these various points, allowing for exaggeration of slopes (Section 7-1), and adjusting the stream and ridge elevations as necessary; (7) transfer all geologic features at or near the section line, using a stereoscope as needed, and project structures into the section line as described in Section 6-5; and (8) complete the sections as described in Section 6-5.

If the vertical positions of points must be controlled more exactly than by the methods just described, any number of elevations can be measured by walking the line of section and using the methods of Sections 2-6 or 2-7. If still greater control is needed, as for the final cross sections of an area of gently dipping units or structures, a survey along the section line can be made with a plane table and alidade (Chapter 8).

### 7-10. Compiling a Map by the Radial Line Method

If neither a base map nor small-scale photographs are available for compiling geologic data, a planimetric map can be constructed from the photographs by a method that corrects for distortions due to relief. This is the *radial line method*, based on the fact that the true position of any image in an untilted photograph lies along the radial line passing from the center of the photograph through the image (Section 7-1). Equipment and materials that are desirable but not essential are a drop-circle (spinner) compass with an ink tip, and two or three colored inks into which a small amount of water-tempera paint or china-white pigment are mixed so that the ink will be visible on dark photo images.

The instructions for the full procedure that follows are based in part on a variation of the method kindly provided by Herbert E. Hendriks:

1. Locate the center (principal point) of each photograph (Fig. 7-4A).

2. Place the pivot pin of a drop-circle compass exactly at each center point and spin a circle approximately 4 mm in diameter, using colored ink.

3. Select several points that should become located accurately (as stream intersections, points where contacts cross ridges, hilltops with structure data, etc.) and spin color circles around each; these *pass points* should lie in the central band of a given photograph and at least one should be in each sidelap area (Fig. 7-9A).

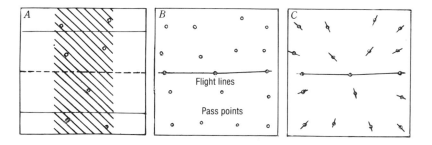

**Fig. 7-9.** Preparations for a radial-line plot. *A.* Locations of pass points, showing central band (lined) and areas of sidelap with adjoining flight strips. *B.* Points and lines traced from photograph to overlay. *C.* Overlay with radial lines added.

4. Transfer each center and pass point to the adjoining two photographs by inspection of the photo image, and spin circles over each. If the image is not distinctive enough to do this, view the photographs stereoscopically and place the drop-circle compass on the unmarked photograph exactly at the pinhole that seems to be in this photograph; spin a circle the same size as the initial circle. If the location is correct, the two circles will merge into one in the stereo image and it will seem to lie exactly at the level of the pinhole. If it lies above or below the pinhole, wipe the ink off quickly and try again.

5. Cut sheets of plastic drawing film that are slightly smaller than the photographs and fasten a sheet to each photograph, matte side up, using several small pieces of drafting tape.

6. Write the photograph number on each overlay and trace all pass points and center points appearing on that photograph.

7. Draw lines (flight lines) on each overlay from the center point through each transferred center point and somewhat beyond (Fig. 7-9*B*).

8. Add lines that radiate from the center point through each pass point, drawing only a 3-cm segment centered on the pass point (Fig. 7-9*C*).

9. Remove the overlays from the photographs and lay both sets out in the order of the flight lines.

10. Select two adjoining photographs from one flight line that have their centers at about the same elevation and preferably in an area where the local photograph scale has been determined (Section 7-5). Superimpose the two overlays of these photographs by bringing their flight lines into exact coincidence. If one flight line is longer than the other, the scales of the two photographs differ. Either length (or any length) can be used; however, *it will set the scale for the entire compilation.*

11. Add the next consecutive overlay to the first two by superimposing flight lines and moving the overlay until the pass points in the overlap area

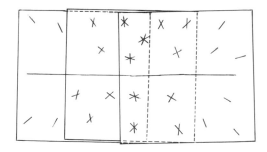

**Fig. 7-10.** Three adjoining overlays with flight lines superimposed and radial lines forming three-line intersections.

of the three photographs form 3-line intersections (Fig. 7-10). Tape the three overlays down lightly.

12. Repeat step 11 with the next overlay and continue similarly throughout that flight line. If small triangles result at some intersections, check the accuracy of the radial lines by placing the overlay on its photograph. If all intersections on one overlay are off consistently, they can sometimes be improved by shifting the overlay laterally or by rotating it slightly, suggesting that the photograph was tilted.

13. When the overlays of the first flight line are assembled as satisfactorily as possible, select three overlays in an adjacent flight line, preferably in an area of relatively low relief. There will be an advantage if the photograph scales along the flight lines are about the same as those used in step 10. Tentatively assemble the three overlays as in steps 10 and 11.

14. Place this tentative assembly over the first assembly and bring all pass-point intersections in the sidelap area as nearly as possible into coincidence. The scales of the two assemblies will typically differ somewhat, so that the new assembly of three overlays may have to be expanded or contracted parallel to its flight lines to achieve a fit.

15. Continue by adding overlays to the second flight line and adjusting them as necessary to the pass points located in the first assembly of overlays. Tape them together lightly so that small readjustments can be made easily. The final result will typically include intersection triangles that cannot be resolved and will have to be accepted.

16. When all flight lines have been assembled and joined laterally as in steps 13, 14, and 15, place a sheet of drawing film over the entire assembly of overlays and trace all center points and pass points with colored ink.

17. Place this master sheet over one of the photographs, preferably starting in an area of low relief and where the scales of the photograph and the compilation are about the same.

18. Bring the traced center into coincidence with the actual photograph center and align the overlay by means of the flight lines and the various pass points. Using a pencil, draw all geologic lines, drainage lines, and cultural features around the center. Then shift the overlay sheet so that the closest

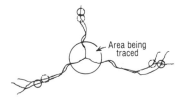

**Fig. 7-11.** Tracing detail in a small area by first positioning pass points of overlay (*outer circles*) proportional to pass points of photograph (*inner circles.*)

traced pass point lies over the corresponding point on the photograph and draw the features around it. Continue similarly with all neighboring pass points.

19. To complete lines between these areas, shift the sheet so that its pass-point marks are positioned symmetrically with respect to two or three pass points on the photograph and draw the lines in that area (Fig. 7-11). This procedure changes the scale of the photograph images to the scale of the compilation, and in areas of high relief it may be necessary to view the stereo image or to locate additional pass points.

20. When the data from all the photographs have been penciled onto the compilation, a second sheet is placed over it and the lines and other features are traced in ink. The map can then be completed to whatever stage is desirable, as by the suggestions given in Section 16-2.

**Controlled radial line plots** can be compiled more firmly and are more accurate, but may or may not be worth the effort spent in surveying control stations or assembling existing survey data (Section 8-7). The control stations are plotted on a single sheet at the scale most suitable for the photograph compilation (generally close to the average scale of the photographs), and the stations are also located exactly on all photographs on which they appear. The compilation will be simpler mechanically if the appropriate control stations are transferred to elongate plastic sheets, which just cover each of the assembled flight lines of photographs. When the radial line assembly is started for any one flight line (as described in step 10 above), radial lines through the control points are used to hold each photograph in position and thus to set the scale of the assembly. This is done by placing the plastic strip with control points traced on it over the three-overlay assembly of step 10, and expanding or contracting the assembly until the three-line intersections fit the corresponding control points of the overlay. Two or so control points per photograph, located in side-lap areas, are an ideal number; however, even a few points for the entire map will help control the compilation.

### References Cited

Colwell, R. N., editor, 1983, *Manual of remote sensing* (in 2 volumes), 2nd edition: Falls Church, VA, American Society of Photogrammetry, 2440 p.

Ellis, M. Y., editor, 1978, *Coastal mapping handbook*: U.S. Geological Survey and National Ocean Survey, Washington, Government Printing Office, 200 p.

Lillesand, T. M., and Kiefer, R. W., 1979, *Remote sensing and image interpretation:* New York, John Wiley & Sons, 612 p.

Miller, V. C., and Miller, C. F., 1961, *Photogeology*: New York, McGraw-Hill Book Co., 248 p.

Mollard, J. D., 1973?, *Landforms and surface materials of Canada, a stereoscopic airphoto atlas and glossary*, 3rd edition: Regina, Sask., J. D. Mollard, 56 p. and 336 p. of stereo views.

Ray, R. G., 1960, *Aerial photographs in geologic interpretation and mapping*: U.S. Geological Survey Professional Paper 373, 230 p.

Sabins, F. F., Jr., 1978, *Remote sensing: principles and interpretation*: San Francisco, W. H. Freeman and Co., 426 p.

Smith, J. T., and Anson, A., editors, 1968, *Manual of color aerial photography*: Falls Church, VA, American Society of Photogrammetry, 550 p.

von Bandat, H. F., 1962, *Aerogeology*: Houston, Gulf Publishing Co., 350 p.

Way, D. S., 1978, *Terrain analysis: a guide to site selection using aerial photographic interpretation,* 2nd edition: Stroudsburg, PA, Dowden, Hutchinson, & Ross, 438 p.

# ■ Mapping with the Plane Table and Alidade

## 8-1. The Alidade

Mapping with the plane table and alidade is traditional for geologists because the map is constructed directly as measurements are made in the field. Geologic features can be related precisely to topography, and relations among complex geologic data can be compared and checked immediately. Projects requiring map-scales greater than 1:10,000 typically employ this method, especially where exact vertical measurements are required.

A variety of modern alidades are obtainable, including exceptionally compact ones with optical devices for making rapid readings, self-indexing level-units, and extending rulers for plotting lines of sight (see a supplier's catalog, Section 2-1). These instruments and their use are described in literature available from manufacturers. The descriptions and instructions in this chapter apply specifically to the *standard alidade* manufactured by several companies; the instructions may be applied in principle to other instruments.

*The standard alidade* (Fig. 8-1) consists of a basal *blade* (1) aligned parallel to a telescope (2) by means of a rigidly mounted pedestal (3) and a horizontal axis and bearing (4) on which the telescope may be elevated and depressed in the vertical plane of the instrument. This movement can be constrained by setting the *axis clamp screw* (5), after which a gradual motion is effected by turning the *tangent screw* (6). The telescope can be rotated on its axis through 180° when the *retaining ring* (7) is loosened; during ordinary measurements,

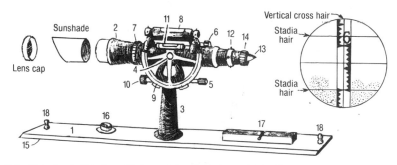

**Fig. 8-1.** Standard alidade, with insert showing view of a stadia rod as seen through the telescope. Numbered parts are identified in the text.

the telescope should be turned firmly against a stop, such that the focusing knob is on the right side (and thus out of sight in the figure). The telescope is brought to a horizontal sight by referring to the *striding level* (8), which lies loosely on two bearing surfaces of the telescope and may be removed after opening a spring clamp or, in some models, by loosening a screw.

Vertical angles are read from a *calibrated arc* (9) that is adjusted by means of a *vernier tangent screw* (10) and a *vernier level* (11). Most arcs are also calibrated with the Beaman arc scale, which is used for correcting stadia readings (Section 8-4). *Stadia hairs* and adjustable *cross hairs* (12) are mounted in front of the eyepiece (13), which is focused by turning a ring (14).

The blade's *fiducial edge* (15) is parallel to the axis of the telescope, and the blade has a level (16) for leveling the plane table, a compass (17) for alignments to magnetic north, and two small knobs (18) for moving the alidade on the plane table.

## 8-2. Care and Adjustments in the Field

If cleaned and adjusted before the field season, an alidade will generally stay in good working order if:

1. It is lifted by the pedestal, never the telescope.

2. It is placed in its carrying box when not in use and not left on the plane table unattended.

3. The lenses are cleaned when necessary with a soft (camel's hair) brush.

4. The shield of the striding level is turned up and the level is removed before placing the alidade in its case.

5. The magnetic needle is lifted off its bearing when not in use.

6. The case is cushioned when being transported in a vehicle.

The adjustments that follow are made routinely in the field.

*Parallax correction.* If a sighted point shifts slightly in relation to the cross hairs when the eye is moved, the cross hairs must be focused precisely by the eyepiece ring after pointing the telescope to the sky. The telescope is then depressed to the original sight and is refocused as necessary.

*Striding level.* To adjust the level so that it is parallel with the telescope axis; (1) center the bubble by moving the tangent screw; (2) reverse the level end for end; (3) if the bubble moves off center, turn the capstan screws on the level to bring the bubble halfway back to center; (4) center it again with the tangent screw; and (5) reverse the level, repeating the procedure until the bubble stays on center when the level is reversed.

*Magnetic needle.* If the compass needle dips so much that it touches the window of the compass box, slide the balancing weight along the needle until it remains level.

*If an alidade has been dropped*, the set of the cross hairs should be checked by: (1) placing the alidade on a solid level surface; (2) releasing the retaining ring and rotating the telescope firmly against its stop; (3) sighting on a distant point and adjusting for parallax; and (4) elevating and depressing the telescope with the tangent screw to see if the point sighted stays on the vertical cross hair. If it does not, the cross hairs are rotated by loosening the four capstan screws of the cross hair mount and rotating the mount by very light taps. The line of collimation (the axial line through the telescope) is then checked by releasing the retaining ring and rotating the telescope through 180° to determine if the cross hairs stay on the point sighted. If they do not, each pair of capstan screws is turned until the cross hairs stay on the point sighted when the telescope is rotated.

### 8-3. The Plane Table, Mapping Sheets, and Tripod

The plane table is a firmly constructed drawing board with a *base plate* for attaching it to the tripod. The standard board is small enough (18 X 24 in.) to permit steep downhill sights from most parts of its surface.

Plane table sheets can be purchased ready-made or can be made easily from a variety of materials. Factors in selecting materials are durability, smoothness, resistance to moisture (especially with respect to dimensional changes), thermal coefficient of expansion, and reflectance (sun-glare from white or metal sheets may be a problem). An inexpensive sheet may be made from medium-gauge matte-faced clear plastic (as Mylar) mounted over a separate sheet of light-brown paper. Expansion and contraction should not change the map scale seriously, except possibly during triangulation over

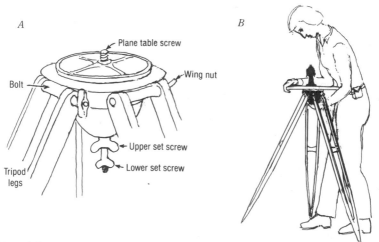

**Fig. 8-2.** *A.* Johnson head and upper part of plane table tripod. The cup shape of the head permits tilting the plane table moderately in any direction when the set screws are loose. *B.* Leveling the plane table by observing the level on the blade of the alidade.

long distances (Section 8-7). The sheet may be kept clean by an overlay of brown paper that is cut away as the mapping progresses.

Plane table tripods have a *Johnson head* that is attached to wooden legs by bolts and wing nuts (Fig. 8-2*A*). The wing nuts should be set just firmly enough to hold the tripod leg unsupported at an angle of about 45°. The Johnson head is designed so that the plane table can be leveled and oriented easily, as by the steps that follow:

1. Set the tripod up, with its legs inclined about 30° from vertical; loosen both set screws of the Johnson head, and move the upper (main) part of the head until it is evenly upright.

2. Tighten both set screws and screw the plane table down so that it fits firmly against the head.

3. Place the assembly over the station occupied such that the map is oriented approximately and the ground marker of the station lies under the corresponding point on the map.

4. Level the board approximately by moving the tripod legs; on steep slopes, place two of the legs downhill.

5. Press the tripod feet into the ground or brace them with chunks of rock so that they will stay solidly in place.

6. Place the alidade without its striding level in the center of the board, parallel to the board's length.

7. Grip the far side of the board with one hand, hold the blade down under that arm, and press the near edge of the board firmly against the waist (Fig. 8-2*B*).

8. Loosen both set screws of the Johnson head with the free hand and level the board by referring to the level on the blade. Tighten the upper set screw of the Johnson head.

The board can be rotated on what is now the vertical axis of the Johnson head, and it is oriented for mapping by the following steps:

1. Place the fiducial edge so that it bisects the map point of the station occupied and also the map point of the station that will be sighted.

2. Rotate the board on its vertical axis until the alidade is pointing at the station to be sighted; look through the telescope in order to bring the station sighted exactly on the vertical cross hair.

3. Tighten the lower set screw of the Johnson head, making sure that the cross hair remains on the station sighted.

4. Check for parallax, and make sure the board and tripod are set firmly enough so that they will not be disoriented during mapping.

The board can also be oriented by reference to magnetic north; however, small local variations in the magnetic field tend to make this method less accurate (see *Magnetic anomalies*, Section 2-5).

### 8-4. Stadia Measurements

The stadia hairs of the eyepiece are spaced so that their intercept on a graduated rod viewed through the alidade is 1/100th of the distance from the alidade to the rod. Thus if the intercept on the rod is 1.23 (m or ft) (Fig. 8-1), the sighted distance is 123 (m or ft). If the line of sight is not level, however, the sighted distance must be corrected to a horizontal (map) distance. The sighted distance must also be converted to a vertical distance in order to map contours or plot elevations. These conversions can be made by using the Beaman arc attached to most alidades, as described in principle below and in detail in a subsection that follows. The conversions may also be made by measuring the vertical angle to the point sighted and using trigonometric functions to calculate the horizontal and vertical distances, a method that will be described after the Beaman arc procedure. Finally, stadia (slope) distances can be converted to horizontal and vertical distances by using a stadia slide rule, or the stadia reduction tables included in some surveying books.

*The Beaman arc* has two scales scribed next to the vertical-angle arc and read by a separate index (Fig. 8-3). The H scale is used for the horizontal conversion and is read by visually projecting the index line across the V scale. The H reading is multiplied by the stadia intercept to obtain the horizontal correction. If, for example, the stadia intercept is 3.53 and the H reading is 3.2, their product, 11.3, is subtracted from the stadia distance, 353, to give the horizontal distance, 342.

As shown in Fig. 8-3, the V scale reads 50 for a horizontal sight. This number is subtracted from the V reading in order to obtain the vertical correction

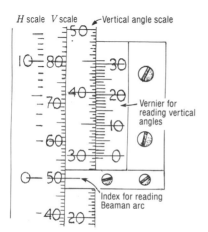

**Fig. 8-3.** Central part of a Beaman arc and vertical-angle arc, with indexes indicating a horizontal line of sight. The arcs illustrated are *edge-mounted* and read from the rear of the alidade; those on the alidade in Fig. 8-1 are *side-mounted* and read from the left side of the instrument.

factor, giving negative factors for inclined sights and positive factors for depressed sights. The vertical correction factor is then multiplied by the stadia intercept to give the vertical distance above or below the alidade. For example, a V reading of 47 and a stadia intercept of 3.53 would give a vertical distance of (47 − 50) × 3.53, or −10.6.

The vertical distance from the alidade to the point read on the rod must then be modified to obtain the elevation on the ground where the rod is held (Fig. 8-4). The height of the alidade above the ground (the H.I.) is first added to the elevation of the station occupied to obtain an elevation of the instrument (sometimes abbreviated E.I.). The cross-hair reading on the rod is then given a − sign and added algebraically to the vertical distance computed from the stadia distance, as just described. For example, if the vertical distance is −10.5 and the cross-hair reading is 6.5, the elevation difference is −10.5 − 6.5 or −17. The elevation on the ground where the rod is held is thus the elevation of the instrument less 17.

**Rods for stadia surveying.** A standard leveling rod (Philadelphia rod) is suitable for large-scale mapping by sights of 100 m or less, and a *stadia rod* graduated in feet and tenths, or meters and tenths, for surveys requiring longer sights. Stadia rods are typically 10 to 15 ft (3 to 5 m) long, the longer ones being hinged or separable at the center. The brightness and distinctness of the graduated patterns affect the accuracy of instrument-readings, so that rods should be handled carefully, cleaned occasionally, and touched up with paint as necessary. When making stadia sights, the rod should be held so that its marked face is in the sunlight if possible, which may require turning it somewhat from the line of sight.

The person holding the rod can keep it plumb (vertical) by balancing it between his or her fingertips or, in winds, holding a clinometer against it. The person with the rod also describes each station in a notebook, number-

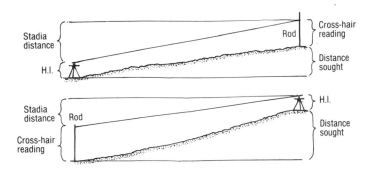

**Fig. 8-4.** Relations of the vertical stadia distance to the H.I., the cross-hair reading, and the vertical distance between two points.

ing the entries consecutively so that they can be matched with the consecutive numbers recorded in the instrument notebook and on the map.

**The Beaman-arc procedure** described here is designed for mapping at scales larger than 1:10,000 and for stadia distances less than 250 m (800 ft). Persons who have gained even moderate experience in plane table surveying can easily modify the steps to suit special needs. If a calculator is not available, the lower part of each notebook page (Fig. 8-5) should be reserved for computations. The brief *remarks* help key the stations to the corresponding numbers and descriptions in the notebook kept at the rod. In the procedure described here, the person with the rod records the main geologic observations.

After the plane table has been set up and oriented (Section 8-3), the height of the instrument is measured with the stadia rod and added to the elevation of the station to give the elevation of the instrument. This can be recorded as in Fig. 8-5. Mapping then commences as follows:

1. Point the alidade at the rod by sighting along the top of the telescope, making sure that the point marking the station occupied is next to the fiducial edge. Using reasonable care, the rod should be visible through the telescope, so that the alidade need be moved only slightly to bring the vertical cross hair onto the rod. As this is done, the fiducial edge must be positioned a pencil-line away from the point occupied.

2. Draw a ray with a sharp 8H or 9H pencil, stopping it short of the point occupied in order not to obscure the needle hole.

3. Tighten the axis clamp screw, then move the telescope with the tangent screw until either stadia hair is exactly on a whole foot or meter division. Check for parallax (Section 8-2) and correct the focus as necessary, because this will be the most important reading taken. *Count* the full units and tenths, and estimate the hundredths, to the other stadia hair—a procedure that prevents errors that might result from subtracting. Check the focus and the count, and record the intercept at once. If too little of the rod is visible to read a full intercept, count the intercept between a stadia hair and the horizontal cross hair and double this number before recording it.

4. Level the telescope by referring to the striding level and use a 2X or 3X magnifying glass to bring the Beaman arc index to 50 on the V scale. Do this by turning the *vernier* tangent screw (10 in Fig. 8-1). This adjustment corrects for any tilt of the board caused by moving the alidade from one position to another.

5. Look through the telescope to see if the horizontal cross hair is on the rod or is close enough to it so that a cross-hair reading can be estimated to the nearest 0.2 m (0.5 ft). If so, H is 0, V is 50, and all that is needed to complete the sight is to record the cross-hair reading (see step 7).

6. If the sight is inclined (and thus step 5 cannot be completed), unclamp and tilt the telescope until the cross hair is near the center of the rod. Then

| Party, date | Plane table at | Rod pt. | Stadia intercept | Beaman V | Beaman H | H x stad. intercept | Horiz. dist. | V factor |
|---|---|---|---|---|---|---|---|---|
| Emory, | P; Elev = 510.2 | 1 | 2.85 | 54 | 0.5 | 1 | 284 | +4 |
| Willis | H.I = 4.3 | 2 | 3.32 | 40 | 1.0 | 3 | 329 | -10 |
| 4 June 83 | Elev. inst = 514.5 | 3 | 4.89 | 73 | 5.5 | 27 | 462 | +23 |
|  |  |  |  |  |  |  |  |  |
|  |  |  |  |  |  |  |  |  |
|  |  |  |  |  |  |  |  |  |

**Fig. 8-5.** (and see opposite page) Two facing pages of notebook used for stadia surveying with a Beaman arc. Products are rounded for typical precision when mapping at a scale of approximately 1:2400.

clamp the telescope and turn the tangent screw (no. 6 in Fig. 8-1) until the V scale graduation closest to the Beaman arc index is exactly at the index. Use a magnifying glass to check this setting. The whole number is used because it can be read precisely from the V scale and simplifies computations. Read the V and H scales and record the numbers.

7. Observe where the horizontal cross hair crosses the rod. This number, the cross-hair reading, generally need be no more exact than the nearest 0.2 m (0.5 ft). It is recorded with a minus sign.

8. Signal the person with the rod to move to the next point.

9. Find the product of the stadia intercept and the H reading and record it (Fig. 8-5). Subtract this number from the stadia distance (stadia intercept x 100) to obtain the horizontal distance. Record the distance and scale it along the pencil ray. Mark the point with a small needle hole and number it.

10. Subtract 50 from the V reading and record the resulting *V factor*; then multiply it by the stadia intercept and record the product. If a calculator is not available, do the multiplication on the lower part of the notebook page so that it can be checked.

11. Add this number algebraically to the cross-hair reading and record the result (the *Net diff. in elev.* in Fig. 8-5).

12. Add the latter number algebraically to the elevation of the instrument to obtain the elevation of the point sighted. Record this elevation in the notebook and next to the point on the map.

In surveys where the person with the rod is moving rapidly through a series of closely spaced points, the procedure will be more efficient if a third person records stadia data at the instrument and computes results as the work proceeds. If only one person is at the instrument, he or she should look up occasionally while computing, because the person with the rod may be difficult to spot when standing still at a new point. A pair of field glasses

| V factor X stad. int. | Cross-hair reading | Net diff. in elev. | Elev. at rod | Remarks |
|---|---|---|---|---|
| + 11.4 | - 5.5 | +6 | 520.5 | Crossing of road and stream |
| - 33.2 | - 10.5 | - 44 | 470.5 | Base of Lyle Ss in steep gully |
| + 112.5 | - 3.0 | + 109.5 | 624 | Lg otcp at ridge crest |
|  |  |  |  |  |
|  |  |  |  |  |
|  |  |  |  |  |

will be helpful, especially in brushy country.

*A vertical-angle procedure* must be used if the instrument has no Beaman arc. If a hand calculator with tangent and cosine functions is available, this procedure may be preferable to the Beaman arc method. The procedure is based on the vertical-angle arc and its vernier scale, which will therefore be described first.

The vertical-angle arc encompasses only 60°, because standard alidades can be inclined or depressed only a little more than 30°. The scale is marked at 30' intervals, and its "0" or horizontal mark has been given the number 30 in order to eliminate mistakes caused by reading and recording the sign of angles close to 0 (Fig. 8-3). Thus, the number 30 is subtracted from the number read on the vertical-angle scale, giving + signs for inclined angles and – signs for depressed angles.

The vernier scale is marked in 1' divisions that are somewhat smaller than the marks on the vertical-angle scale (30 vernier marks is exactly equal to 29 vertical-angle marks). The vernier is read by noting which of its marks exactly matches a mark on the vertical-angle scale. In Fig. 8-3, for example, the 0 mark of the vernier exactly matches the 30 mark on the vertical-angle scale, giving a reading of 30 (and no minutes) and thus an angle of 30 – 30, or 0. If the telescope were inclined 4°12', the vernier mark would lie somewhat above the 34 mark on the vertical-angle scale, and the vernier 12' mark (and only that mark) would match a mark on the vertical-angle scale. Vertical angles are thus read in these steps: (1) note where the 0 mark of the vernier scale meets the vertical-angle arc, which will give a reading to the nearest 30'; (2) using a 2X or 3X magnifying glass, find which of the vernier marks exactly matches a mark on the vertical-angle scale; and (3) add the latter vernier reading, which will be in minutes, to the number observed in step (1). As an example, if the vernier 0 mark lies between the 20° 30' and 21° marks of the vertical-angle scale, and the 22' mark on the vernier exactly matches a vertical-angle mark, the reading is 20° 52' and the angle is 20° 52'– 30°, or –9° 08'. Finally, in cases where no one vernier mark *exactly* matches a vertical-angle mark, the positions of the two vernier marks closest to two vertical-angle marks can be used to esti-mate the nearest 30″, or perhaps 20″, of arc. The latter degree of precision is usually unnecessary in stadia mapping but may be required in control sur-veys (Section 8-7).

Most of the steps in the vertical-angle stadia procedure are the same, or similar, to those in the Beaman-arc procedure. The instrument notebook may be like that in Fig. 8-5, except that the columns for recording and calculating Beaman-arc data must be replaced by columns for recording (1) the *vertical angle*, (2) the *horizontal distance*, and (3) the *vertical distance*.

The first three steps in the stadia procedure are exactly the same as those given above for the Beaman-arc procedure. Step 4 is the same except that the vernier *0* mark is brought to the *30* mark on the vertical-angle scale. Step 5 is the same except that the vertical angle is recorded as 0, and in step 6 the vertical angle is read (as just described) and recorded. Steps 7 and 8 are identical in the two procedures, but the steps involving stadia calculations (9 and 10) are replaced by these three steps:

1. Multiply the stadia intercept by 100 to obtain the stadia distance.

2. If the sight is not horizontal, multiply the stadia distance by the square of the cosine of the vertical angle, thus obtaining (and recording) the hori-zontal distance to the rod (the derivation of this calculation is given in most surveying texts).

3. Multiply the horizontal distance by the tangent of the vertical angle to obtain (and record) the vertical distance.

The procedure is then completed by steps 11 and 12 of the Beaman-arc procedure.

**The stadia constant and stadia interval.** The point of origin for a stadia mea-surement is not at the axis (pedestal) of external-focus alidades but generally about 1 ft in front of it. This short distance, called the stadia constant, is inscribed in the instrument box by the manufacturer. The constant need be

| Station occupied | Point sighted | Cross-hair readings | Grad. readings | Stadia distance | H | .01 H × stad. dist. | Horiz. dist. |
|---|---|---|---|---|---|---|---|
| M | 15 | 2) 10.0 | 2) 0.62 | 4.5 × 100 | 2 | .02 × 882 | |
| | | 1) 5.5 | 1) 0.11 | .51 | | = 18 | 864 |
| | | 4.5 | 0.51 | = 882 | | | |

**Fig. 8-6.** Notebook and set of data for obtaining horizontal distance from gradienter readings.

applied only in very large-scale mapping, as 1:1000 or more. It must be used, however, in determining the stadia interval factor.

The stadia hairs in many alidades are not set to give a stadia ratio of exactly 1:100. The difference is typically so small that it can be ignored, but it may be large enough in some alidades to affect large-scale mapping that must be unusually precise. The actual stadia interval can be determined by observing a graduated scale held exactly at right angles to the line of sight and at a distance of 100 ft (or 100 m), as measured exactly with a tape. The proximal end of the tape must be held at the point prescribed by the stadia constant just described. If the stadia interval proves to be significantly different from a ratio of 1:100, it is used in all computations of stadia distances.

### 8-5. Methods for Long Sights

In cases where sights are so long or brush and trees so thick that even a half-stadia interval cannot be read directly, the methods that follow will generally give adequate results.

*Gradienter or Stebinger method.* The tangent screw on most alidades is calibrated so that one complete turn moves the telescope through one stadia interval. A gradienter or Stebinger drum attached to the tangent screw is graduated into 100 divisions; one division is thus equal to 1/100th of a stadia interval. The calibrated drum is used to make measurements as follows:

1. Turn the tangent screw in a clockwise direction (the telescope may not move smoothly in the opposite direction) until the horizontal cross hair is on the lowest visible division on the rod.

2. Record this number and also the gradienter reading (opposite the index by the gradienter drum) (Fig. 8-6).

3. Elevate the telescope by the tangent screw until the horizontal cross hair is on the highest division on the rod; record that number and the gradienter reading.

4. Calculate and record the difference between the two cross-hair readings and the difference between the two gradienter readings (Fig. 8-6). Divide the former by the latter and multiply the quotient by 100; the product is the stadia (slope) distance.

5. For sights shorter than 300 m, read the Beaman H scale and use that number to correct the stadia distance to a horizontal distance (Fig. 8-6). Read the V scale and corresponding cross-hair reading; obtain the V factor as described in Section 8-4 and multiply it by 1/100th of the stadia distance. This result is the vertical distance between the alidade and the point of the cross hair on the rod. Finally, add the cross-hair reading algebraically to obtain the net difference in elevation to the point on the ground where the rod was held.

6. For sights longer than 300 m, Beaman arc methods (see the preceding

step) will not be precise enough. If the sight is at a low angle, use the method described in the following subsection. If the angle is too steep for that method, read a vertical angle and calculate the horizontal and vertical distances trigonometrically as described in Section 8-4.

**Vertical distances by stadia stepping** is an especially useful method for long sights, because they tend to be more nearly horizontal than short sights. In addition, the vertical-angle arcs of alidades can be read only to the nearest 1 or ½ minute, introducing a low precision in long sights. The stepping method proceeds as follows:

1. Level the telescope by reference to the striding level.

2. Look in the telescope and select a distinct object that is just at the level of the upper stadia hair, if the rod is above the elevation of the alidade, or the lower stadia hair if it is below the alidade (Fig. 8-7A).

3. Elevate or depress the telescope with the tangent screw until the other stadia hair lies on the object selected (Fig. 8-7B).

4. Repeat steps 2 and 3 after selecting another object in the view, and continue similarly until either the upper stadia hair (if the rod is above the alidade) or the lower one (if it is below the alidade) is on the rod (Fig. 8-7C).

5. Record the rod reading for the stadia hair just noted, giving it a – sign; also record the number of steps used in moving the stadia hair onto the rod.

6. Bring the horizontal cross hair to the nearest even division on the rod and read the half-stadia intercept.

7. Multiply the stadia intercept (twice the half stadia intercept) by the number of steps, and to this product algebraically add the rod reading recorded in step 5. The result is the difference in elevation between the alidade and the point on the ground where the rod was held. This difference in elevation is then added to the elevation of the instrument to obtain the elevation at the point sighted.

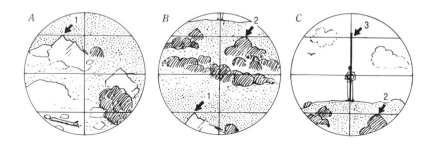

**Fig. 8-7.** Measuring vertical distance from instrument to rod by successive steps of the upper stadia hair, keyed here to the numbered points.

## 8-6. Preparations for a Plane Table Project

Compared to geologic mapping on a topographic base or on aerial photographs, plane table mapping requires expensive equipment, extra personnel, and routines that are time-consuming and can be tedious. The purpose of the project should therefore be considered thoroughly. Is an accurate, large-scale, geologic and topographic map actually needed?

If so, a field reconnaissance is essential (Section 1-4). Particulars to determine include: (1) the degree to which exposures permit detailed mapping of rocks and structures; (2) the effect of trees, brush, and steep slopes on visibility from typical instrument stations; (3) the degree of vehicle access within the area; (4) local wind and weather conditions during the field season; and (5) the locations of preexisting survey data that can be used for control points.

The map scale may be selected during the reconnaissance, and it is determined by the smallest features that must be mapped accurately and shown routinely to scale. It is generally difficult to work with map points spaced more closely than 4 mm (0.16 in.), and geologic features smaller than that are tedious to represent to scale. Thus if the smallest features to be

**Table 8–1.** Equipment for Plane Table Mapping, Including a Control Survey

| | |
|---|---|
| Aerial photographs | Pencils, 2H to 9H |
| Alidade, with pins and screwdriver | Pencil pointer (file or sandpaper) |
|     for adjustments | Pens, fine, medium, for inking map |
| Altimeter | Pen, ruling |
| Beam compass | Plane table, in case |
| Brush, sable, for dusting lenses | Plane table sheets |
| Calculator, pocket | Pliers with wire cutters |
| Colored cloth for station signals | Plumb bobs for taping |
| Colored tape for marking stadia points | Reading glass, about 3X |
| Color pencils | Rod, Philadelphia |
| Crayons (keel) for marking points | Rod, stadia |
| Dark glasses | Scale, precise, for plotting points |
| Ephemeris, solar | Stereoscope |
| Erasers | Straightedge, steel, 24 in. or more |
| Field glasses | Surveying text |
| Hatchet | Tables, mathematical |
| Ink, black, green, blue, red, brown | Tape, 6-ft roll-up |
| Lumber for signals and stakes | Tape, precise, 100-300 ft |
| Manufacturer's instruction books | Taping pins |
| Nails | Taping scale for judging pull |
| Needles for pricking points | Tracing material |
| Notebooks | Triangulation and level survey data |
| Paper, brown, for covering map sheet | Tripod, Johnson, with cap |
| Paper pads for computations | Wire for guying station signals |

mapped accurately are 10 m across on the ground, the map scale should be about 1:2500 (1 cm = 25 m or 1 in. = 200 ft).

The main factor in selecting the contour interval is the vertical control required by the geologic purpose of the project. Additional factors are steepness of slopes, scale of the map, and spacing of stadia points permitted in the time available for the survey. For typical areas that will be mapped at scales of 1:1000 to 1:4000, intervals of 5 or 10 m (20, 25, or 40 ft) are typical; however, contour intervals of 2 or 3 m (5 or 10 ft) may be needed where low topographic forms must be defined or in cases where the map will be used to determine vertical dimensions in many places, as in construction or quarrying operations.

A checklist of equipment for plane table mapping is given in Table 8-1. Sections 8-1, 8-3, and 8-4 include suggestions for selecting and preparing equipment. Enlarged aerial photographs might be considered as plane table sheets if the area has low relief (Section 8-9).

## 8-7. Horizontal and Vertical Control

Accurate maps of areas more than a kilometer or so across typically require a system of control points that must be surveyed two or three times more precisely than the other points used to make the map. The control points keep the scale, orientations, and elevations of the map acceptably accurate and consistent throughout. Systems of map control should be considered before field reconnaissance and planned specifically during the reconnaissance.

Days of field time can be gained by basing a control system on existing triangulation stations and bench marks. If, for example, two or three triangulation stations lie within the area to be mapped, there will be no need to measure a base line and the stations can be plotted on a plane table sheet before the field season. Inquiries for control data should be started a month or so before the field reconnaissance. The National Cartographic Information Center of the U.S. Geological Survey can supply some control data and suggest sources for data held by other federal agencies (see Section 6-1 for addresses). Control data may also be available from state geodetic and land survey offices, state highway departments, and county surveyors' offices.

Triangulation data generally include geographic positions (precise latitudes and longitudes) of stations, azimuths and distances between pairs of stations, elevations and descriptions of stations, and state plane coordinates of stations. Level data consist of elevations and positions of bench marks and maps showing survey lines of leveling stations. Inquiries must cite the boundaries of the area of interest, and an index map showing latitude and longitude lines should be submitted, if possible.

Nongeodetic data that may be available include highway route surveys, railroad surveys, and plats of property lines and corners. Most surveyors' blueprint maps are not accurate enough for direct scaling of distances or azimuths, but the survey coordinates can be used to compute distances and bearings between any two survey points, as described in surveying texts.

**General scheme of triangulation.** A triangulation survey locates preselected stations by lines of sight that are treated as sides of joined triangles (Fig. 8-8). If the angles at each apex are measured with a transit or theodolite, the accuracy of the work can be checked as the survey progresses, because the internal angles of each triangle must sum to 180°. Triangulation with a plane table and alidade consists of drawing a ray from any station occupied to each nearby station and thus locating stations by line intersections. Accuracy in this procedure is checked by reverse sights and by sights to and from additional stations, for example the sights $BD$ and $DB$ in the two triangles $ABC$ and $ACD$ (Fig. 8-8).

If the locations of two or more of the stations have been obtained from a government agency, the length and azimuth of at least one of the triangle legs will be known; if not, one side will have to be taped in the field as described in *Selecting and measuring a base line*. If a transit or theodolite are used for triangulation, lengths of triangle sides are computed trigonometrically from the known side (base line) and the measured angles. In plane table triangulation, the base line is plotted to scale at the outset and all other triangle sides are measured from the plane table map when intersection is completed. Vertical distances between stations are generally determined in both kinds of surveys by measuring the vertical angles among stations and computing trigonometrically from the horizontal distances.

**Selecting and marking stations.** Control stations can be planned tentatively by examining aerial photographs and are selected during the field reconnaissance and during the early part of the field season. The stations must be visible at instrument height (about 1.3 m) from surrounding stations and must give a clear view of the surrounding terrain for stadia surveying. Although government triangulation stations are commonly on high peaks, control stations for mapping may be more usefully located on low ground or on low hills or spurs. Convenience to roads is a consideration because a full

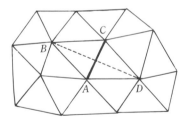

**Fig. 8-8.** Triangulation network expanded from a baseline *(AC)* to a quadrilateral *(ABCD)* and thence to adjoining triangles.

set of plane table equipment is heavy to carry. The spacing of stations should be such that at least two will fall on each plane table sheet used for stadia mapping. Internal angles of the triangulation network should be between about 40° and 90° if the stations will be intersected with a plane table and alidade.

As locations are selected, they are marked with a signal that can be seen from all nearby stations. A flag of brightly colored cloth 0.5 m long can be seen with field glasses or an alidade for several kilometers. The pole can be mounted over a permanent stake by driving a finishing nail partway into the stake and drilling a hole in the base of the pole, to fit over the nail. The pole may be guyed by three pieces of baling wire attached to heavy stones or vegetation. The signal can thus be taken down and reset easily during triangulation. Computations of vertical distances will be simplified by making all flagpoles the same height.

**Selecting and measuring a base line.** An ideal base line is located on reasonably level and open ground. The two ends must be visible from all points along the line, and all nearby triangulation stations must be visible from both ends. If the control system will be intersected with a plane table and alidade, the base line should be at least half as long as the other triangle sides in the system. The precision of measuring the base line need be no greater than the precision with which it can be plotted on a plane table sheet. The latter is determined by the thickness of graduation lines on scales of good quality, which permit reading lengths of lines to approximately the nearest 0.05 mm (0.002 in.). A plotted base line 12.5 cm (5 in.) long can thus be scaled with a precision of 1 part in about 2500.

This same degree of precision would require that a base line 500 m long must be measured to the nearest 0.2 m, which should not be difficult by using a steel tape of good quality. Additional equipment needed are two plumb bobs, a set of chaining pins (or long nails flagged with brightly colored tape), and a hand level or Brunton compass. The precision just noted will not require that a temperature correction be applied, but the manufacturer's description of the tape should be checked to determine whether it was standardized fully supported or for being held unsupported at a 10-lb pull. If it is standardized fully supported, sag is likely to introduce a systematic error of approximately 1 part in 1000, and the actual sag factor can be measured by laying the tape on a floor, marking its length, and then comparing its length when held at both ends under a 10-lb pull.

The taping is done by three persons and will be easiest going downslope, as by the steps that follow:

1. Examine the markings at the two ends of the tape (some tapes are graduated beyond their end measures).

2. One person carries the 0 end of the tape ahead for the first full measure

or to the point that the tape is level when held at chest height. The second person (the recorder) sights along the tape with a hand level, thus helping the person holding the 0 end to keep the tape level within 0.2 m or so. The third person sits over the end of the base line and sights along it to guide the others into line with the marker at the far end (Fig. 8-9).

3. The person holding the 0 end places the string of the plumb bob against the 0 graduation and lets the plumb bob hang just above the ground (Fig. 8-9). The tape is pulled with a force of 10 lb.

4. The person at the other end of the tape holds an even division over the station marker and calls out the measure to the recorder who sticks a pin into the ground at right angles to the taped line and under the end of the plumb bob.

5. The recorder records the measure and the procedure is checked by repetition.

6. As the tape is moved ahead for the second measure, the person holding the measure-end of the tape uses the pin as the marker and collects it when the second measure is completed. The total number of pins thus collected serves to check the number of measures that should have been recorded at any stage of the taping.

7. Measures are taken and recorded similarly until the other end of the base line is reached. The entire measurement should then be repeated as a check of major mistakes and as a means of ascertaining the precision of the total measurement.

If the control system will be surveyed with a transit or theodolite, a shorter base line can be used if it is taped with greater precision, as by taping over stakes set on slope, described in surveying texts.

***Intersecting stations.*** Although triangulation with an alidade is less precise than with a transit or theodolite, it takes less time and is precise enough for

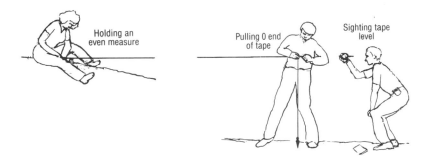

Holding an even measure

Pulling 0 end of tape

Sighting tape level

**Fig. 8-9.** Taping downhill with tape held level.

most plane table projects (see *Horizontal precision*, below). The plane table sheet should be smooth, matte-surfaced, and waterproof (Section 8-3). Stations for which geodetic data have been obtained may be plotted in the office before the field season. The stations are plotted by their latitudes and longitudes after a polyconic grid is constructed on the control-survey sheet. The methods are described in surveying texts, and the linear distances subtended by arcs of latitude and longitude are given in feet in Birdseye (1929) and in meters by the Coast and Geodetic Survey (1935). A steel straightedge, a beam compass, and an accurate scale are needed for the construction.

If a base line is being used as a starting reference for the survey, it is plotted on the control-survey sheet by using a steel straightedge, an accurate scale, and a fine needle to prick the station points at each end. The plotted base line should be remeasured occasionally over a period of several days to see how expansion and contraction of the sheet will affect the scale. The survey can then be planned for times of day when changes in scale are acceptably small (generally less than 1 part in 1000).

The survey is started at one end of the base line. The alidade is placed on the sheet so that the fiducial edge bisects the two needle holes marking the base line stations, and the plane table is oriented by sighting on the other end of the base line (Section 8-3). The alidade is then moved so as to sight a control station signal, and a thin pencil ray is drawn from close to the point occupied to the full length of the fiducial edge (so that the plane table can be oriented accurately by a backsight). The vertical distance is determined by the stepping method, if possible (Section 8-5), or from a vertical angle (see *Measuring vertical angles*, below). The pencil ray is lettered lightly with the letter of the station sighted, and the alidade is turned to sight on the next station signal. The far end of the base line should be sighted after every two or three stations to be sure that the plane table has remained correctly oriented.

When all visible stations have been surveyed, the plane table is moved to the other end of the base line and the procedures are repeated, resulting in a pencil-line intersection for each station. The intersected stations are then occupied in turn, and the plane table is oriented by backsights to either base line station. The vertical distance is determined again by stepping or measurement of a vertical angle, and rays are drawn to all other visible stations. The order of occupying the triangulation stations should be based on the probable accuracy of the initial intersection, which will decrease with horizontal distance, vertical distance, and difference between the intersected angle and 90°.

**Horizontal precision** of such surveying varies with the factors just mentioned and also with plotting precision. As each person may determine for him or herself, a thin line can be measured to about the nearest 0.05 mm (0.002 in.), but it is difficult to prick a needle hole with that degree of precision (the size of the hole is not so much a problem as its placement). Plotting precision is thus approximately 0.08 mm (0.003 in), equivalent to 0.5 m on the ground at a map-scale of 1:6000 (1 cm = 60 m or 1 in. = 500 ft.). The pencil ray itself, however, cannot generally be drawn along the fiducial edge with more than half that precision. The lines can usually be improved by: (1) using a hard, sharp pencil; (2) holding it at right angles to the fiducial edge, with the upper end tilted slightly in the direction in which the line will be drawn; and (3) drawing the line in one steady stroke. In spite of care, however, errors perpendicular to plotted lines are likely to be 0.1 mm in many cases.

An average plotting error for locations made by plane table intersection is thus likely to be between 0.05 and 0.1 mm on the map. For a triangle side 10 cm (4 in.) long this gives a precision of approximately 1 part in 1500. Although this precision is considerably less than can be obtained by surveying the control stations with a transit or theodolite, it is acceptable for controlling subsequent stadia locations, which can be located, at best, about half that precisely.

***Measuring vertical angles*** with the alidade requires particular care because the vertical-angle arc cannot be read as precisely as the arc on a transit, and sideways tilt of the plane table cannot be detected during the measurement. Use of the vertical-angle arc and its vernier are described in Section 8-4 (in the first part of *A vertical-angle procedure*). Vertical angles are measured as follows:

1. Level the plane table as exactly as possible and make sure the tripod is set firmly.

2. Record the vertical distance from the station marker to the horizontal axis of the alidade (the H.I.) and enter the height of the station signal being sighted (H.F. in Fig. 8-10).

| Station occupied | Station sighted | Readings level | Readings to signal | Vertical angles |
|---|---|---|---|---|
| A M | R R | | 38° 26′ 00″ | |
| H.I. = 4.2 | H.F. = 6.0 | 30° 21′ 00″ ⎫ 30° 21′ 30″ ⎭ | 30° 21′ 15″ | 8° 04′ 45″ |
| | | | 41° 25′ 00″ | |
| | | 33° 20′ 00″ ⎫ 33° 20′ 20″ ⎭ | 33° 20′ 10″ | 8° 04′ 50″ |

**Fig. 8-10.** Notebook record and calculation of a vertical angle based on two sets of readings.

3. For maximum stability, place the alidade in the center of the plane table.

4. Align the telescope on the station and bring the horizontal cross hair onto the top of the signal. Make the final adjustment by turning the tangent screw (No. 6 in Fig. 8-1) clockwise. Check to be sure that the cross hair rather than a stadia hair has been used in the sighting. Do not touch the plane table or walk around it until the measurement has been completed.

5. Read the vertical-angle vernier with the help of a magnifying glass; estimate the reading to the nearest half or quarter minute if possible, and record it.

6. Level the telescope, again by making the final movement with a clockwise turn of the tangent screw.

7. Read the vertical-angle vernier and record the reading; then reverse the striding level, center the bubble again by turning the tangent screw, and take a second reading. Averaging the two readings (Fig. 8-10) compensates for an unadjusted striding level.

8. To catch mistakes, repeat steps 4 through 7 after shifting the vernier a few degrees by turning the vernier tangent screw (No. 10 in Fig. 8-1).

9. Subtract each set of readings, and repeat the entire procedure if the results do not agree within acceptable limits.

The difference in elevation is computed from the trigonometric relation:

$$vert.\ dist. = horiz.\ dist. \times tan\ vert.\ angle$$

and the + or – sign of the angle is carried throughout the computation. The H.I. is always given a + sign, the height of the signal (H.F.) a – sign, and the net difference in elevation is found by adding algebraically (Fig. 8-4). Sights over horizontal distances greater than 1 km can be corrected for curvature of the earth and for refraction of light by the atmosphere by multiplying the square of the distance in kilometers by 0.175, the result being in meters.

The elevation of each station should be determined by sights to or from all nearby stations. The final elevation is then taken as an average of these determinations and may be weighted according to the estimated reliability of each.

*Elevation control* can generally be brought to the control network by reference to a bench mark or other established station within the area. If elevations cannot be obtained locally, they can usually be surveyed from a neighboring area by a plane table traverse (see *Control traverses*, below). An approximate elevation can be introduced with an altimeter, allowing for the variations described in Section 6-3.

*A north arrow* can be based on any azimuth supplied for triangulation surveys or other prior surveys of the area. It can also be obtained directly by a sight on Polaris, as described in surveying textbooks; however, the star is too high in the sky at some seasons and in some areas to be observed with an alidade. An approximate north arrow can be transferred from a topographic map covering the same area or can be based on the magnetic needle of the alidade, corrected for local declination (Section 2-4).

**Control traverses** may have to be used to survey a control network where trees or brush preclude triangulation. The traverse should be started at a geodetic station or other well-controlled point and must be closed on that point or completed at another well-controlled point. Traverse legs should be measured with a somewhat greater precision than can generally be obtained in plotting the legs, such as 1 part in 2000 for lines longer than 200 m. Careful stadia measurements are adequate for measuring legs shorter than 200 m. Longer legs must be taped as described in the subsection *Selecting and measuring a base line*, above. The legs should also slope as little as possible so that horizontal distances can be measured more precisely and vertical distances can be measured by the stepping method (Section 8-5).

A traverse is started at a point of known location (as a geodetic station) by orienting the plane table on another known station or by using the magnetic needle to orient on north. The first traverse station is then selected and marked clearly so that it can be found and used again during stadia mapping. A ray is drawn to it and the vertical distance is determined by stepping (if possible) or by measuring a vertical angle. The distance is then measured by taping or stadia, and the plane table is moved to the station, where it is oriented by a backsight. The vertical distance is measured in a backsight to serve as a check. A second traverse station is then selected, and the procedure is repeated similarly until the traverse is closed. Errors of closure too large to be plotting errors must be corrected by resurveying all or part of the traverse (see Fig. 6-2 and related text).

### 8-8. Locating Stations for Stadia Mapping

If the area is to be mapped on several plane table sheets and the control stations were surveyed on one sheet, control stations must first be plotted on each sheet. The plotting is based on distances and bearings from the control survey, and is done by: (1) drawing a triangle side on a new plane table sheet (as *AB* in Fig. 8-11*A*); (2) setting a compass for an adjoining triangle side, as *AC*; (3) striking an arc in the general location of *C*; and (4) doing the same with *BC*, so that *C* is plotted by intersection.

Additional stations will be needed for stadia mapping, and they should be spaced no farther apart than twice the longest stadia sight acceptable for the project. A typical spacing for mapping at scales of 1:2000 to 1:6000 is 600 to 700 m. Other factors in selecting stadia stations are visibility of the terrain and control signals, steepness of slopes, and weather (especially wind, which may decrease the precision of long stadia sights greatly).

In relatively open areas, the stations should be selected and surveyed onto

the plane table sheets before stadia mapping is started, using either intersection or traverse methods (Section 8-7). Where visibility is limited or
accessibility difficult, the stations can usually be surveyed more efficiently
during stadia mapping. For example, when visibility is such that control
stations are being surveyed by traverses, stadia mapping around each traverse station might be completed as the survey proceeds. The two subsections that follow describe additional methods for locating stations during the
course of stadia mapping.

   **Resection** is a method of selecting and surveying a new station after stadia
work at an occupied station has been completed. Generally, the person with
the rod selects the new station and holds the rod on it, while the person at
the alidade draws a ray toward it and determines the vertical distance by
stepping or by measuring a vertical angle. The plane table is then set up at
the new site and oriented by a backsight on the station just occupied. The
alidade is placed with the fiducial edge next to the point of any other station
that will provide a strong intersection with the ray just drawn. That station
is sighted, and a line is drawn back along the fiducial edge to intersect the
station occupied. When other stations have been sighted as a check, stadia
mapping can commence from the new station.

   **Three-point methods** must be used when the plane table has been moved to
a new site to which no rays have been drawn but from which at least three
control stations can be seen. The quickest method requires a piece of tracing
paper (or tracing film) and some drafting tape, and proceeds as follows:

   1. Prick a needle hole near the center of the tracing paper and tape it
anywhere on the plane table sheet.

   2. Lay the fiducial against the hole and sight each of three (preferably four)
signals, drawing pencil rays from near the hole toward each station.

   3. Untape the tracing paper and shift it until each pencil ray passes over
the corresponding point on the plane table sheet (Fig. 8-11*B*).

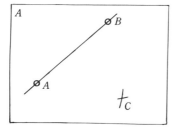
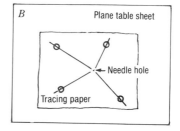

**Fig. 8-11.** *A.* Plotting control stations on a plane table sheet by intersection. *B.* Using
tracing paper to locate an unmapped instrument station. The lines represent pencil rays
on the tracing paper, and the circles are stations plotted on the plane table sheet.

4. Insert a needle through the hole in the tracing paper to prick a point in the sheet.

5. Remove the tracing paper and place the alidade so that the fiducial edge bisects the new hole and that of the farthest station sighted.

6. Loosen the lower screw of the Johnson head and orient the plane table on the corresponding signal; tighten the head and check the orientation by sights to other signals.

The only points that cannot be located by three rays are those lying on or near the circle that passes through the three points to be sighted (the *great circle* of Fig. 8-12). These cases become apparent during step 3, and either a fourth point must be used or the plane table must be moved to a new location.

Lacking tracing paper, the plane table is first oriented by the magnetic needle or by estimation. Three signals are then sighted and a ray from each is drawn back along the fiducial edge. The resulting intersection will typically form a small triangle, and the following rules from Birdseye (1928, p. 203) will help in estimating the true position of the point relative to the small triangle (Fig. 8-12).

1. If the small triangle is within the great triangle, the true position of the point is within the small triangle.

2. If the small triangle falls between the great triangle and the great circle, the true position of the point is outside the small triangle and opposite the side formed by the ray from the middle station sighted.

3. If the small triangle lies outside the great circle, the point lies outside

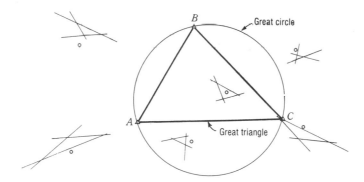

**Fig. 8-12.** Three stations sighted *(A, B* and *C)* used for orienting the plane table by estimation, and the small triangles that might result upon drawing rays to intersect the station occupied. The actual positions of the point occupied are shown by the small circles. The use of the great circle and great triangle are explained in the text. After Birdseye (1928, p. 203).

the small triangle and on the same side of the ray from the most distant station sighted as does the intersection of the other two rays.

In addition, the distances between the true position of the point and the three rays drawn are proportional to the distances from the point occupied to the three stations sighted. After a new position is estimated and marked, the plane table is reoriented by sighting on the farthest station, and rays are drawn back from each station point as before. If this does not resolve the location, the procedure is repeated until it does, typically once or twice more.

### 8-9. Stadia Mapping

Stadia mapping of geology and topography may be started at any convenient station and should be completed there before being extended to an adjoining station. Mapping complex areas will generally be more efficient if the survey party walks out the geologic features first in order to plan the mapping. Contacts, faults, and key outcrops may be flagged with strips of colored tape where relations are crucial or complex. Sketch maps of such areas, made by pace and compass methods, may also be helpful.

As the stadia mapping starts, the person with the rod should describe the first series of points to be visited so that the person at the plane table can concentrate on computing and plotting points and still know approximately where the person with the rod will be. The person with the rod examines the geology and makes notes that will be used in drawing the map. The notes may include large-scale sketch-maps of features that have been paced or measured with the rod, with the stadia points providing a framework for these details (Fig. 8-13). Most stadia points are marked by writing the point number on a smooth stone, outcrop, or piece of colored tape. They can thus be found again when the map is being checked or refined.

After about 10 to 20 points are mapped, the person with the rod should return to the plane table to help plot geology and topography. Corresponding notes in the two notebooks are compared, and the terrain is studied visually. Drainage lines, roads, and buildings are drawn first; then geologic features; and finally contours. Drawing is done lightly in pencil in order to permit adjustments. The first step in drawing contours is to interpolate short contour segments between pairs of stadia points on more-or-less even slopes. The segments are then connected by using other points or by visual estimation (Fig. 8-14). Where geologic features are spaced so widely that stadia points are taken solely for contours, the person with the rod can use a hand level to follow a given contour. After this has been done for every fourth contour or so, the other contours can be interpolated.

In an efficient survey, neither the person at the plane table nor the one with the rod has to wait long while the other person completes his or her part of the work. Small-scale mapping of large geologic features may thus

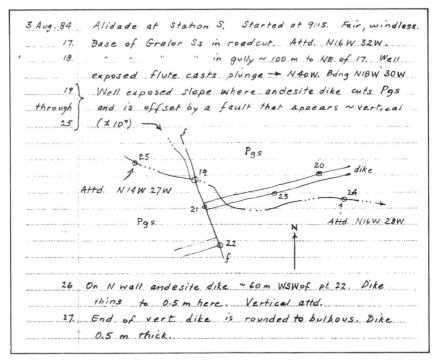

**Fig. 8-13.** Notes taken per stadia point by the person at the rod.

require two or three persons with rods, and detailed large-scale mapping of complex relations may require two persons at one rod.

*Mapping based on aerial photographs* is well-suited to areas of low relief and may save a good deal of time. An enlarged photograph can be used as a

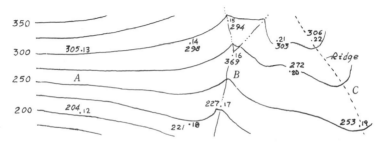

**Fig. 8-14.** Drawing contours by (1) interpolating evenly spaced contours between two points on an even slope *(A)*; (2) using points (as 14 and 18) to extend the contours; (3) interpolating again along a stream of even gradient *(B)* and along a ridge *(C)*, with local variations keyed to points such as 20 and 21 by visual estimation.

plane table sheet after control stations have been transferred to it by intersection or by inspection of the photograph image. Stadia mapping from such
points is the same as that on a plane table sheet. Many lines and points may
be located by inspection of the photograph image, so that the principal surveying procedure is determining elevations. Contours may be drawn more
easily on the photograph by using a stereo image than by visualization in the
field, and thus fewer elevations are needed for contouring.

Data can also be transferred from photographs to a plane table sheet by
the method described in Fig. 8-15.

***Moving the plane table instead of the rod*** is a stadia method by which one
person can map alone. Mapping is based on a rod made from square lumber
(as 2 X 2 in.) painted with graduations on all sides and fixed with guy wires
at any station that has been located on the plane table sheet. Such a rod
station is equivalent to one of the instrument stations of Section 8-8. The
geologist carries the plane table to an outcrop or other point of value, orients
the table by using the magnetic needle of the alidade, sights on the rod, and
makes the usual stadia readings and computations. A ray is drawn back
from the point representing the rod station, and the place occupied is plotted
with a scale. After geology and topography around the plane table are
mapped by direct observation, the plane table is moved to the next point of
interest.

***Office work on the map.*** Stadia stations may be inked with very small
numbers from time to time, and structure symbols and any resolved contacts
or faults should also be inked. All lettering should read from the south edge
of the map. Elevation numbers, contours, and any geologic features that
remain questionable should be left in pencil until the final field check of the
map. They can be protected in the meantime by an overlay of brown paper.
Additional routine office work is described in Section 1-6.

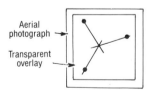
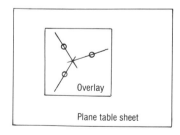

Aerial
photograph

Transparent
overlay

Overlay

Plane table sheet

**Fig. 8-15.** A point can be transferred from an aerial photograph to a map by these steps:
(1) on a transparent overlay, draw rays from any three located stations through the point
*(left)*; (2) place the overlay on the map so that the rays pass over the corresponding station
points (the circles on the diagram to the right); and (3) prick a needle hole through the
intersection and into the map.

***Completing a plane table map*** requires a field check, during which questionable relations are reexamined and additional data plotted. The data are located by means of the topography, by the stadia point markers already described, and by additional measurements made with a tape or by pacing. The sheet is carried into the field on the plane table but without the tripod or the alidade. Rock units might first be colored lightly to clarify their distribution. Samples may be collected systematically during the field check, and section lines should be selected and walked out to be sure that geology and topography along them are as complete as possible. The cross sections should be completed in pencil before leaving the field (Section 6-5).

At the close of the mapping project, the map should be inked, traditionally with geology in black, contours in brown, and drainage in blue. An explanation is required so that outcrop areas of rock units can be identified by their colors or map symbols (Section 16-2). Each sheet must be labeled with a geographic title, the location of the sheet, the party members, and the dates of the survey. A north arrow and bar scale must be added.

### References Cited

Birdseye, C. H., 1928, *Topographic instructions of the United States Geological Survey*: U.S. Geological Survey Bulletin 788, 432 p.

Birdseye, C. H., 1929, *Formulas and tables for the construction of polyconic projections*: U.S. Geological Survey Bulletin 809, 126 p.

U.S. Coast and Geodetic Survey, 1935, *Tables for a polyconic projection of maps and lengths of terrestrial arcs of meridian and parallels based upon Clark's reference spheroid of 1896*, 6th edition: Special Publication 8, 101 p.

# 9
# Primary Features of
# Marine Sedimentary Rocks ∎

### 9-1. Beds and Bedding

Sections 9-1 through 9-6 describe sedimentary structures, many of which may be either marine or nonmarine. The remainder of the chapter presents structures and rock associations typical of certain marine environments, and Chapter 10 describes associations in nonmarine environments. The descriptions are brief and might well be supplemented by other sources, such as the exceptionally clear introductory book by Leeder (1982) or the more specific ones by Reineck and Singh (1980), Harms and others (1982), Scholle and Spearing (1982), and Scholle, Bebout, and Moore (1983).

The basic structure in all environments is the sedimentary *bed (stratum)*, which is a distinct layer of sediment or rock that may differ in a variety of ways from overlying and underlying layers. Genetically, the base of a bed represents an abrupt change in depositional conditions or sediment supply; the bed represents more or less uniform conditions; and the top represents another abrupt change. The changes may include a period of erosion or a pause in deposition, and thus some bed surfaces are minor unconformities or omission surfaces (Section 9-6). When erosion surfaces between beds of closely similar materials are so faint that the composite unit looks like a single bed, the beds are said to be *amalgamated*. For example, a dense turbidity current might sweep away all but the basal sand division of a preceding turbidity-current deposit, so that two sand beds become amalgamated into a single thick unit. Amalgamated deposits should be considered as separate beds, no matter how obscure their internal contacts may be.

Beds are studied and described according to (1) how they differ from adjoining beds (by grain size, fabric, composition, or primary color); (2) their shape (Fig. 9-1); (3) their thickness (actual thickness is most useful); (4) their lateral extent, noting the degree to which it can be determined; (5) their

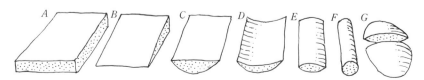

**Fig. 9-1.** Some bed shapes: *A*, tablet; *B*, wedge; *C*, trough; *D*, lunate trough; *E*, lenticular spar; *F*, cylinder; and *G*, lens.

internal structures; and (6) the nature of their contacts with adjacent beds. The most common internal structure, *lamination*, consists of layering within beds. In *cross-bedding* (cross-stratification), the laminations are oblique to the bed containing them, a common relation in current-deposited sediments. Cross-bedding is described according to the overall geometry of the cross-bedded units, the shape of the inclined surfaces, and the angles they make with the top and bottom of the bed (Fig. 9-2).

*Grading* in beds may be by grain size or by composition, and often is noticeable due to a color gradation. Some size-graded beds have a matrix of fine sediment throughout and some consist of relatively well-sorted coarse and fine materials. Size grading is normally from coarser up to finer sizes, but may be inverse, complex, or inconsistent.

Similar beds may be grouped in *bedsets* (Campbell, 1967), and specific kinds of beds or bedsets may be repeated cyclically, as ABCABC···, or rhythmically, as ABCBABCB···. Cycles of beds that become broadly coarser grained upward or finer grained upward are of particular value in interpreting cyclic changes in the energy of transporting currents. Beds in coarsening-upward and fining-upward cycles tend also to thicken upward and thin upward, respectively; however, cycles based solely on changes in thickness may be due to other causes, such as variation in sediment supply.

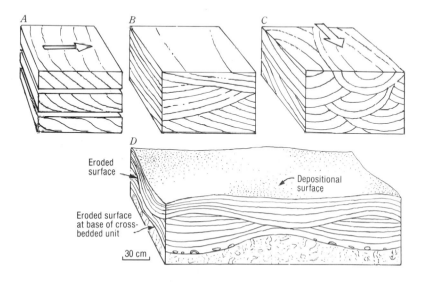

**Fig. 9-2.** Cross-bedding, with arrows indicating current direction where appropriate. *A.* Three tabular sets with *(from top)* angular, parabolic, and sinusoidal cross-lamination. *B.* Wedge sets indicating variable current direction. *C.* Trough sets or festoon cross-bedding. *D.* Hummocky cross-bedding, thought to be formed by storm waves on the lower shoreface (Hunter and Clifton, 1982).

In addition to their genetic value, specific kinds of cycles may be used to identify formations or stratigraphic tops of beds in areas of complex structure. A number of kinds of cycles and rhythms are described in this chapter, and Einsele and Seilacher (1982) have presented additional kinds.

**Bedding** differs from beds in being the general aspect of bedded rocks, as might be shown by a succession of beds in a large exposure, by lamination in a single bed, or by the grain fabric within a small sample (Fig. 9-3). Cross-bedding has already been described.

### 9-2. Depositional Bed Forms and Structures

These features vary with the nature of the transporting current and with the concentration and texture of the sediment. Currents that transport sediment can be classified into two categories, one being *fluid gravity currents*, which are generated by gravity acting on a fluid. Rivers and tidal currents are examples. *Sediment gravity currents*, the other category, are generated by gravity acting on a mass of suspended sediment that entrains intergranular fluid. Turbidity currents and debris flows are the best known of this category, which also includes fluidized sediment flows, liquefied sediment flows, and grain flows. The mechanics of sediment gravity flows have been reviewed by Lowe (1982), and deposition from fluid gravity currents has been described by Harms and others (1982).

**Fluid gravity currents** deposit sediment mainly from the traction load (grains that slide, roll, or skip along the bed). The sediment is cast into bed forms that vary in size and shape with the depth and power of the current and the grain size of the sediment. Where the forms are not preserved on the upper surface of a bed, they can often be identified by the bed's internal lamination and fabric.

At low velocities (as between 20 and 40 cm/sec, averaged through the lower 35 cm of a stream), silt and sand finer than 0.6 mm are deposited as *current ripples*, which migrate slowly downstream as grains are deposited on their steeper downstream slopes, finally constructing thin cross-laminated beds (Fig. 9-4*A,B*). When the sediment supply is moderate, the

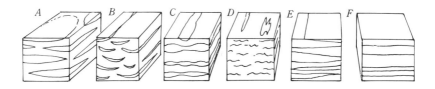

**Fig. 9-3.** Bedding patterns: *A,* lenticular; *B,* trough-flaser; *C,* linear pinch and swell; *D,* wavy discontinuous; *E,* nonparallel planar; and *F,* parallel planar.

beds dip at low angles upstream, and these angles increase as the sediment supply increases, such that steeply *climbing ripples* are produced when sediment is deposited over the entire ripple bedform (Fig. 9-4*D*). Oscillating bottom-currents produced by waves commonly cast sediment into *oscillation ripples*, which are symmetrical if waves are acting alone and asymmetrical if a unidirectional current is superimposed. Oscillation ripples can be distinguished from current ripples by straighter crests that branch here and there (Fig. 9-4*C*).

At moderate current velocities (up to about 90 cm/sec averaged through the lower 35 cm of a stream), current ripples coalesce and grow, and sediment coarser than about 0.2 mm is cast into larger forms called *dunes, megaripples*, or *large ripples*. Dunes may initially be ridges transverse to the current but generally change at higher velocities to mounds or cuspate forms with the internal structure of festoon cross-bedding (Fig. 9-5*A*). As velocity increases, the dunes become lower and spaced farther apart, and at high velocities (as at 120 cm/sec averaged through the lower 15-20 cm of a stream) the bed becomes planar. Deposited sand is laminated parallel to the plane bed, and the upper surface is commonly lineated by thin stripes of grains aligned parallel to the current. At higher velocities the bed may be cast into a series of low sand waves called *antidunes*, some of which migrate slowly upstream as sand is deposited on their upstream slopes and others remain stationary or migrate downstream. The laminations are occasionally disturbed by the shear of the swift current (Fig. 9-5*B*).

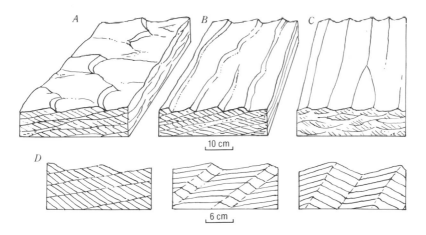

**Fig. 9-4.** Ripple structures, all from currents flowing toward the right: *A*, linguoid current ripples; *B*, transverse current ripples; *C*, oscillation (wave) ripples affected by a superimposed current; *D*, sections through ripple-drift beds, with angle of climb increasing to the right.

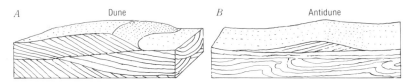

**Fig. 9-5.** *A*. Dune formed by flow from left to right. *B*. Antidune with cross-bedding indicating migration upstream, lying on planar and convoluted laminations.

Bed forms in sediment coarser than about 0.6 mm are exceptions to this progression, in that a planar bed rather than ripples forms at low current velocities, and antidunes may form before a plane bed at high velocities (Harms and others, 1982). Bed forms in associated finer sediments must thus be used to distinguish between low-velocity and high-velocity planar bedding in coarse sandstone.

**Turbidity currents** keep sediment suspended by turbulence and by additional dispersive mechanisms in their lower, denser parts. Beds are deposited in successive stages, generally giving a broadly graded layer consisting of

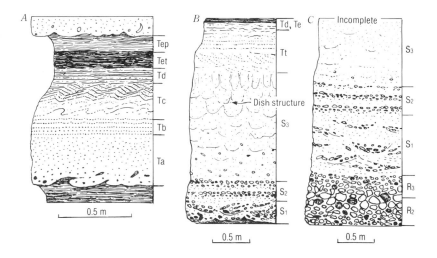

**Fig. 9-6.** *A*. Section through an idealized deposit of a sand-rich turbidity current, with divisions lettered as by Bouma (1962). *B*. Idealized deposit of a high-density turbidity current that carried abundant sand and pebbles. Division $S_1$ was deposited in traction; $S_2$ was transported as a coherent traction carpet; and $S_3$ was deposited from suspension. *Tt* is a traction-laminated division that commonly includes *Tc* and *Tb* of the Bouma sequence as well as a basal division with large-scale cross-bedding. *C*. Deposit of a high-density turbidity current that carried abundant sand, pebbles, and cobbles. Division $R_2$ was a traction carpet and $R_3$ was deposited from suspension. *B* and *C* are after Lowe (1982), © The Society of Economic Paleontologists and Mineralogists, copied with permission.

divisions that may be obvious or may require close examination. The relatively coarse divisions (as *Ta* in Fig. 9-6*A*) are absent from deposits formed from low-density turbidity currents, and the coarse divisions formed from high-density currents vary with the range of grain sizes being transported. These coarse divisions are not completely understood, but the classification proposed by Lowe (1982) provides a basis for analysis and comparisons. Briefly, and for idealized cases, high-density flows carrying sediment finer than about 0.5 mm deposit the *Ta* division by rapid settling of large amounts of sand. The deposit may be normally size-graded and is unlaminated because it is not reworked in traction. If the sand is fairly well-sized, upward escape of intergranular water may redistribute fine sediment so as to form *elutriation pillars* and *dish structures* (Fig. 9-7).

In contrast, high-density flows carrying abundant coarse sand and fine pebbles may form coarse basal deposits in three stages: (1) settling of sand and pebbles that are reworked in traction, forming cross-bedding and cut-and-fill structures (division $S_1$ in Fig. 9-6*B*); (2) rapid settling of a carpet of sand and pebbles, which is swept along bodily beneath the main current and becomes reverse size-sorted due to intergrain collisions and the buoyant effects of fine sediment (division $S_2$ in Fig. 9-6*B*); and (3) rapid settling of a thick layer of sand that may develop water-escape structures as already described (division $S_3$ in Fig. 9-6*B*). More than one size-graded layer (as in $S_2$ of Fig 9-6*B*) may result if the current surges (undergoes an increase in velocity).

Finally, high-density flows containing abundant cobbles as well as smaller grains generally deposit the cobbles in a basal sandy conglomerate. Deposition of this conglomerate may start with a traction-worked (cross-bedded or imbricated) division, but more commonly the first division is deposited bodily

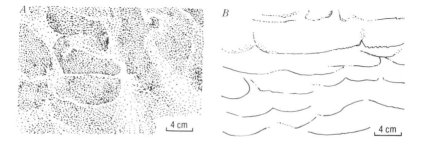

**Fig. 9-7.** *A.* Vertical sections through elutriation pillars (light-toned) and adjacent dish structures, in dark sandstone. *B.* Vertical section through dish structures in light sandstone. From photographs by Wentworth (1966). Dish structures tend to become narrower and taller in the upper parts of thick beds, as indicated diagrammatically in Fig. 9-6*B*.

and is structureless except for reverse grading (division $R_2$ in Fig. 9-6C). It is followed by a division ($R_3$) with broad normal grading. The graded divisions are then overlain by divisions similar to the pebbly and sandy divisions in Fig. 9-6B.

The finer-grained, laminated divisions of ideal turbidites ($Tb$ and higher in Fig. 9-6A) tend to be eroded away by subsequent high-density turbidity currents, but they are commonly preserved in sequences deposited by low-density currents. The laminated division $Tb$ is formed when sand that settles from the current is swept along in traction. Then, as velocity and sediment concentration of the current decrease, fallout and traction produce: (1) cross-laminated fine sand and silt typified by climbing ripples ($Tc$); (2) very fine sand, silt, and clay sorted into more or less parallel thin layers ($Td$); and (3) silty clay deposited from the last of the turbid cloud, largely after it stops ($Tet$). The turbidite bed may then be overlain by silt, clay, and microskeletons carried to the site by other processes ($Tep$). Turbidity currents that are especially diffuse (typically because they have traveled long distances) may deposit beds that start with the $Tc$ or the $Td$ division.

A variety of scour casts and load casts are typical of the bottoms of sandy or conglomeratic turbidite beds (Section 9-3), and scour commonly emplaces fragments of $Te$ or $Td$ division mud into these beds. Lamination in the $Tc$ division is commonly convoluted (Fig. 9-6A and Section 9-3).

**Debris flows** suspend sediment by the cohesive strength of their mud matrix, by its buoyant effect on larger clasts, and by its lubrication of contacts between large clasts (Lowe, 1982). Their deposits are unsorted except for a reverse-graded layer that may form at the base due mainly to the dispersive forces produced by collisions of the clasts. Large floating clasts may project partly above the upper surface of the deposit, and the deposit may be overlain by a fining-upward layer of sediment that settled from suspension after the flow was emplaced. Coarse sand needs no more than 20% of clay matrix to move as a debris flow, and fine sand only 4% (Hampton, 1975). Thus debris flow deposits resulting from mobilization and resedimentation of somewhat clayey fine sand may be recognizable only by the fine graded layer that caps them. Marine debris-flow deposits, also called *olistostromes*, must be distinguished from tectonic mélanges, as described in Section 12-7.

**Other kinds of sediment gravity flows** are much less widespread than the two already described; however, the mechanisms that support grains in them contribute to the support of grains in all high-density sediment gravity flows. The grains in *liquefied flows* are partly supported by upward flow of intergranular fluid, and those in *fluidized flows* are totally supported in this way. Either kind of flow may be generated by slumping of sorted sand on a moderate slope, followed by liquefaction and perhaps by fluidization. The deposits are likely to have water-escape features (Fig. 9-7) but otherwise should be structureless. They may be overlain by a thin division of sand with cross-bedding or cut-and-fill structure.

The sediment in *grain flows* is supported by upward-directed forces (dispersive forces) resulting from grain collisions. The flows should be thin, should form on slopes as steep as the angle of repose, and should travel only short distances beyond the slopes. Sand flows on the steep faces of eolian dunes are grain flows of well-sorted sediment (Fig. 10-4). In grain flows of gravelly sand, dispersive forces should sort the larger clasts toward the top of the flow.

## 9-3. Postdepositional Structures

Marks caused by scour or by objects moved across the substrate are most sharply defined in clay-rich mud. They are commonly made by sediment gravity flows and become filled by the basal sand or gravel of the flow that made them. It is the casts on the bottoms of the latter beds that are used in most studies, because mudstone usually disintegrates more rapidly than sandstone or conglomerate in surface exposures. The casts are valuable in determining paleocurrent directions (Section 9-4) and stratigraphic tops of beds (Section 9-10). The prominent narrow ends of scour casts called *flute casts* point up-current (Fig. 9-8*A*, 8*B*), and the beaklike ends of the more prominent forms indicate erosion by small eddy currents (Fig. 9-8*D*). Marks

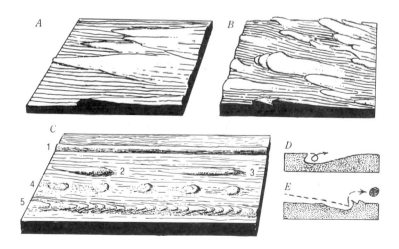

**Fig. 9-8.** Casts on the bottom of beds, filling marks made by currents flowing toward the right. *A.* Pointed flute casts (in turbidites, typical of divisions *Tb* and *Tc*). *B.* Bulbous flute casts (in turbidites, typical of divisions *Ta* and *S₁*). *C.* Groove casts (1), prod or impact cast (2), grazing impact cast (3), casts of bounce marks (4), and cast of chevron marks (5) (in turbidites, all are common at the base of divisions *Tb* and *Tc*). *D.* Longitudinal section of a flute mark (in mud). *E.* Longitudinal section of an impact mark.

with asymmetry opposite to flutes are produced when objects carried by the current prod the bottom sediment, saltate (bounce) across it, or are moved over it so closely as to create eddies that scour rhythmically spaced chevron-shaped forms (Fig. 9-8C, E). *Groove casts* fill linear grooves formed by objects dragged along the bottom, and thus give only an either-or current direction (Fig. 9-8C). Dzulynski and Walton (1965) described additional kinds of scour forms and bed markings and illustrated many forms with superb photographs.

**Soft-sediment deformation.** A number of structures may result when loose sediment is disturbed: (1) *convolution* due to liquefaction within a sediment layer or to shear of a loose substrate by currents (Fig. 9-9A); (2) *slumps* caused by creep or failure on slopes, whereby the upper ends of slope deposits are faulted and extended, and the lower parts buckled or converted into slides and perhaps debris flows (Fig. 9-9B); (3) *load structures*, due to deposition of relatively dense sediment on less dense sediment, most commonly sand over

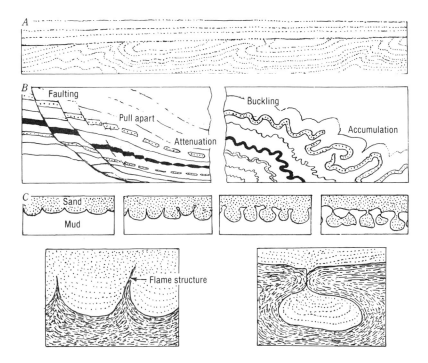

**Fig. 9-9.** *A*. Convolution caused by a current flowing toward the right. *B*. Schematic structural facies of a subaqueous slump, which may pass into a debris flow to the right. *C*. Progressive loading of sand into watery mud, with enlarged views showing idealized patterns of bedding lamination.

water-rich mud (Fig. 9-9C); and (4) *flame structures,* which are upward narrowing wedges and wisps of mud typically produced by loading (Fig. 9-9C). That all these structures are sedimentary rather than tectonic may be proven by lack of deformation of overlying beds, or by undeformed burrows that cut through the structures.

Deformational features are also caused directly or indirectly by the transmission of pore fluid, which flows most readily through well-sorted sand and gravel. Some sands are thus liquefied and convoluted in place, and both fine and coarse sediments may be mobilized and injected by lurching during earthquakes (Fig. 9-10). Soft or partly consolidated sediments may also be deformed tectonically, as indicated by structural relations described in Section 12-1.

***Diagenetic structures due to compaction.*** Structural relations may give a measure of compaction where laminations pass through concretions (Fig. 9-11A). Delicate fossils that have been broken and rotated toward bedding planes may also indicate compaction (Fig. 9-11B), as may buckle-folding of steeply inclined dikes or tubes filled with loose sand, and elliptical sections of originally cylindrical burrows (Fig. 9-11C).

To determine the actual amount of compaction, thicknesses of beds or laminae must be compared with original sediment thicknesses, or bulk densities and porosites must be compared with original values. Original thicknesses, bulk densities, or porosities can be obtained from data on modern sediments accumulating in an environment identical to that interpreted from the rocks being studied (Hamilton, 1976).

### 9-4. Paleocurrent Direction and Paleoslope Direction

A number of the primary structures described in this chapter and Chapter 10 indicate current direction, and some of the deformational features suggest slope direction at the time of deposition. Structures that initially dip in the direction of the current are cross-bedding and cross-lamination (Figs. 9-2 and 9-4). Planar structures that dip opposite to the direction of the current are cross-laminated beds formed by climbing ripples (Fig. 9-4D) and planar

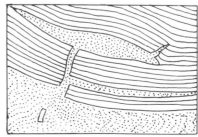

**Fig. 9-10.** Bed of liquefied sand *(below)* convoluted and injected into overlying shale as a dike, a sill, and a laccolith.

clast-imbrication (Fig. 4-4). Convolutions with consistently asymmetric fold forms suggest drag under a current that flowed opposite to the dip of their axial planes (Fig. 9-9A); however, similar forms might be generated by downslope flow in a liquefied sediment. Linear structures that give a unidirectional sense of a current are the troughs of festoon cross-bedding (Fig. 9-2C); flute casts, prod or impact casts, bounce casts, and chevron casts (Fig. 9-8C); and plunging linear imbrication (Fig. 4-4). Linear structures that give only an either-or sense of current direction are groove casts (Fig. 9-8C), cut-and-filled channels, and linear fabrics (e.g., of elongate fossils) that are parallel to bedding surfaces.

Paleoslope direction can be determined from: (1) consistently overturned load structures; (2) the dip direction of early formed faults (Fig. 9-9B); (3) the direction in which beds are pulled apart (Fig. 9-9B); (4) rotated concretions or other solid bodies; and (5) sheared-out trace fossils or primary structures. Asymmetric convolutions are suggestive of paleoslope direction but may also form by drag of a bottom current, as already noted. Postdepositional convolutions and other deformational effects can be identified in cases where trace fossils, flute casts, cross-bedding, and other depositional features have become deformed.

**Procedures.** Outcrops must be examined in three dimensions to make certain of measuring true dips and plunges rather than apparent ones. Sev-

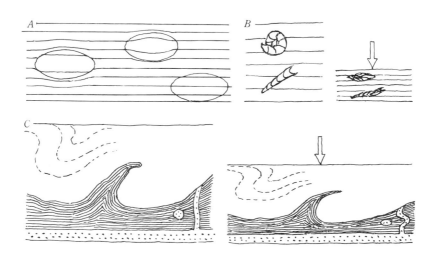

**Fig. 9-11.** *A.* Concretions formed before *(left),* during *(middle),* and after compaction of muddy sediment. *B.* Foraminifer shells crushed by compaction. *C.* Compaction of flame structure and sand-filled burrows.

eral measurements should be made at a given locality, because all currents tend to be variable in orientation.

If beds dip less than 15° and are known to be unstrained and undeformed, paleocurrent directions can be read directly with a compass. For planar structures in nearly horizontal beds, the paleocurrent direction is at right

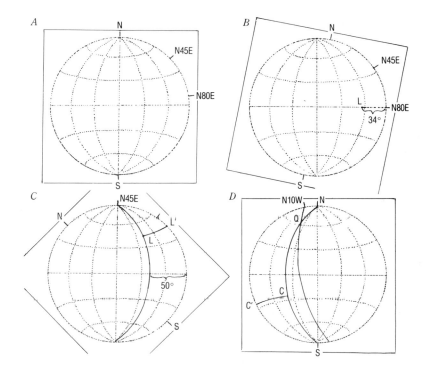

**Fig. 9-12.** Rotating current indicators to original orientation; stereo net is dotted and overlay and construction lines solid. Data for *A, B,* and *C* are: attitude of bedding, N 45°E, 50°E; plunge of linear current indicator, 34° → N 80°E. *A.* With overlay as shown, plot ticks at due N, due S, N 45°E, and N 80°E. *B.* Rotate overlay so that tick of current indicator is at E-W diameter and mark the point *(L)* 34° inward from outer circle. *C.* Rotate overlay so that tick for N 45°E is at N pole of net, and draw trace of bedding along the meridian 50° inward from outer circle. This line will pass through *L* if the data were measured accurately. Find the stereo net small circle that passes under *L;* follow it to the outer circle, and mark point *L'.* Finally, rotate the overlay into the position shown in *A,* in order to read the arc *SL',* which is the original orientation of the current indicator, here, S 88°E. *D* illustrates the correction of a planar current indicator for these data: bedding is due N, 50°W; cross-bedding is N 10°W, 70°W. First, plot ticks for due N, due S, and N 10°W and construct planes for each bedding surface as in step *C* above. Next, find the current direction *(C)* in the bedding plane by counting an arc of 90° from the intersection of the bedding and cross-bedding planes (point *Q*). Finally, follow the stereo net small circle from *C* to the outer circle and tick the point *C'.* The arc *C'S* gives the bearing of the current (here, S 64°W).

angles to the strike of the planar features. Linear features are measured by a compass bearing (Section 3-6). Neither the amount of plunge of linear structures nor the dip of planar ones is required; in fact, linear structures in moderately tilted beds may plunge at a low angle opposite to the paleocurrent direction. It is thus crucial to record the current direction correctly, which is worth a double check of the outcrop and systematic note-taking (e.g., a symbol such as →N46E could be used for paleocurrent directions and a two-ended arrow for either-or paleocurrent indicators such as groove casts).

Where beds dip more steeply than 15°, or where gently dipping beds are likely to be parts of large folds or strained in other ways, the structure of the area must be determined as completely as possible. The current structures may be measured at any time, however, and these data are required: (1) the strike and dip of the bed in which, or on which, the current indicators lie; (2) the strike and dip of planar indicators, or the trend (bearing) and plunge of linear ones; and (3) the sense of the paleocurrent direction relative to the dip or plunge, which is generally recorded as (+) if the paleocurrent direction is downward in the same direction as the dip or plunge, and (-) if it is the opposite (e.g., if flute marks plunging 40° N 65° E have an opposite paleocurrent direction, the reading would be recorded 40→N65E (-), and the direction would point 40° above the horizon in the direction S 65° W).

The structural data needed to correct the measurements just described are: (1) strike and dip of beds; (2) bearing and plunge of fold hinge lines (fold axes) in the area; (3) strike and dip of faults along which the rocks have been rotated; and (4) orientation and amount of strain in the rocks at sites of measurements.

In the commonest case, beds have been tilted by one episode of cylindroidal folding on a more or less horizontal axis. This kind of folding is indicated by hinge lines that are approximately horizontal and are approximately parallel from one fold to the next. Any linear current-indicator may be plotted on a stereographic net and rotated back to its original orientation by the steps

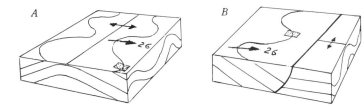

**Fig. 9-13.** A major fold *(A)* and a fault *(B)* that have tilted preexisting folds from which current data have been collected. The stippled patches represent possible sites for the data used in Fig. 9-14.

described and illustrated in Fig. 9-12*A*, *B*, and *C*. For a planar current-indicator, a somewhat different construction is used, as described in Fig. 9-12*D*.

Another common structural relation is that in which cylindroidal folds plunge, either because they have been rotated by later folding or faulting, or because strata inclined by folding or faulting have later been folded. Mapping or use of pre-existing regional maps should disclose both sets of folds (Fig. 9-13*A*) or perhaps a major fault along which the folded rocks were rotated (Fig. 9-13*B*). The age relations are not needed to correct current indicators measured in such areas, but it is essential to know the trend and plunge of the hinge lines of the local folds as well as the trend of the major folds or the strike of the major fault. Figure 9-14 describes a method for such cases.

Deformation that is still more complex can generally be recognized in the

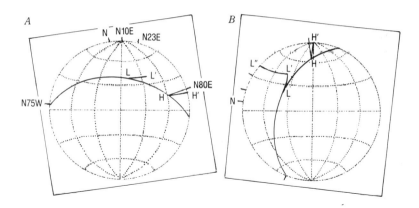

**Fig. 9-14.** Correcting a current direction for rocks folded or faulted as in Fig. 9-13. Data for the case illustrated are: hinge of greater fold or strike of fault, 0° → N10°E; hinge of lesser fold, 26° → N 80°E; attitude of bed containing current indicator, N 75°W, 50°NE; plunge of current indicator, 49° → N 23°E. The steps are: (1) with the overlay as in Fig. 9-12*A*, mark ticks for due N, N 10°E, N 80°E, N 75°W, and N 23°E; (2) rotate the overlay so that the N 80°E tick is at the E-W diameter and mark the point *H* 26°inward from the outer circle; (3) rotate the overlay so that the N 23°E tick is at the E-W diameter and mark the point *L* 49° inward from the outer circle; (4) with the N 75°W tick at the N pole, draw a line along the meridian 50° inward from the outer circle (this line will pass through *H* and *L*); (5) with the N 10°E tick at the N pole (as in part *A* of this figure), draw a line from *H* along the underlying small circle to the outer circle of the net (point *H'*); (6) count the degrees in this arc (here, 28°) and project *L* for the same number of degrees along the underlying small circle, thus locating the point *L'* (these operations correct for the effects of the greater fold or for the fault); (7) bring *H'* to the N pole of the net (as in part *B* of the figure) and project *L'* along the underlying small circle to the outer circle of the net (point *L"*), an operation that unfolds the lesser fold; (8) finally, rotate the overlay so that the tick for N is at the N pole and read the arc *NL"*, the corrected current direction (here, N 34°E).

field by the unusual forms of folded beds or other deformed primary structures (Sections 12-2 and 12-3). In cases where several episodes of folding are all approximately cylindroidal, it may be possible to obtain approximate current-directions by extending the method just described. If strains in complexly deformed rocks cannot be resolved, mapping the distribution of depth-related or slope-related sedimentary facies may indicate paleoslope directions (see, e.g., *Deep-sea fans*, Section 9-9).

### 9-5. Trace Fossils; Bioturbation

Trace fossils (ichnofossils) are often abundant where true fossils are scarce or missing, and provide clues to depositional environments as well as to sedimentation rates and processes (Crimes and Harper, 1970; Frey, 1975). Animals make traces in six principal ways: (1) by touching or resting on the bottom; (2) by moving across the bottom; (3) by feeding on the sediment surface; (4) by probing or excavating sediment for food; (5) by digging or boring a living space; and (6) by escape from such a dwelling. Trace fossil assemblages commonly include several of these kinds of structures, so that the animal itself can sometimes be deduced from the traces (Seilacher, 1970). Most traces, however, have not been matched conclusively with specific animals, and therefore trace fossils have been classified independently by their shapes and structural details. A large number have been named, described, and illustrated (Häntzschel, 1975). Trace fossils that are invisible in fresh rock may contain enough iron sulfide to develop iron-oxide stains where weathered. Farrow (1975) has described the use of ink and Alizaron Red or Methylene Blue dye to make trace fossils more visible.

Because traces are records of animal behavior, different animals that behave similarly may leave nearly identical traces. Behavior, however, is connected broadly to environment, and therefore trace fossils are valuable in interpreting conditions of deposition (Fig. 9-15). Substrate alone, however, may lead to similar trace fossils at greatly different water depths. Burrows common in shallow marine sands, for example, may occur here and there in similar sands deposited on deep-sea fans.

*Bioturbation* is the sum effect of animal movements that destroy primary fabrics and structures. The degree of bioturbation depends on: (1) the concentration of animals; (2) the rate of accumulation of sediment; (3) the food content of the sediment; and (4) the kind of feeder—for example, burrowing animals that feed on particles suspended in water bioturbate far less sediment than animals that ingest deposited sediment. Mud generally contains more food and becomes more bioturbated than sand, except where anaerobic organic-rich muds preclude most life. In fact, *lack* of bioturbation of laminated mudstones containing organic materials is evidence of an oxygen-depleted environment (see *Dysaerobic environments*, Section 9-9).

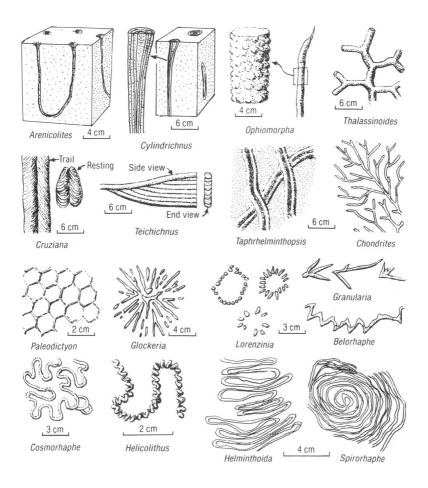

**Fig. 9-15.** Some distinctive trace fossils indicative of sedimentary facies. *Top row:* burrows in sand beds 15 cm or more thick and typical of the littoral and nearshore environments. *Second row: Cruziana* and *Taphrhelminthopsis* are trails cast on the bottoms of sandstone and siltstone beds and the others are burrows; all are typical of interbedded sandstone, siltstone, and mudstone of the shelf and somewhat deeper slopes and basins. *Third row:* casts of surface traces and shallow burrows on the bottom of thin beds (typically turbidites) of sandstone and siltstone of the slope and ocean basin. *Bottom row:* delicate casts of surface traces on the bottoms of turbidite beds of very fine sandstone and siltstone of the same environment. *Cruziana* is after Seilacher (1970) (©Liverpool Geological Society; copied with permission). *Arenicolites, Cylindrichnus*, and *Teichichnus* are after Howard and Frey (1984) (©National Research Council of Canada; copied with permission). *Taphrhelminthopsis* and all the drawings in the third and bottom rows are after Ksiazkiewicz (1970) (© Liverpool Geological Society; copied with permission).

Degrees of bioturbation may be quantified as the percentage of bioturbated material per bed or locality (Fig. 9-16). Fine-grained sediments must be examined closely with a hand lens (and possibly with a microscope), because some mud-dwellers are so small as to bioturbate thin layers without mixing them with other layers. Degrees of bioturbation have been used together with kinds of fossils and trace fossils to evaluate the details of depositional environments (Dörjes and Howard, 1975).

## 9-6. Unconformities; Rates of Deposition

Unconformities result from erosion or nondeposition, followed by renewed deposition. When the older rocks are arched or otherwise deformed before being eroded, the resulting *angular unconformity* is easily recognized in exposures or by geologic mapping. An unconformity parallel to the underlying beds (a *disconformity*) may be indicated or suggested by: (1) datable fossils or volcanic rocks lying just above and just below the surface; (2) relics of a soil profile in the underlying rocks; (3) truncation of root tubes, burrows, joints, faults, dikes, or alteration zones in the underlying rocks; (4) concentrations (lags) of resistant clasts or fossils on the surface; (5) accumulations of clay or chert nodules on an irregular surface between two limestone beds; (6) the underlying rock being distinctly more compacted, cemented, or crystallized than the overlying one; (7) irregularities or cracks in the older rock filled by the younger; (8) relics of encrusting organisms or holes bored by animals at the top of the underlying rock; and (9) fragments of the underlying rock (preferably rounded) included in the overlying deposit.

Of these relations, the first is by far the most valuable because it can be used to determine the time-equivalence of the missing strata; however, it is the least likely to be of help at the outcrop. The last relation has little signifi-

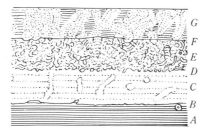

**Fig. 9-16.** Some kinds and degrees of bioturbation: *A,* laminated kerogen-rich shale, 0% bioturbated; *B,* diastem crossed by burrow; *C,* laminated sandstone with burrows, 15% bioturbated; *D,* contact 100% bioturbated; *E,* laminated siltstone mottled by 100% bioturbation; *F,* diastem cutting underlying trace fossils, locally cut by overlying burrows; and *G,* laminated muddy sandstone partly homogenized at large burrows, 75% bioturbated.

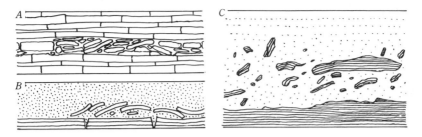

**Fig. 9-17.** *A.* Intraformational conglomerate in platy limestone. *B.* Curled mud plates accumulated with windblown sand over cracked mud. *C.* Mud fragments caught up by a turbidity current while they were still soft.

cance if the fragments are of clay-rich rocks, because they may become coherent shortly after being deposited and may thus be ripped up by storm waves, wind, or other strong currents (Fig. 9-17).

Many of the relations enumerated above can result from geologically brief scour and nondeposition, which result in minor unconformities (*diastems* or *omission surfaces*). Whether local or extensive, these surfaces are important in interpreting the depositional environment as well as specific events. Traced laterally, they may pass into patches or thin wedges of very slowly accumulated sediment, recognizable by concentrations of glauconite, phosphatized fossils or coprolites, teeth and bones, nodules of manganese oxides, or silicified fossils.

Cementation and alteration during a long pause in sedimentation may cast carbonate sediments into a surficial *hardground* or oxidize them to yellow or red tints. A hardground may be indicated by trace fossils that include: (1) an early suite made when the sediment was soft; (2) a suite of

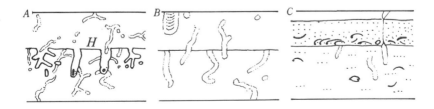

**Fig. 9-18.** *A.* Burrowed limestone with hardground *(H)* drilled by organisms and later covered by lime mud that was burrowed much like the older deposit. *B.* Burrow traces in mudstone, some cut off at an unhardened diastem and younger ones crossing it. *C.* Storm-sorted lag of shelly sand crossed by an escape trace of animal that was in underlying muddy sand.

borings in the hardground; and (3) a still younger suite in the soft sediment above the hardground (Bromley, 1975) (Fig. 9-18A). Brief erosion is shown where unlithified burrows are cut off at a diastem and similar ones cross it and occur above it (Fig. 9-18B). Distinct layers consisting of the larger clasts in the underlying sediment indicate a period of erosion and thus an omission surface (Fig. 9-18C). Brief erosion is also indicated by burrow fillings with concave-downward laminations made by animals that were evidently burrowing downward to maintain an optimal depth in the sediment.

**Rates of deposition** must generally be computed as averages (as in cm/1000 yr) based on ages derived from fossils or radioactive isotopes. Exceptions are sequences that consist of well preserved annual cyclic beds, such as the varved sediments of glacial lakes (Section 10-5) and of some dysaerobic deposits (Section 9-9). Rates of accumulation of other kinds of deposits must be judged from analogous modern or late Pleistocene deposits, which can be observed in formation or dated by [14]C content.

Trace fossils help in evaluating accumulation rates. Slow deposition of mud and fine sand rich in organic detritus typically results in complete bioturbation of primary features, producing mottled sandy mudstone in which traces are truncated and crossed by younger ones (Fig. 9-16). As rates of deposition increase, traces become less numerous, and burrow fillings may have concave-upward laminations made by animals that were burrowing upward in order to maintain an optimal depth. Finally, traces are typically absent from the main parts of thick sandy layers formed by a single storm or sediment flow. The upper parts of such layers generally contain trace fossils, however, and layers less than 30 cm thick may be crossed by escape traces made by animals living in the underlying sediment at the time the layer was emplaced (Howard, 1975) (Fig. 9-18C).

### 9-7. Environments Affected by the Tides

Tidal currents are a major cause of transportation and deposition in estuaries, bays, lagoons, and over certain parts of the seafloor. Where ebb and flood currents are moderate and roughly equal, tidal deposits may be recognized by the opposite facings of cross-lamination in ripple-deposited mud and fine sand, and locally in coarser cross-bedded sand (Fig. 9-19A). If sediment is supplied abundantly, ripple forms may be preserved between linear crescents of fine mud accumulated in ripple troughs (Fig. 9-19C). Tidal currents swift enough to transport medium and coarse sand are typically divided into domains dominated by either ebb or flood flow (Terwindt, 1981). Thus dunes at an ebb-dominated locality will have sets of cross laminations that all dip in the ebb direction but are separated by *reactivation surfaces* eroded by flood-tide currents (Fig. 9-19B). At nearby localities, the cross lamination will dip in the opposite direction.

**Estuaries** are coastal waterways in which salinities vary measurably from a maximum at their mouths to a minimum where they are joined by a river. Tidal currents are concentrated in central channels, which are characterized by sorted sand or shelly gravel. Dunes and bars will have reversing structure (Fig. 9-19*A*) or complex crescentic shapes with alternate crescents dominated by one or the other tidal currents (Fig. 9-19*B*). Sandy or silty mud is deposited on the gently sloping bottoms near channels, and fine silt and clay are caught up in the bordering marshes, which lie just below the highest tide level. A

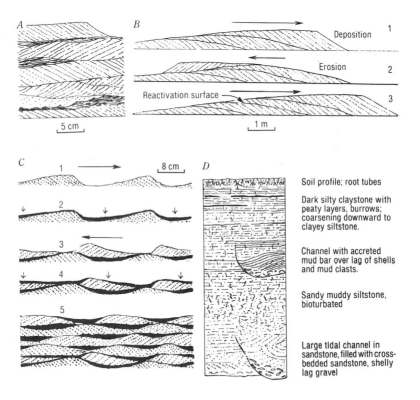

**Fig. 9-19.** *A.* Herringbone cross-lamination due to sedimentation of sandy mud by tidal currents. *B.* Evolution of cross-bedding and erosion surfaces *(reactivation surfaces)* where one tidal current is dominant. *C.* Depositional sequence leading to flaser-bedded mud and sand in a tidal environment: (1) fine sand transported in traction during peak tidal flow and deposited as cross-laminated ripples; (2) mud deposited from suspension during slack current; (3) partial erosion of mud followed by deposition of more sand ripples during reverse tidal flow; (4) deposition of another mud layer during slack current; and (5) sequence resulting from additional tidal cycles. After Reineck (1960); ©*Geologische Rundschau,* copied with permission. *D.* Generalized sequence deposited during the filling of an estuary.

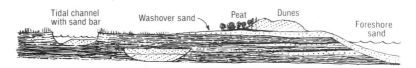

**Fig. 9-20.** Transverse section through deposits associated with a tidal marsh behind a sand barrier that has migrated slowly landward (to the left). Diagrammatic, with vertical scale exaggerated at least 4X.

filled estuary thus shows a sedimentary sequence that becomes finer grained upward, with local channel facies at the thickest sections (Fig. 9-19*D*). Tidal distributaries in marshes develop steep-sided sinuous forms that migrate laterally and show cutbank and point-bar arrangements of mud (Fig. 9-19*D*). Bioturbation is strong in the lower (seaward) parts of estuaries and decreases away from the sea.

*Barrier-bounded lagoons and bays* have sedimentary facies similar to those of estuaries but generally have more sand fed in from the sea. The barriers consist mainly of dune ridges with beach deposits on their seaward side (Section 9-8). Tidal channels cut deep transects through them. Sandy washover fans form in the lagoon where low parts of the sand barrier are topped at high tide, and broad sand wedges are deposited over lagoonal muds when storm waves cross the barrier (Fig. 9-20). Commonly, subsidence and landward erosion cause the barrier to migrate slowly landward, placing eolian dunes, beach, and nearshore marine structures over eroded marsh and bay deposits (Fig. 9-20). Unlike true estuaries, bioturbation is abundant throughout bays, but it decreases near bay mouths because of rapid reworking of sandy sediments by waves and tidal currents.

*Carbonate associations.* Where lagoons or banks are rimmed by organic reefs or by partially cemented beach or submarine ridges, carbonate production is rapid throughout an array of subenvironments (Fig. 9-21). The

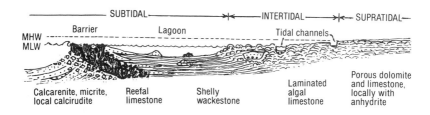

**Fig. 9-21.** Diagrammatic transverse section showing the principal subenvironments and deposits of a carbonate-producing reef and lagoon. *MHW* = mean high water, and *MLW* = mean low water.

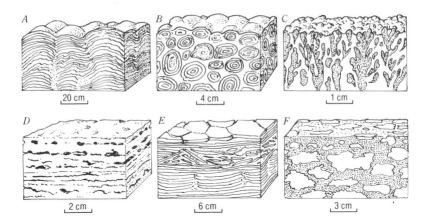

**Fig. 9-22.** Carbonate structures typical of intertidal and supratidal zones: *A,* stromatolites; *B,* oncalites; *C,* upward-branching small stromatolites; *D,* crinkled algal layers with open spaces *(fenestra* or *birdseye structure)*; *E,* flat algal layers, cracked and curled *(below),* and locally broken and buckled into *tepee structure;* and *F,* anhydrite nodules.

reef typically comprises a large variety of organic and detrital structures (James, 1983). It is flanked on its steep, seaward side by distinctly sloping proximal beds of breccia and calcarenite, and by distal beds of lime mudstone and occasional coarse deposits of sediment gravity flows. The lagoonal sediments are mainly bioturbated fossiliferous lime mud, with sand or shells forming local deposits near the reef or as isolated dunes or mounds.

Sediments of the intertidal zone, which is alternately submerged and exposed daily, can be recognized by filled tidal channels, which consist mainly of fossiliferous lime mud but have a basal layer of gravel composed of shells and intraclasts. The intertidal sediments may also have algal struc-

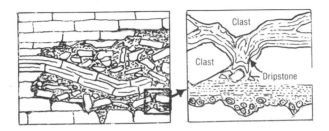

**Fig. 9-23.** Breccia formed by solution and piecemeal collapse, with detail showing cave pearls (with concentric structure) overlain by laminated sediment and by banded deposits of dripstone.

tures (Fig. 9-22*A*, *B*, and *C*), abundant burrows, reversed (tidal) cross-bedding, and tracks of wading animals.

The supratidal zone is flooded only during spring tides and strong onshore (storm) winds, and thus is exposed most of the time (Shinn, 1983). As a result, the laminae of washover lime mud are commonly dried, cracked, and locally recast into breccia and conglomerate (Fig. 9-22*E*). Shrinkage and the expansion of gases commonly produce a variety of openings in the laminated rocks (Fig. 9-22*D*). Dolomite is common, and evaporation in arid climates results in nodular anhydrite (Fig. 9-22*F*), bedded gypsum, and other evaporite minerals, any of which may later be dissolved away and perhaps replaced by chalcedony. Prolonged solution and collapse may form large breccia-filled openings (Fig. 9-23).

If the lagoon fills with sediment faster than the deposits subside, the tidal zone will migrate seaward, placing the subfacies in a distinctive vertical sequence (Fig. 9-24). Where well exposed, the oxidized supratidal rocks are noticeably lighter in tone than the darker gray (reduced environment) lagoonal limestone beneath.

## 9-8. Beach and Shelf Deposits

Beach sands are characterized by simple planar laminae that are 2 to 40 mm thick and inversely size-graded (coarsen upward) (Clifton, 1969). Heavy

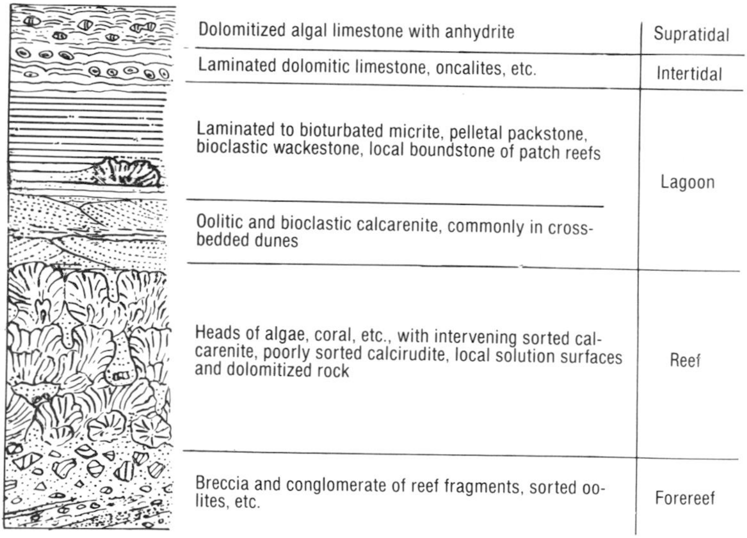

| | |
|---|---|
| Dolomitized algal limestone with anhydrite | Supratidal |
| Laminated dolomitic limestone, oncalites, etc. | Intertidal |
| Laminated to bioturbated micrite, pelletal packstone, bioclastic wackestone, local boundstone of patch reefs | Lagoon |
| Oolitic and bioclastic calcarenite, commonly in cross-bedded dunes | |
| Heads of algae, coral, etc., with intervening sorted calcarenite, poorly sorted calcirudite, local solution surfaces and dolomitized rock | Reef |
| Breccia and conglomerate of reef fragments, sorted oolites, etc. | Forereef |

**Fig. 9-24.** Vertical sequence of carbonate rocks formed by seaward migration of the subtital, intertidal, and supratidal environments.

minerals are more abundant near the base of each lamina and may form distinctive dark sheets where waves temporarily scoured and winnowed the foreshore. The set of inclined laminae is capped by a layer of sand deposited by overrun of waves on the berm (Fig. 9-25*A*). Beach deposits are particularly recognizable where deposited in sequence between nearshore marine sands and sands of wind dunes (Fig. 9-25*B*). Deformation and bioturbation are scarce above the strandline, where vertical burrows occur locally. Cross-laminae in marine dunes outside the breaker zone may dip alongshore rather than onshore. Another feature of the seaward facies, particularly on high-energy coasts, are *rip channels*, which are typically a meter deep, several meters across, and aligned perpendicular to the shore. They can be recognized by crosscutting relations, by gravelly (often shelly) basal lags, and, at their outer ends, by sand bars with seaward-dipping cross-lamination.

Beach gravels may be recognized by strongly aligned, locally vertical, or even reversed imbrication, and by abrupt lateral and vertical changes in grain size and fabric. Large isolated clasts in beach sands tend to be discoid and lie either parallel to sand laminae or to dip more steeply seaward. Gravel may form a winter lag that becomes overlain by summer deposits of sand (Fig. 9-25*A*). These gravels are often spaced rhythmically in mounds that

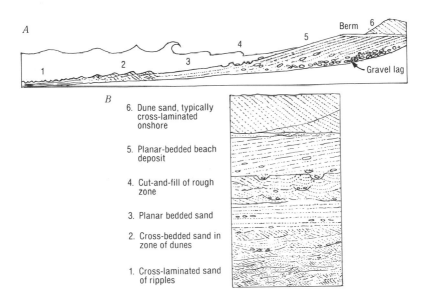

**Fig. 9-25.** *A*. Diagrammatic section through nearshore and beach deposits of a high-energy coast. See also Clifton and others (1971). *B*. Stratigraphic sequence resulting from falling sea level or seaward aggradation, such that the numbered features in *A* become superimposed.

were gravel cusps when the winter beach was exposed.

Wave energy is suggested by the dip of beach laminae, by grain size, and by the height of the berm (thickness of the deposit). Low-energy beaches along low-lying coasts have foreshore dips of less than 3°, and have abundant trace fossils near the strandline. Between major storms, large sand bars migrate onshore, creating runnels that become filled in part by rippled sand with cross-laminations dipping alongshore (Fig. 9-26). Because beaches on low-lying coasts are commonly backed by dune-ridges and marshes, they are transected here and there by tidal channels that are 10 to 30 m deep and extend for kilometers offshore. Where filled, these channels can be recognized by a coarse lag at their base, by sands with alternately facing (tidal) cross-lamination (Section 9-7), and by a complex of sand bars with alongshore-directed cross-lamination resulting from filling at the end of a spit (Hine, 1979).

**Pre-Quaternary shelf deposits** may differ from modern ones, which tend to reflect major changes in sea level during the Pleistocene glaciations. Older deposits should be well-sorted cross-laminated sandstone near shore and grade broadly seaward to finer deposits, typically interbedded silty mudstone and laminated sandstone, and then clayey mudstone, all bioturbated to various degrees. Perhaps the best indicators of environment are indigenous fossils, which are likely to be abundant locally. Hummocky cross-bedding (Fig. 9-2D) is thought to result from erosion and resedimentation of sand by major storm waves, and would be preserved if below fair-weather wave base, thus indicating a shelf environment. A storm and its waning effects are indicated by a fining-upward sequence that ideally shows: (1) a base scoured in burrowed sand or mud, often with a coarse lag; (2) sand in hummocky cross-beds as in Fig. 9-2D passing up into (3) flat-laminated sand, (4) ripple-laminated sand, and (5) siltstone or mudstone that typically is burrowed (Dott and Bourgeois, 1982). Hummocky cross-bedded sandstone may also be part of large sandstone lenses that are many meters thick and formed offshore as shoaling sand bodies. These bodies may be distinguished by a capping of shelly gravel. Opposed (tidal current) cross-bedding may develop locally on shelves, and long periods of slow sedimentation may be marked by

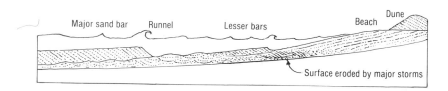

**Fig. 9-26.** Diagrammatic section through nearshore and beach deposits of a low-energy coast. Vertical scale exaggerated.

concentrations of glauconite, chamosite, or phosphate pellets.

**Carbonate deposits** of shallow, high-energy shelves and barrier-free banks can be recognized by an abundance of oolites and sand-sized skeletal grains, which are typically well sorted (grainstone and packstone) and locally cast into tidal cross-bedding structures (Section 9-7). In great contrast to these calcarenites are the pelagic chalk and marl formed in shelf seas that were so broad as to receive little terrigenous detritus. These deposits can be recognized by: (1) abundant fossils of shelf-environment organisms; (2) chalk-marl (light-dark) rhythms that are laterally persistent and may be anywhere from centimeters to meters thick; and (3) numerous hardgrounds or omission surfaces, commonly associated with nodular flint, glauconite, and phosphatic nodules and grains. Scholle, Arthur, and Ekdale (1983) also noted the common occurrence of the trace fossil *Thalassinoides* in sediments of this environment (Fig. 9-15). As described by Garrison and Kennedy (1977), nodules of lithified chalk may form during early diagenesis, and subsequent compaction and pressure solution of interstitial carbonate may lead to lenticular (flaser) structure (Fig. 9-27). The initial chalk nodules may also become concentrated by erosion of the loose sediment around them, forming a residual nodule gravel, locally with eroded-out *Thalassinoides* fillings. With further lithification, these gravels become striking hardgrounds (Kennedy and Garrison, 1975).

**Marine deltas** consist of seaward-aggraded deposits of marine, salt marsh, tidal channel, and alluvial deposits (Fig. 9-28A). The thickness and extent of each deposit varies with the size and sediment load of the river system, the tidal range, and the rate at which the delta has subsided (Morgan, 1970). At a given locality, the delta association can be recognized by a coarsening-upward sequence of marine sediments overlain by marsh and alluvial deposits, which may be repeated cyclically due to periodic subsidence (Fig. 9-28B).

Distinctive parts of this sequence in a river-dominated delta have been

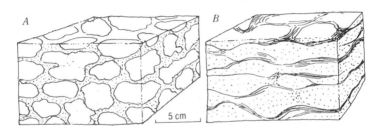

**Fig. 9-27.** Nodular chalk *(A)* modified by pressure solution into lenticular structures *(B)*. After photographs supplied by R. E. Garrison.

described by Coleman and Prior (1982). Briefly, bioturbation in the thinly laminated prodelta clays and silts decreases markedly upward, and these sediments locally grade upward to deposits of distal bars, in which the fine sediments are interlayered with thin beds of cross-laminated silt and sand. The thick sands of the distributary mouth bar are well sorted and cross-laminated, and are commonly distorted by slumping; they may grade upward into coarse sands reworked by tidal currents, or into marsh deposits associated with overspill (splay) sands of the distributary system. Distinctive sedimentary facies of the subaerial delta below high-tide level are: (1) numerous channels filled with woody detritus and organic-rich clay and silt in which bedding is distorted during compaction; and (2) bay fills, which coarsen upward from marsh clay with much organic debris to increasingly sandy sediments deposited as small deltas. Deposits of the delta plain not reached by high tides can be recognized by: (1) dark organic clays of swamps and lakes, commonly containing pyrite, vivianite, and nodules cemented by siderite, and locally broken perpendicular to bedding by contraction (syneresis) cracks; (2) shells of fresh water mollusks; (3) a capping of oxidized (slightly reddened) mud with root tubes; and (4) cross-cutting braided river channels that fine upward as described in Section 10-2.

## 9-9. Marginal and Basinal Deposits of the Deep Sea

Of the environments of the deep sea, those near its margin show the great-

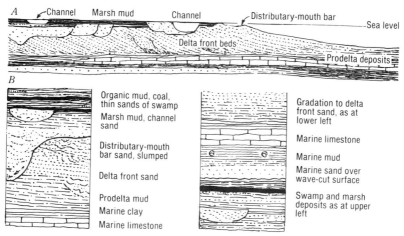

**Fig. 9-28.** *A.* Diagrammatic longitudinal section through a large delta, with vertical scale exaggerated greatly. The marsh and swamp deposits thicken landward due to gradual subsidence of the delta. *B.* Vertical sequence with base at the lower left and top at the upper right, illustrating a complete deltaic cycle closed by marine erosion due to subsidence and followed by the beginnings of another cycle.

est variety of sedimentary associations. The relatively smooth slope and rise of the western Atlantic is covered by mud and sand transported by the contour-parallel geostrophic boundary current (Heezen and Hollister, 1971). These *contourites* can be recognized by widespread distribution, by persistent thin laminations without size grading, by moderate numbers of trace fossils (Fig. 9-15), by cross-lamination in thin beds of coarse silt and fine sand, and by consistently oriented linear sediment tails and ripple sets which indicate a current that flowed equatorward. In contrast, the continental slopes of tectonically active margins develop thick local sequences of mud, sand, and local gravel caught behind fault or slide blocks. Slumping forms convolute strata, breccias, and debris flows in these sequences.

**Dysaerobic environments** are fairly common where the oxygen-minimum zone of the ocean column (for dysaerobic conditions, water with 2 to 0.1 ml/l of dissolved oxygen) impinges on slopes and basins not flushed by oxygen-rich currents. Typical indicators in sedimentary rocks are: (1) lack of benthic invertebrate fossils except for certain protozoans; (2) unscavenged remains of fish or delicate foraminifers, such as arenaceous forms; (3) laterally continuous planar lamination, commonly consisting of annual varves of dark/light pairs less than 1 mm thick; (4) scarcity of trace fossils, which are typically thin forms parallel to bedding, such as *Chondrites* (Fig. 9-15); (5) abundant organic matter; and (6) drab and commonly dark colors (gray, gray-green, brown). These rocks may alternate with beds deposited during aerobic episodes, during which benthic animals bioturbate them into structureless layers or into sequences with irregular and discontinuous lamination (Govean and Garrison, 1981; Isaacs, 1983). The other characteristics of the aerobic beds are generally opposite those enumerated above. Truly anaerobic (oxygen-free) depositional environments are probably rare; however, pore water in deposits containing abundant organic matter may become anaerobic, as indicated by formation of sulfides.

**Deep-sea fans** are forming today where sediment is funneled through submarine canyons to the floor of the deep ocean, and some turbidite sequences appear to represent similar pre-Quaternary fans. Other turbidites may represent fans in shallower or more restricted basins, and some are problematical (Normark and others, 1983/1984). Ancient fan (or turbidite) associations are studied by mapping paleocurrent directions and sedimentary facies, and the facies that have been especially useful are those of Mutti and Ricci Lucchi (1978). The brief descriptions of them that follow are modified somewhat after Howell and Normark (1982).

A. *Sandy-conglomeratic facies* — medium to very coarse sandstone and conglomerate (the latter either clast- or matrix-supported) in beds more than 1 m thick; beds may be amalgamated or separated by thin beds of mudstone or by zones of ripped-up mudstone clasts, some clasts more than 1 m in diameter; beds vary in thickness and thin abruptly at channel margins; some beds are ungraded, some have reverse grading, and many are normally graded and may be laminated in their upper parts; scour and load casts are common and may be deeper than 1 m.

B. *Sandy facies* — coarse to fine sandstone in beds typically less than 3 m thick; beds more tabular and with more numerous and persistent mudstone interbeds than in *A*; laminations may be common and are often broadly wavy through most of bed; dish structures and elutriation pillars may be present; scour marks scarce but filled channels, mudstone clasts, and soft-sediment deformation common.

C. *Sandy-muddy facies* — sandstone-mudstone turbidite beds like those of Fig. 9-6*A* dominant; coarsest sand usually medium-grained but some *Ta* divisions contain coarse sand or pebbles, usually near their base; beds typically 0.5 to 1.5 m thick and more tabular than in *B*; scour casts abundant; intraformational mudstone clasts usually less than 3 cm in diameter.

D. *Muddy-sandy facies I* — turbidite beds consisting of divisions *Tc* to *Te* typical, and otherwise lowest division is *Tb*; all beds laminated; sandstone:-mudstone ratios from 1:2 to 1:9; sandy parts of beds 3 to 40 cm thick and composed of fine sand to coarse silt.

E. *Muddy-sandy facies II* — turbidites similar to those of *D* but beds are thinner (generally less than 0.3m thick), more lenticular, have sandstone to mudstone ratios about 1:1 or more, have sand coarser and less sorted, have

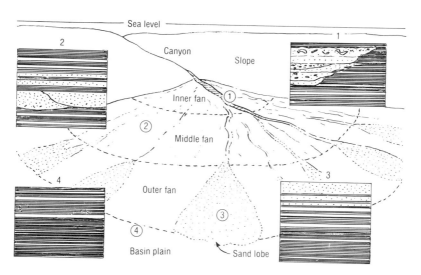

**Fig. 9-29.** Oblique view of an idealized deep-sea fan, with inner, middle, and outer environments delimited by dashed lines. Typical locations of the four columnar sections are shown by circled numbers, and the sedimentary sequences are described in the text.

basal divisions unlaminated and with mudstone clasts, and have sandstone and mudstone distinctly separated in a given bed.

F. *Chaotic facies* — slumped beds, contorted or broken strata, or debris flows (Figs. 9-9*B* and 12-29).

G. *Hemipelagic and (or) pelagic facies* — mudstone (siltstone to claystone) with varying content of fine sand, detrital mica, and carbonates; structureless to laminated; may consist of thin separate deposits on *Tet* divisions or as thicker sequences; generally with more microfossils than *Tet* mudstone and thus lighter colored.

These facies form typical associations and sequences in the fan environments of Fig. 9-29. The *inner fan* is characterized by channel deposits of sandy-conglomeratic facies with subordinate sandy facies and occasional chaotic facies. Hemipelagic or pelagic facies is typical of the interchannel parts of the inner fan, locally being associated with muddy-sandy facies II. The *middle fan* is dominated by channel deposits of sandy facies, locally with sandy-conglomeratic facies and sandy-muddy facies. Muddy-sandy facies II forms levees and grades outward to muddy-sandy facies I and perhaps to pelagic facies in areas farthest from channels. Laterally shifting channels result in thinning-and-fining-upward sequences over large areas (Fig. 9-29, column 2). The *outer fan* is generally defined as fan-areas beyond the ends of the distributary channels, and sandy muddy facies and muddy-sandy facies I are characteristic, commonly being associated in thickening-upward deposits of sand lobes, with sandy facies forming the uppermost part of these deposits (Fig. 9-29, column 3). The outer fringe of the fan may show thickening-upward sequences starting with hemipelagic or pelagic facies whereas the *basin plain* is composed of the latter facies, with occasional beds of muddy-sandy facies I and perhaps sandy-muddy facies (Fig. 9-29, column 4).

A paleofan is proven by the distribution of these facies associations and by a radiating pattern of channels and of flow casts (Section 9-3). The latter are well developed in many fan sediments and are oriented fairly consistently at any one locality. Convolute bedding is common in the coarser deposits. Sandy and conglomeratic facies are likely to contain displaced shallow water fossils without accompanying burrows, whereas adjoining mudstones contain pelagic fossils and trace fossils typical of deep water (Fig. 9-15).

**Basinal sediments** in areas adjacent to deep-sea fans are typically laminated clayey siltstones. Farther basinward, clays and planktonic skeletons become the dominant components, forming biogenic calcareous deposits and, in deep cold waters, siliceous deposits and clays that are residual to the solution of all carbonate. Bedding in most of these deposits is parallel, thin, and highly persistent laterally. Cross-lamination is fairly common but is so thin and often at such low angles to bedding as to require sawing and etching in dilute HCl in order to measure paleocurrent direction. Well-preserved trace fossils are scarce and are mainly grazing trails or shallow probing marks and burrows. Sediments are generally oxidized and may thus be reddish or brownish. Association of vesicle-poor basalt with calcilutite, bedded chert, or porcelanite is indicative of a deep-sea environment. In addition to occasional far-traveled turbidity currents, silt and coarser sediments may be introduced by floating ice, by winds from arid continental areas, and by major volcanic explosions.

**Deep-sea carbonates** formed from calcareous plankton are preserved at depths less than around 4 km, forming chalks and marls (and their derivative limestones). These deposits can be distinguished from shelf-sea chalks (Section 9-8) by scarcity of fossils of benthic organisms and by the absence of hardgrounds and the trace fossil *Thalassinoides* (Scholle, Arthur, and Ekdale, 1983). Beds are typically tabular and highly persistent. Where amounts of carbonate decrease, due either to dissolution or to local scarcity of plankton, chalk and marl (or limestone and calcareous shale) grade laterally to carbonate-free clay or organic deposits of silica. As noted by Scholle, Arthur, and Ekdale (1983), these fine-grained deposits accumulate so slowly that delicate trace fossils (as the bottom two rows in Fig. 9-15) are destroyed by bioturbation, and the only preserved traces are *Planolites,* a small flattened burrow-filling that is variously curved and lies about parallel to bedding (see Häntzschel, 1975).

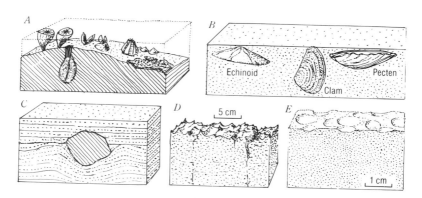

**Fig. 9-30.** Miscellaneous structures indicating stratigraphic sequence. *A.* Organisms attached to the top (free face) of rock substrate or boulder; from left to right, solitary corals, drilling pelecypod, brachipods, barnacle, and bryozoan. *B.* Living positions of certain organisms. *C.* Dimple under dropstone (ice-rafted or kelp-rafted) overlain by planar beds. *D.* Fluted top of exposed limestone. *E.* Prints of raindrops or hail. Partly after "Criteria for determining top and bottom of beds" by S. W. Muller (1958, Stanford University), reproduced in full by Dietrich and others (1982).

## 9-10. Structures Indicating Stratigraphic Facing (Tops) of Beds

Primary sedimentary structures can often be used to determine stratigraphic tops in areas where strata are deformed. Although a number of these features have already been described, a list may be helpful: cross-bedding with cut-and-fill relations or parabolic foreset lamination (Fig. 9-2); ripple forms preserved at the top of cross-laminated sequences (Fig. 9-4); turbidite beds (Fig. 9-6); elutriation pillars and dish structures (Fig. 9-7); scour forms (Fig. 9-8); load casts (Fig. 9-9C); trace fossils consisting of surficial tracks, trails, or grazing marks (Fig. 9-15); truncated burrows, bored hardgrounds, and escape traces (Fig. 9-18); all relations indicating unconformity (Section 9-6); inverse size-grading in beach sands (Section 9-8); cut-and-fill relations in reversing (tidal) cross-bedding (Fig. 9-19B); sequences of aggraded estuarine and bay deposits (Fig. 9-19D); gravel or shell lags at the base of cut forms (Fig. 9-19D); growth forms of stromatolites (Fig. 9-22), dessication plates curled upward at their edges (Fig. 9-22E); collapse breccia (Fig. 9-23); sequence in barrier-related carbonate deposits (Fig. 9-24); sequence in aggraded near-shore and beach deposits (Fig. 9-25); channel forms in turbidite fan deposits (Fig. 9-29). Additional structures include stratified fillings in vesicles and fossils (Fig. 4-7C), as well as the features illustrated in Figure 9-30.

Structures indicating stratigraphic tops in nonmarine deposits (Chapter 10) are: fining-upward sequences in alluvial sediments (Fig. 10-1B); cut-and-fill in alluvial fan deposits (Fig. 10-2); forms and cross-bedding of eolian dunes (Fig. 10-3); slump-and-fill in dune foreset beds (Fig. 10-4); inverse size-grading in eolian ripple-drift layers (Fig. 10-5); gravel lags at the base of eolian sand accumulations (Section 10-4); sequence in composite till sheets (Fig. 10-6); size-grading in talus deposits (Fig. 10-11); upward-opening shapes of filled ice wedges (Fig. 10-12); facing of involutions in Vertisols (Fig. 10-16); and almost all soil profiles (Section 10-9).

Where volcanic and sedimentary rocks are intercalated, stratigraphic sequence can be determined from these features in basalts: (1) position of pipe vesicles, pahoehoe toes, and pahoehoe tops (Fig. 13-1); (2) stratified filling and stalactites in lava tubes (Fig. 13-3B); (3) patterns of columnar joints (Fig. 13-4); and (4) cuspate forms at the base of some lava pillows (Fig. 13-5). In viscous lavas, stratigraphic tops may be indicated by thrust faults and vertical distribution of flow structures (Fig. 13-7). Indicators of sequence in pyroclastic deposits include bomb sags (Fig. 13-9B), cross-bedding in surge deposits (Fig. 13-9), original textural sequences in ash-flow deposits (Fig. 13-10A and B), and the facing of gradations due to welding and compaction (Fig. 13-10C).

Finally, in sedimentary sequences that have been folded once strongly, tops of beds in isolated outcrops can be based on the geometric relations between bedding and cleavage (Figs. 12-18 and 19).

## References Cited

Bouma, A. H., 1962, *Sedimentology of some flysch deposits: a graphic approach to facies interpretation*: Amsterdam, Elsevier Scientific Publishing Co., 168 p.

Bromley, R. G., 1975, Trace fossils at omission surfaces, p. 399-428 *in* Frey, R.W.

Campbell, C. V., 1967, Lamina, laminaset, bed and bedset: *Sedimentology*, v. 8, p. 7-26.

Clifton, H. E., 1969, Beach lamination: nature and origin: *Marine Geology*, v. 7, p. 553-559.

Clifton, H. E., Hunter, R. E., and Phillips, R. L., 1971, Depositional structures and processes in the non-barred high-energy nearshore: *Journal of Sedimentary Petrology*, v. 41, p. 651-670.

Coleman, J. M., and Prior, D. B., 1982, Deltaic environments of deposition, p. 139-178 *in* Scholle, P. A., and Spearing, D.

Crimes, T. P., and Harper, J. C., editors, 1970, *Trace fossils*: Liverpool, Seel House Press, 547 p.

Dietrich, R. V., Dutro, J. T., Jr., and Foose, R. M., compilers, 1982. *AGI data sheets*, second edition:Falls Church, VA, American Geological Institute.

Dörjes, J., and Howard, J. D., 1975, Estuaries of the Georgia coast, U.S.A.: Sedimentology and biology, IV. Fluvial-marine transition indicators in an estuarine environment, Ogeechee River-Ossabaw Sound: *Senckenbergiana Maritima*, v. 7, p. 137-179.

Dott, R. H., Jr., and Bourgeois, J., 1982, Hummocky stratification: significance of its variable bedding sequences: *Geological Society of America Bulletin*, v. 93, p. 663-680.

Dzulynski, S., and Walton, E. K., 1965, *Sedimentary features of flysch and greywackes*: Amsterdam, Elsevier Scientific Publishing Co., 274 p.

Einsele, G., and Seilacher, A., 1982, *Cyclic and event stratification*: New York, Springer-Verlag, 536 p.

Farrow, G. E., 1975, Techniques for the study of fossil and recent traces, p. 537-554 *in* Frey, R. W.

Frey, R. W., editor, 1975, *The study of trace fossils: a synthesis of principles, problems, and procedures in ichnology*: New York, Springer-Verlag, 562 p.

Garrison, R.E., and Kennedy, W.J., 1977, Origin of solution seams and flaser structure in Upper Cretaceous chalks of southern England: *Sedimentology*, v. 19, p. 107-137.

Govean, F. M., and Garrison, R. E. 1981, Significance of laminated and massive diatomites in the upper part of the Monterey Formation, California, p. 181-198 *in* Garrison, R. E., and Douglas, R. G., editors, *The Monterey Formation and related siliceous rocks of California*: Society of Economic Paleontologists and Mineralogists Pacific Section, Special Publication.

Hamilton, E. L., 1976, Variations of density and porosity with depth in deep-sea sediments: *Journal of Sedimentary Petrology*, v. 46, p. 280-300.

Hampton, M. A., 1975, Competence of fine-grained debris flows: *Journal of Sedimentary Petrology*, v. 45, p. 834-844.

Häntzschel, W., 1975, Trace fossils and problematica, Part W, Supplement 1, *in* Teichert, C., editor, *Treatise on invertebrate paleontology*, Lawrence, University of Kansas, and Boulder, CO, Geological Society of America, 269 p.

Harms, J. C., Southard, J. B., and Walker, R. G., 1982, *Structures and sequences in clastic rocks*: Society of Economic Paleontologists and Mineralogists, lectures for short course no. 9, 8 chapters paged separately.

Heezen, B. C., and Hollister, C. D., 1971, *The face of the deep*: New York, Oxford Univ. Press, 659 p.

Hine, A. C., 1979, Mechanisms of berm development and resulting beach growth along a barrier spit complex: *Sedimentology*, v. 26, p. 333-351.

Howard, J. D., 1975, The sedimentological significance of trace fossils, p. 131-146 *in* Frey, R. W.

Howard, J. D., and Frey, R. W., 1984, Characteristic trace fossils in nearshore to offshore sequences, Upper Cretaceous of east-central Utah: *Canadian Journal of Earth Sciences*, v. 21, p. 200-219.

Howell, D. G., and Normark, W. R., 1982, Sedimentology of submarine fans, p. 365-404 *in* Scholle, P. A., and Spearing, D.

Hunter, R. E., and Clifton, H. E., 1982, Cyclic deposits and hummocky cross-stratification of probable storm origin in Upper Cretaceous rocks of the Cape Sebastian area, southwestern Oregon: *Journal of Sedimentary Petrology*, v. 52, p. 127-143.

Isaacs, C. M., 1983, Compositional variation and sequence in the Miocene Monterey Formation, Santa Barbara coastal area, California, p. 117-132 *in* Larue, D. K., and Steel, R. J., editors, *Cenozoic marine sedimentation, Pacific margin, U.S.A.*: Society of Economic Paleontologists and Mineralogists Pacific Section, Special Publication.

James, N. P., 1983, Reef environment, p. 345-440 *in* Scholle, P. A., Bebout, D. G., and Moore, C. H.

Kennedy, W. J., and Garrison, R. E., 1975, Morphology and genesis of nodular chalks and hardgrounds in the Upper Cretaceous of southern England: *Sedimentology*, v. 22, p. 311-386.

Ksiażkiewicz, M., 1970, Observations on the ichnofauna of the Polish Carpathians, p. 283-322 *in* Crimes, T. P., and Harper, J. C.

Leeder, M. R., 1982, *Sedimentology, process and product*: London, Geo. Allen & Unwin, 344 p.

Lowe, D. R., 1982, Sediment gravity flows: II. Depositional models with special reference to the deposits of high-density turbidity currents: *Journal of Sedimentary Petrology*, v. 52, p. 279-297.

Morgan, J. P., 1970, Depositional processes and products in the deltaic environment, p. 31-47 *in* Morgan, J. P., editor, *Deltaic sedimentation modern and ancient*: Society of Economic Paleontologists and Mineralogists Special Publication 15.

Mutti, E., and Ricci Lucchi, F. (translated by T. H. Nilsen), 1978, Turbidites of the northern Apennines: introduction to facies analysis: *International Geology Review*, vol. 20, p. 127-166 (reprinted in AGI Reprint Series 3).

Normark, W. R., Mutti, E., and Bouma, A. H., 1983/84, Problems in turbidite research: a need for COMFAN: *Geo-Marine Letters*, v. 3, p. 53-56.

Reineck, H. E., 1960, Über Zeitlücken in rezenten Flachsee-Sedimenten: *Geologische Rundschau*, v. 49, p. 149-161.

Reineck, H. E., and Singh, I. B., 1980, *Depositional sedimentary environments, with reference to terrigenous clastics*, 2nd edition: New York, Springer-Verlag, 549 p.

Scholle, P. A., Arthur, M. A., and Ekdale, A. A., 1983, Pelagic environment, p. 619-691 *in* Scholle, P. A., Bebout, D. G., and Moore, C. H.

Scholle, P. A., Bebout, D. G., and Moore, C. H., editors, 1983, *Carbonate depositional environments*: American Association of Petroleum Geologists Memoir 33, 700 p.

Scholle, P. A., and Spearing, D., editors, 1982, *Sandstone depositional environments*: American Association of Petroleum Geologists Memoir 31, 410 p.

Seilacher, A., 1970, *Cruziana* stratigraphy of "non-fossiliferous" Paleozoic sandstones, p. 447-476 *in* Crimes, T. P., and Harper, J. C.

Shinn, E. A., 1983, Tidal flat environment, p. 171-210 *in* Scholle, P. A., Bebout, D. G., and Moore, C. H.

Terwindt, J.H.J., 1981, Origin and sequences of sedimentary structures in inshore mesotidal deposits of the North Sea, p. 4-26 *in* Nio, S.-D., Shuttenhelm, R.T.E., and van Weering, T.C.E., editors, *Holocene marine sedimentation in the North Sea basin*, International Association of Sedimentologists Special Publication no. 5.

Wentworth, C. M., Jr., 1966, *The Upper Cretaceous and lower Tertiary rocks of the Gualala area, northern Coast Ranges, California*:   Stanford University, CA, PhD dissertation, 197 p.

# ■ Surficial Sediments; Continental Environments

### 10-1. The Quaternary Record

The descriptions in this chapter pertain to nonmarine deposits of all ages but are applied especially to Quaternary deposits because of the areal extent and importance of that System. Field studies of Quaternary deposits are somewhat special for several reasons. The cyclic nature of the System makes it necessary to recognize specific subunits within sequences of similar units, which requires careful attention to small differences in color, texture, induration, and primary structures. Soils and paleosols are among the most important units, and they are alteration zones rather than deposits. In addition, all kinds of surficial deposits tend to be loose, which makes for infrequent exposure and the need for picks and shovels, soil augers, coring devices, and entrenching machines. Aerial photographs are typically essential. Study of landforms and their relations to soils, plant communities, and underlying deposits is one of the more basic approaches (Ruhe, 1975). Geomorphological techniques (Goudie, 1981; Dackombe and Gardiner, 1983) may be useful. Intercalated volcanic deposits provide means of numerical dating (Chapter 13; Self and Sparks, 1980), and active faults are important (Section 12-5). Most of the basic terms and relations described in the first six sections of Chapter 9 apply to continental deposits as well.

*Quaternary stratigraphy* is based on cyclic climatic variations, because the period was too brief to evolve sets of index fossils or faunal-zone assemblages. The acknowledged subdivisions are four glacial periods and three interglacials, based on glacial drifts separated by warm-climate soils or other materials. A number of studies have now demonstrated, however, that there were additional cold-warm cycles during the Quaternary Period. As examples, sequences of loess layers and climate-related fauna and flora in central Europe indicate at least 17 cold-warm cycles in the past 1.7 m.y. (Fink and Kukla, 1977), and unusually complete pollen records (Woillard, 1978) and thick sedimentary sections that can be dated paleomagnetically (Cooke, 1981) also indicate many climatic cycles.

One must thus proceed with caution in assigning local indications of a cold-warm change to one of the classical subdivisions. Before beginning a study, the local and regional Quaternary stratigraphy should be reviewed thoroughly, with special note of all horizons dated by [14]C or other methods. Valuable unpublished data are commonly available through local geology or engineering offices. Regional stratigraphic data for North America have been presented in books edited by Mahaney (1976, 1981, 1984), Black and others (1973), and Wright (1983), as well as in many articles in the journal *Quaternary Research*. These sources also provide descriptions of the principal methods used in recognizing and identifying cold-warm variations in sediments, rocks, and plant materials.

## 10-2. Alluvial Deposits

Rivers that have developed a floodplain and are deepening their valleys very slowly deposit a single fining-upward sequence as they erode their banks at bends (Fig. 10-1A and B). Cross-bedding is common and dips, on

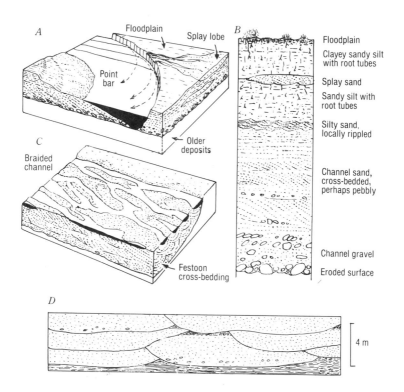

**Fig. 10-1.** *A*. Floodplain deposit made by a stream that is deepening its valley very slowly. The solid arrows in the stream indicate surface currents, which turn under at the bank and flow along the bottom toward the point bar (*dashed arrows*). *B*. Vertical section through fining-upward deposit of floodplain like that of *A*. *C*. Trough sets of sand with sparse interbeds of finer sediment, the result of continuous rapid aggradation. Stream is at low (braided) stage and flowing from left to right. *D*. Transverse section of channel fills made by a rapidly aggrading braided river.

the average, downstream in straight reaches and toward the outside of bends. Laminae in sands are locally convoluted during antidune-stage (high velocity) flow (Fig. 9-5B). Generally, alluvial conglomerate is crudely bedded and has imbrication in which the longest axes of clasts lie perpendicular to the current. Gravels in straight reaches have the usual upstream imbrication and those in bends may be imbricated toward the bend. Levee deposits are typically sandy and outlying floodplain deposits silty. Both kinds of overbank deposits are commonly rippled in arid regions but may be structureless in temperate regions due to deposition among plants or to later bioturbation. Clay, organic muck, and peat are deposited in abandoned channels and other low parts of floodplains, which may merge downstream with deltaic plains (Section 9-8). Floods locally erode channels (*crevasses*) in levees and deposit sand or gravel as fan-shaped splays near the levee and as thin channel deposits intercalated with floodplain siltstone downvalley (Fig. 10-1A and B). Crevasse splays near the riverbank may form upward-thinning sets of sand beds in which the thicker beds have basal lags of mud clasts ripped up from the levee.

Rivers that aggrade their valleys slowly tend to deposit a number of fining-upward cycles like that of Fig. 10-1A and B. In contrast, rapid aggradation results in braided channels that together cover the entire valley floor, constructing broad sand or gravel deposits (braid-plains) characterized by closely spaced sets of trough cross-strata (Figs. 10-1C and 9-2C). In many braid-plain deposits, stacked channel fills are the most clearly visible structure (Fig. 10-1D). Fine sediments are deposited at receding flood stage and are thin and discontinuous compared with the fine-grained deposits of more slowly aggrading rivers. They are likely to be covered so quickly that ripple cross-laminations are not bioturbated.

***Alluvial fans*** are best known from the many examples studied in arid and semiarid regions (Nilsen, 1982), but they may form in many other environments; for example, in Arctic Canada, where organic mud forms the greater part of some fans (Legget and others, 1966). The essential features of fan deposits must usually be determined by mapping: (1) a radiating pattern of channels and current indicators; (2) a concentric pattern of textural facies; and (3) abrupt termination against an upland slope (Fig. 10-2A and B). Debris-flow deposits are typically abundant in the upper fan, where the sediments generally are coarser than those in the mid and lower fan. Margins of debris flows may be nearly vertical, and the tops of mud-rich flows commonly crack as they dry. Clasts in gravels are subangular and those in arid and semiarid regions include rocks susceptible to chemical decomposition. Exposures parallel to any fan radius show few channel cross sections and those normal to any radius show many (Fig. 10-2C and D). In many fans, large exposures show a great variety of grain size and sorting among the many tongues and layers, which commonly have thicknesses in the range of

0.1 to 2 m. Bedding and textural facies may be obscure, however, in fans that had their sources in fine-grained friable rocks.

In arid and semiarid regions, the more permeable deposits commonly become oxidized. Soils on these deposits contain soil carbonate (*caliche*), and the lower fan deposits may grade laterally to gypsiferous or salty playa deposits. Soils form on the stratigraphically upper parts of any fan sequence that has become entrenched due to degradation of the source area and slowing of fan growth. Deposition thus shifts to the foot of the trench, on the lower part of the fan.

### 10-3. Lake Deposits

Lakes were particularly numerous during the Pleistocene glacial periods (Feth, 1964). Deposits in large lakes with gently sloping sides typically consist of beach sand or gravel grading basinward to fine rippled sand and thence to extensive fine sediments that range from silty or tuffaceous clay to calcitic and dolomitic mud or diatomite. Episodic drying may lead to cracked mud and evaporite intercalated in the finer facies. In the deeper parts of lakes in which oxygenated water is not mixed downward, dysaerobic conditions may lead to preservation of fine organic-rich (sapropelic) sediment or to kerogen-rich calcilutite and dolomite ("oil shale"). At steep lake margins formed by faults or landslide dams, fine sediments may interfinger with

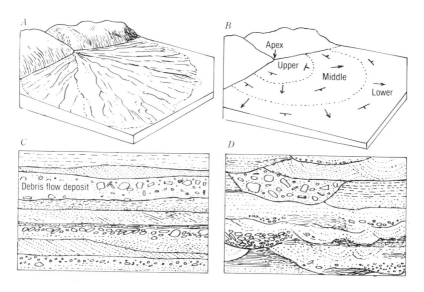

**Fig. 10-2.** Alluvial fan (*A*); its facies, bedding attitudes, and current indicators (*B*); and vertical sections parallel to the radial channels (*C*) and transverse to them (*D*).

talus or coarse gravel. Elongate lakes formed by damming of rivers are filled mainly by river-borne sediment, partly by the growth of deltas and partly by turbidity currents and the settling of sediment from surficial turbid plumes (Pharo and Carmack, 1979).

Where specific facies cannot be recognized, lake deposits may be indicated by: (1) freshwater fossils in fine-grained partly laminated deposits; (2) abundance of oscillation ripples rather than current-formed structures; (3) absence of tidal structures (Section 9-7); (4) well-preserved fossil leaves and other delicate land-based organisms, notably insects and larvae; (5) laminated or structureless clays and other fine deposits that pass upward into plant-rich deposits or evaporites; (6) stromatolites not associated with tidal features (Section 9-7); and (7) abundant beds of tuff.

### 10-4. Eolian Deposits

Eolian sand accumulations can generally be recognized by their relict dune forms and cross-bedding (Fig. 10-3). Dune forms are the surest indicators of wind direction, because cross-lamination dips downwind only in transverse dunes or in the transverse parts of barchans and parabolic dunes. The distinctive cross-bed patterns of seifs and other elongate dunes suggest winds in at least two directions, giving an average (resultant) direction parallel to the dunes (Fig. 10-3). Determination of wind direction from ancient deposits thus requires considering dune forms as well as making many measurements of cross-bedding, which must be corrected for tectonic rotations (Section 9-4).

In ancient deposits where only the cross-bedded rocks can be observed, eolian origin is indicated by: (1) foreset beds that dip consistently more than 28°, commonly 30 to 34°, and which are many meters high; (2) thin slump and sand-flow structures aligned downdip and cutting foreset beds (Fig. 10-4); (3) absence of pebbles except in lags under cross-sets or in well-sorted pebble dunes; (4) gently inclined inversely sized-graded laminations, formed

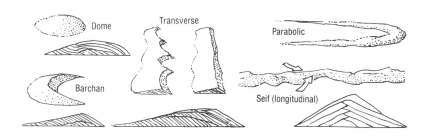

**Fig. 10-3.** Eolian dunes, with transverse sections beneath (a section through the parabolic dune at its apex would be like that of the transverse dune). Based on McKee (1979).

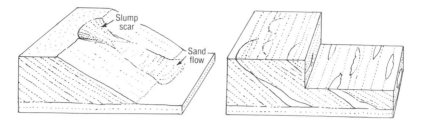

**Fig. 10-4.** Recent slump and flow on frontal slope of dune, and (*right*) slump scars and slump accumulations in cross-bedded dune sand.

by sets of climbing ripples (ripples are otherwise rarely discernible in eolian sand) (Fig. 10-5); (5) animal tracks, which may be sharply delineated in fine sand; and (6) collapsed lamination around large animal burrows. Eolian dune sand that has advanced over vegetation is easily recognized because it covers plants and loose soil without disturbing them. Major slumps in dune sands may produce soft-sediment faults, pull-apart of beds, pinch-and-swell of beds, convolutions, flame structures, and breccias of plastically deformed fragments (McKee, 1979, p. 113). Obscurely laminated parts of dune deposits suggest deposition of sand that was wetted during storms (Hunter and others, 1983).

Major wind-eroded surfaces in dune accumulations can be recognized by a sheet of sand, typically less than 1 m thick, that may be poorly sorted, may show thin parallel lamination, or may be inversely size-graded due to the development of large ripples. The sheet commonly includes a basal lag of sand-faceted and silt-polished pebbles or cobbles.

**Loess** (chiefly wind-deposited silt) can be recognized by its uniform texture, lack of lamination, blanketing of topography, vertical joint systems, internal weathering or soil profiles, buried plant materials, shells of land snails, and other remains of nonmarine organisms. Uncemented loess is

Wind ———▶

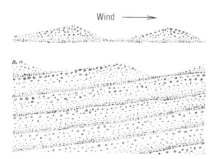

**Fig. 10-5.** Eolian ripples have larger, less dense grains at crests and smaller, denser grains in troughs (*top*); thus, climbing ripples form inversely graded layers, typically without ripple-form cross lamination (*bottom*). See Hunter (1977) for additional varieties.

highly porous (40-50%) and thus hydrocompactible. Loess deposits become thinner and finer away from the river floodplains from which the grains were deflated (Péwé, 1975).

### 10-5. Glacial Deposits

All deposits formed directly or indirectly by glaciers are called *drift*, including several kinds of till as well as associated stream, wind, lake, and glaciomarine deposits. A given drift sheet or complex may be mapped as a composite unit, but interpretations of glaciation may require delineating some or all of the following subunits (see Dreimanis, 1976, for a more complete list):

*Lodgment till* — deposited directly from the base of the glacier as it flows; typified by a firm matrix of sand, silt, and clay that may develop subhorizontal fissility upon exposure; color is due to original materials rather than to weathering; sand laminations, if present, tend to dip up-glacier; locally derived rock fragments are angular, whereas far-traveled ones are rounded and with fresh uneroded striations; many flat and elongate stones are oriented so as to dip or plunge 5 to 30° up-glacier (Fig 10-6); till is locally shaped into drumlins and very locally underlain by sorted sand and gravel of subglacial meltwater tunnels.

*Ablation till* — formed as a residuum on top of the glacier during flow and afterward (Fig. 10-6); generally loose, oxidized, and without consistent stone fabric; textures vary both vertically and laterally; locally moderately sorted to coarse sizes due to winnowing by water and wind; stones commonly broken by frost rifting and striations blurred by abrasion; contains a smaller proportion of locally derived rock debris than lodgment till.

*Meltout till* — englacial moraine deposited over lodgment till when wast-

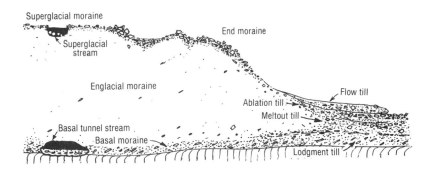

**Fig. 10-6.** Diagrammatic vertical section through terminous of glacier, showing kinds of moraine (rock detritus in and on glacier) compared with kinds of till.

ing of the glacier is so advanced that glacier flow has ceased; may be partly washed of silt and clay in sorted layers intercalated with unsorted material; although not widely studied, stone fabric is relict to englacial moraine and thus oriented parallel to base of deposit and to flow direction of ice (Boulton, 1971); commonly included with ablation till.

*Flow till* — forms when ablation or meltout till becomes water saturated and creeps, slides, or is solifluected down the slopes of ice-cored moraine (Fig. 10-6); typically intercalated with other kinds of till and outwash deposits; fabrics variable, often with blade-shaped stones oriented parallel to flow fronts and sides (Boulton, 1971, p. 54).

*Kame deposits* — washed sand and gravel, often with flow till layers, deposited against an ice face and thus with layers internally slumped during final wasting of glacier; typically forming elongate mounds, irregular ridges, or terrace relics along valley slopes, the latter distinguishable from alluvial terraces because the kame terrace surface and fabric dip toward the valley side (Fig. 10-7).

*Esker deposits* — narrow, long, often sinuous ridges of sorted sand and gravel formed by melt water flowing in tunnels under the glacier or in it (Figs. 10-6 and 10-7); commonly oriented parallel to the direction of glacier flow.

*Outwash apron or plain* — coalesced alluvial fans formed by melt water at glacier terminous; consisting of well-washed sand and gravel cross-bedded in sets about 1 m thick; locally collapsed at sites of kettles; intercalated with flow till and other debris flows at upper end.

*Valley outwash train* — deposited where melt water flows down major river valleys; rapidly aggraded and thus with structures like those of Fig. 10-1C and D.

*Glaciomarine deposits* — may include: (1) marine till, deposited where glaciers were in contact with the sea floor; (2) a variety of deposits from sediment gravity flows (Section 9-2); (3) unsorted but often fossiliferous deposits resulting from the rain of glacial sediments from floating ice; and

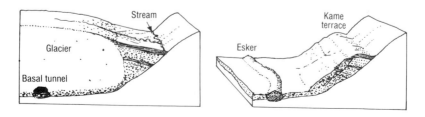

**Fig. 10-7.** A basal melt-water tunnel and ice-contact moraine and outwash (*left*) become an esker and kame terrace (*right*).

(4) sediments reworked and sorted by marine currents but commonly including dropstones from floating ice. All of these deposits may be deformed by overriding glaciers (Eyles and Eyles, 1984).

*Glacial lake deposits* — consist of gravelly or sandy deltas and beach or bar deposits (often with shingled fabrics) and bottom sand, silt, and clay in distinctly planar beds and laminations, the latter commonly less than 0.1 mm thick in clay; lakes with freeze-thaw cycles develop annual *varves*, which are sharply bounded couplets of light-toned silt or fine sand overlain by dark clay, the two layers typically several centimeters thick and internally laminated; occasional stones dropped from floating ice; local folds and faults caused by drag of glacier or ice floes.

*Eolian deposits* — common because of openness of glaciated terrains; deposits include winnowed pavements, sand sheets, dunes, and loess (Section 10-4).

**Extent of glacier and direction of ice flow.** Mapping the deposits just described may show the former extent of a glacier and of its outwash and other marginal deposits. The terminous may also be indicated by relatively high hummocky to ridged ground of the terminal moraine. Aerial photographs show these features especially clearly. Some ice margins are marked by valleys or deposits of rivers that flowed along the ice front. In deep valleys, the upper ice contact can be located by lateral till moraines, tops of kames, height of scour on rock faces, the lower limit (trimline) of soil and trees, and by positions of hanging tributary valleys.

Direction of ice flow in till terrains can be determined from streamlined till hills (drumlins) and from till fabrics, as noted above. Folds in underlying sediments may be overturned in the direction of flow, and slabs of these sediments may be carried upward and down-glacier over till. Eskers are commonly aligned more or less at right angles to glacier terminal margins. In glacially eroded areas, flow direction is indicated by streamlined forms of eroded hills and hillocks that have steep (ice-plucked) slopes on their down-glacier ends; and by grooves, striations, and chatter marks on rock surfaces.

**Drift stratigraphy.** Glaciated regions have generally been affected by more than one glaciation, so that drifts are commonly superimposed. In mapping contacts between them, it is important to anticipate that contacts between the subunits already described may be more apparent than contacts between separate drifts. Another possible relation is the repetition of lodgment till by glaciotectonic thrusting, which is generally indicated by folds formed at the same time or by intercalated slices of underlying deposits (Moran, 1971). These complications should not be problems where the older drift has developed a soil profile (typically with the A horizon eroded), an eolian sand sheet, or a deflated lag, or where the two drifts had such different sources as to have different textures, primary colors, or assemblages of specific kinds of rocks. If the drifts are lithologically similar and cannot be seen in contact, they may be distinguished by characters due mainly to difference in age, specifically:

1. Degree of leaching of carbonates by meteoric water, generally tested by using HCl to determine the depth to which fine matrix-carbonates are leached out completely; soil survey maps showing leached versus unleached soil series are helpful (Section 10-9).

2. Degree of weathering indicated by color changes, as from gray to shades of yellow, orange, and red; determined in the field by use of precisely graduated color charts (Appendix 6).

3. Intensity of leaching based on specific kinds and sizes of fragments (e.g., granite cobbles), as judged by thickness of weathered rinds, depth of surface roughness, or overall firmness, the latter tested by effects of hammer blows (Appendix 4).

4. Degree of erosion and smoothing of surficial forms, such as drumlins, eskers, kettled outwash, morainal ridges, kame terraces, and so on.

5. Degree of compaction and coherence, which increases with age in similar materials (Appendix 4).

6. Fossils of rapidly evolving organisms, as some vertebrates.

7. Numerical dating by $^{14}C$ determinations or measurements of hydration rinds on obsidian.

8. Amounts of displacement along a specific fault.

9. Vertical positions of lateral moraines or outwash terraces along high valley sides, the highest being the oldest.

Matching tills over distances of many kilometers may require especially detailed study where a single till has been supplied from more than one source; for example, a carbonate-rich source and a carbonate-free one (Wright, 1962). Determining the identity and source of older drifts may also be difficult where all rock fragments except chert, quartzite, and vein quartz have been altered to clay or grus and obscured by creep, bioturbation, or frost heaving. Old drifts once rich in carbonate may be thinned greatly by leaching, by factors of 2 or 3, and their thickness may thus differ between places where they were covered by succeeding drift and places where they have remained exposed since they were deposited (Ray, 1974).

## 10-6. Colluvium and Soil Creep

Most materials eroded on slopes are redeposited as *colluvium* where the slopes become more gently inclined. The principal transporting agents are raindrop splash, sheetwash, downslope creep, and occasional small slides, earthflows, and debris flows. The resulting deposits may accumulate to thicknesses of many meters. Colluvium can be recognized by its position along the base of steeper slopes, by its poorly sorted to unsorted texture, and by its crudely bedded to unbedded structure. Bioturbation may destroy colluvial bedding in temperate regions, as indicated by abundant root tubes and burrows.

**Discontinuous creep** probably takes place on all slopes due to diurnal variations in temperature and to seasonal wetting and drying or freezing and thawing. Bioturbation is probably a major cause. The soil at the surface thus moves incrementally downslope at average rates of less than 1 mm to perhaps 2 cm per year (Carson and Kirby, 1972). The rate decreases exponentially with depth, and typically only the upper 10 to 20 cm are affected. The more rapid rates are typical of soils containing abundant clays, especially the expandable clays. Deeply cracked Vertisols (Section 10-9) probably creep episodically in thicknesses equivalent to their crack systems, commonly around a meter. Creep is also suggested by downslope rotation of telephone poles and fence posts; by broken or cracked retaining walls and pavements; and by blocks of rock embedded in soil directly downslope from their sources. Bending of strata and other planar structures in a downhill direction may be caused by currently active creep or by deep freeze-thaw during Pleistocene periglacial periods (Section 10-8). Bending in the lower parts of tree trunks may be caused by several processes besides creep of the soil (Carson and Kirby, 1972, p. 283).

**Continuous creep** is caused by the downhill component of gravity acting on materials that can flow plastically, and is thus probably restricted to moist clay-rich sediments. Relatively rapid rates of creep (1–2 cm per year) measured in clay-rich soils may be due largely to continuous creep. Variations in rates of continuous creep along hillslopes may lead to systematic crack-arrays, as in glaciers, and may be precursors of landslides (Carson and Kirby, 1972, p. 273). *Solifluction* is the downhill flow of water-saturated soil and is described further under periglacial features (Section 10-8).

### 10-7. Landslides, Nonvolcanic Debris Flows, and Rockfalls

Landslides result from failure along distinct shear surfaces or within relatively thin shear zones. The displaced masses move downslope at rates ranging typically from a few millimeters to several meters per day, and in unusual cases at rates up to meters per second. The shear surface is concave

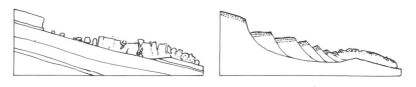

**Fig. 10-8.** Longitudinal sections through translational (*left*) and rotational slides.

upward in most cases, resulting in *rotational slides*; in *translational slides* it is planar (Fig. 10-8). Some displaced masses remain largely coherent but most disintegrate to at least a moderate degree, forming arrays of fault-bounded blocks in their upper parts and rubbly aggregates where they override adjoining slopes (Fig. 10-9*A*). Slides called *earthflows* are more or less loose throughout and move pseudoviscously, developing a fairly smooth concave-upward form in their upper parts and a convex-upward form in their lower parts (Fig. 10-9*B*). *Soil slips* are slides involving only the soil layers and form on slopes steeper than 25°. They are typically generated after prolonged rainfall and commonly produce debris flows (Campbell, 1975).

Most slides are characterized by arcuate main scarps, by straight side scarps (lateral shear surfaces), and by distinctly lobate toes (Fig. 10-9). Because sliding is typical in certain materials and on certain slopes, slides are commonly multiple and may be nested one in another, may form elongate trains of lobate forms, or may lie side-by-side for kilometers along valley slopes. Very large sheets of debris or huge coherent slabs may move kilometers over gently inclined slopes of clay or other ductile material.

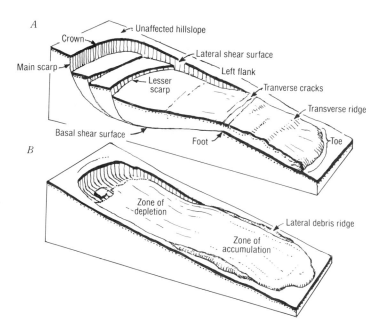

**Fig. 10-9.** Generalized forms of a typical landslide (*A*) and an earthflow (*B*). The form of *B* and the terminology of *A* and *B* are based mainly on a diagram by Karen H. Keefer in Keefer and Johnson (1983).

New slides should be anticipated under any of these conditions: (1) bedding, foliation, joints, or faults dip in the same direction as the slope, and some rock layers are much more ductile than others; (2) relatively impermeable rocks or deposits lie on permeable, porous ones; (3) deposits or rocks contain abundant ductile minerals, as clays, talc, chlorite, micas, serpentine minerals, gypsum, and glauconite; (4) deposits consist of loose sand or gravel in which grains are smoothly rounded; (5) soils or rocks are cracked or closely jointed (they are especially unstable if the cracks have clay skins); (6) sedimentary rocks contain soluble cements, or clays that were deposited in salt water, and are in a position to be leached by fresh water (Hansen, 1965); (7) slopes are occasionally undercut by waves, streams, quarrying, and so on; (8) slopes have no deeply rooted vegetation; (9) slopes are cleared of brush by fire.

Landslide deposits may be recognized by their displaced or jumbled blocks and by fine materials lying on a slickened and grooved surface or, in the overriding toe, on a soil profile. The upper parts of most deposits include displaced soil and vegetation. Deposits of wet earthflows are typically only a meter or so thick (Keefer and Johnson, 1983). Scars above landslide deposits may be subdued and eroded by minor tributary streams that cut headward around the sides of each slide (Fig. 10-10). Ancient slides of more or less coherent material may be recognized by structural attitudes that are rotated or disordered relative to surrounding structures, and by slabs of rock moved out of stratigraphic position.

A common cause of sliding is increase in hydrostatic pressure of pore water, which is typical during or just after heavy rains. Most sliding materials are thus wet, and major parts of earthflows may be so nearly saturated as to have the consistency of thick mud. Further increase in water results in *debris flows*, which have the consistency of a thick mud slurry or, if gravelly, of mixed concrete, and typically move at rates on the order of meters per minute. Large nonvolcanic debris flows are produced when storm-induced landslides move into flooded streams (Williams and Guy, 1973). Debris-flow deposits are described in Sections 9-2 and 12-7.

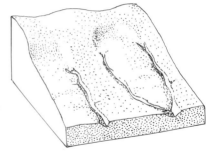

**Fig. 10-10.** Scars of two old slides, with small stream valleys eroded headward along the slide margins.

**Fig. 10-11.** Vertical section through talus, showing typical angle of repose of upper slope, approximately 35°, and the lesser slope of the coarse basal deposit.

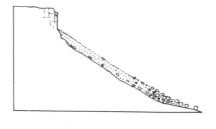

Major dry debris sheets may be produced when large *rockfalls* or landslides partially disintegrate at the base of steep slopes, and spread laterally as partly coherent and partly broken material. Features suggesting that the more coherent parts move on a cushion of compressed air are breccias with fragments unrotated relative to adjoining fragments ("jigsaw puzzle" breccias), and lack of mixing between superposed sheets of differing lithology (Shreve, 1968, Johnson, 1978).

Smaller rockfalls form fans of unsorted angular rubble at the base of cliffs, whereas piecemeal accumulation of small fallen fragments results in a cone or apron of *talus*, characterized by large fragments lying farther downslope than small ones (Fig. 10-11).

### 10-8. Periglacial Deposits and Features

Periglacial conditions result in permafrost (perennially frozen ground) or in surficial processes dominated by freezing and thawing of water. Much of the northern hemisphere above 50° latitude is presently periglacial or has relict permafrost (Péwé, 1983), and similar conditions prevailed much farther south (locally below 35°N latitude) during the last glacial period. Field studies in temperate regions must thus allow for relict periglacial features, which may be modified or may be so fresh as to seem modern.

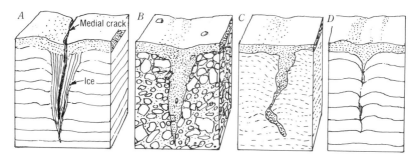

**Fig. 10-12.** Ice wedge (*A*) and three former wedges, partly filled with sandy sediment, as explained in the text.

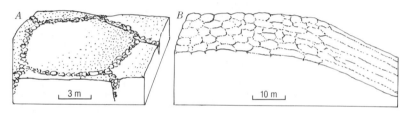

**Fig. 10-13.** Mounded patterned ground with stone lines shown in detail (*A*) and in relation to a sloping surface (*B*).

Studies in periglacial regions have provided criteria for recognizing them (French, 1976; Washburn, 1980; Péwé, 1983).

*Patterned ground, ice-wedge deposits.* Viewed from the air, the surface of the ground in a periglacial environment is commonly divided into regularly repeated polygons, circles, or reticulated forms. The larger (10-30 m per cell) are formed today where permafrost contracts as it becomes colder during the winter. The resulting cracks are filled in the summer, either with melt-water or windblown sand, forming a thin wedge of ice or sand that grows each year and may finally form a thick wedge (Fig. 10-12*A*). Former ice wedges are indicated by top-filled dikes in sediments (Fig. 10-12*B*); by partial fillings that are pinched and deformed when the ice melts (Fig. 10-12*C*); or by traces against which beds are buckled (Fig. 10-12*D*).

Large-scale patterned ground is usually visible on aerial photographs, and smaller cells, typically 1 to 10 m across, can generally be seen on the ground. The latter may develop by freeze-thaw without permafrost and are common relics in temperate regions. Many are accentuated by lines of stones that were sorted upward and outward by freeze-and-thaw of the sediment within the cells, which tend to form low mounds (Fig. 10-13*A*). The patterns vary in shape according to the angle of slope, grading into downhill stone-stripes on slopes steeper than 15° or so (Fig. 10-13*B*).

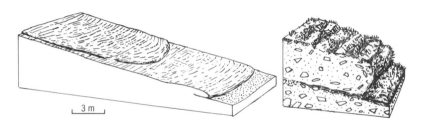

**Fig. 10-14.** Solifluction lobes, with detail showing angular clasts in unsorted soil emplaced over sod.

***Convolutions, involutions.*** Flat-lying sediments in periglacial areas commonly show convolutions (folds) and involutions (underlying materials penetrating overlying materials) that may result from pressure built up during zonal freezing of the annually thawed surficial sediments (French, 1976, p. 42).

***Subsidence hollows, thermokarst.*** Sediments in permafrost areas may develop shallow flat-lying lenses of sediment-free ground ice. When the ice melts, subsidence hollows form at the surface, and they commonly become rounded lakes, or arrays of elliptical ponds. Broadly subsided terrains (thermokarst) result where the ground ice was thick and extensive and the ground has largely been drained.

***Slope forms and deposits.*** Creep and solifluction (flow of surficial water-saturated materials) are widespread in permafrost areas due to frequent freeze and thaw and to the presence, during annual thaw, of a sheet of water-rich sediment lying on permafrost. Periglacial slope deposits are suggested by solifluction lobes or by their relict structure (Fig. 10-14) and by several kinds of terraces (Fig. 10-15). Talus is developed in abundance under high outcrops and may feed block streams on hill slopes or rock glaciers in valleys.

The large amounts of detritus carried downslope accumulate as thick deposits in valleys and swales. Low periglacial temperatures and rapid rates of frost wedging result in clasts being angular and free of weathering rinds or secondary coatings. Rubbly periglacial deposits, especially talus at the base of valley walls, is commonly covered by finer colluvium during interglacial periods.

## 10-9. Soils

A working knowledge of soils is helpful in all types of geologic field work and essential in studies of surficial deposits. A given soil profile may

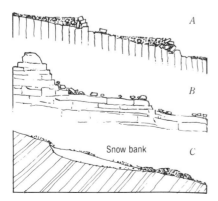

**Fig. 10-15.** Terrace surfaces developed (*A*) where large stones creep downslope, become hung up, and act as barriers; (*B*) where rocks have gently sloping joints; and (*C*) where perennial snowbanks become localized at certain elevations.

encompass several thin stratigraphic units, as ash or loess layers, which must thus be recognized in variously altered states. In addition, such units must be distinguished from layers (horizons) produced by the soil processes themselves. The relations between soil-forming processes and climate, in turn, are used to interpret paleosols, which are indicators of Quaternary cycles (Section 10-1).

**Descriptions of soils.** Soils are customarily studied and described in the field according to characteristics that are here greatly abbreviated from descriptions and procedures of the Soil Conservation Service (Soil Survey Staff, 1951, 1975):

1. *Horizons* (soil sublayers) are identified, measured, and systematized as in Table 10-1.

2. The profile is studied carefully to determine if more than one *parent material* (geologic unit) is represented; any unit beneath the uppermost parent unit is designated by the Roman numeral II (as horizon IIB), and the next underlying parent unit by III, and so on.

**Table 10-1.** Principal Kinds of Soil Horizons, in Normal Vertical Sequence

*O horizons.* Formed by surficial accumulation of organic material; contain > 30% organic matter if the mineral components are mainly clay, or > 20% organic matter if they are mainly coarser than clay (for definitions based on analyzed carbon content, see Soil Survey Staff, 1975, p. 65). $O_1$ *horizon*: original form of most vegetative matter is visible to the naked eye. $O_2$ *horizon*: original form of most vegetative matter cannot be recognized with the naked eye.

*A horizons.* Mineral horizons (defined as containing less organic matter than defined above for O horizons) that occur under an O horizon or at the surface. $A_1$ *horizon*: contains accumulated organic matter and thus typically dark. $A_2$ *horizon*: typically under an O or $A_1$ horizon and pale due to leaching away of iron, aluminum, or clay, and consequent concentration of quartz.

*B horizon.* Mineral horizon normally under an A horizon and characterized by one or more of: (1) concentration of clay minerals, aluminum, iron, or humus; (2) darker or stronger (commonly redder) colors than those of overlying and underlying horizons; or (3) structures of parent material obliterated, clay minerals and oxides formed, and horizon transected by prismatic, blocky, or granular crack systems.

*C horizon.* Mineral horizon developed from material either like or unlike the parent of overlying horizons, lacking characteristics of those horizons (see above), but modified by one or more of: (1) discoloration and mineral alterations below the zone of major biologic activity; (2) conversion to brittle clay; (3) cementation; (4) alteration under reducing conditions to gray tones (gleying); (5) accumulation of carbonates or more soluble salts; or (6) accumulation of silica or iron oxide and silica.

*R horizon.* Underlying consolidated bedrock or unaltered sediment; need not be parent material of overlying horizons.

3. *Colors* are coded by the Munsell Color System (Appendix 6) for each horizon and for dry or moist material as appropriate; color mottles (curving linear to irregular color forms, generally 0.5-3 cm across) and color bands are described.

4. *Texture*, the grain-size constitution per horizon, is estimated by hand-lens examination of intact and washed samples, and by feel of grittiness and smoothness of wet materials. The proportion of outcrops of bedrock, termed the *rockiness* of a soil, is recorded as percentage of area.

5. The soil is broken at a free face and crumbled in the hand to determine its *structure*, which is its tendency to separate along systematic cracks or invisible surfaces of weakness. The resulting separate forms, called *peds*, are described as to size and classified by shape as *prismatic, columnar* (prisms with rounded tops), *angular blocky, subangular blocky, platy, and granular* (the latter are peds less than 1 cm in diameter with shapes not accommodated by adjoining peds). Soils may be *strongly, moderately,* or *weakly* structured, or *structureless* (*massive* if coherent and *single-grain* if not). Thorough treatment of soil structure has been described by Larionov (1982).

6. *Consistence*, which is a measure of a soil's cohesion under pressure, is determined by: (1) testing the degree of friability, firmness, or hardness; (2) squeezing and then releasing a lump of moist soil between the thumb and forefinger to determine its stickiness; and (3) rolling a lump of moist soil in the hands to determine its plasticity (the ease with which it can be rolled into a thin rod).

7. Breaking harder or whitish parts of the soil to determine the *degree of cementation* and the cementing minerals, such as calcite, dolomite, gypsum, opal, iron oxides, manganese oxides, and dark organic complexes with iron or other metals.

8. Examining freshly broken fragments to estimate the volume percentage of roots, root tubes, cracks, and other pores, and also the presence of mineral coatings in pores, especially *clay skins*, which are gelatinous to waxy, generally pale brown to reddish or greenish films of clay transported into that part of the soil.

**Kinds of soil.** The descriptions that follow summarize the ten Orders of the soil classification currently in use in the United States, as it applies to unplowed soil profiles. The Soil Survey Staff (1975) has given complete definitions and keys for identification, and has described classification systems used in other countries. Buol and others (1980) have summarized this same information and presented interpretations and genetic cases in point.

**1. Histosols** — the only soil Order consisting dominantly of organic materials (which are defined under *O horizons* in Table 10-1). These soils represent nonoxidizing, and thus water-saturated conditions; for example, the peat and muck of former bogs and ponds.

**2. Entisols** — soils somewhat paler than parent materials but lacking soil horizons. Their embryonic nature is due to one or more of: (1) youth; (2) slowness of soil formation because of coldness, dryness, or frequent water saturation; (3) inertness of parent materials, such as quartzite; (4) parent materials, such as serpentine, that are poorly supportive of plants; and (5) organic acid complexes becoming bound with fine-grained calcite or opal.

**3. Inceptisols** — *A* horizon distinctly pale or dark gray (especially over volcanic ash); *B* horizon often reddened, bioturbated, and with prismatic structure. The weakly developed horizons in these soils are due to the same causes as listed for Entisols except for dry climate, which is definitive of Aridosols (see below).

**4. Vertisols** — dark, clay-rich soils with cracks 1 cm or more wide at depths of 50 cm, caused by seasonal drying. Expansion and contraction produce numerous slickensided fractures that intersect in wedge-shaped forms, and expansion of materials fallen from the surface into cracks may result in partial inversion of soil layers and upward bulging mounds (Fig. 10-16). Expandable clays (chiefly smectites) make the soil unusually sticky and plastic when wet.

**5. Aridosols** — *A* horizon typically pale (due to low content of organic material), dry, and loose; *B* horizon generally reddish because of oxidizing conditions, bioturbated, or with prismatic structure in younger soils and with accumulated clays in older ones; *C* horizon partly salty or partly cemented by carbonates, gypsum, or opal, due to evaporation of pore water at that depth. These soils of arid climates may develop in semiarid climates where expandable clays in the *A* horizon swell during rains and keep the underlying soil dry.

**6. Mollisols** — more than one-third of combined *A* and *B* horizons organic-rich and thus very dark brown to black (except where locally powdered by fine calcite cement); moderately to strongly structured and quite loose when dry; high availability of cations suggested by unleached silicate minerals and strong vegetative cover in fairly dry climates; may have clay-enriched *B* horizon; in relatively dry climates, *C* horizon has accumulations of carbonates or salt. These soils characterize grasslands, and may be thin or thick

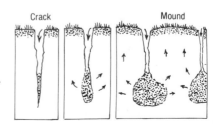

**Fig. 10-16.** Stages of inversion of a Vertisol, starting with a system of deep cracks.

depending on rainfall and thus the luxuriance of the grass cover.

**7. Alfisols** — $A$ horizon pale unless climate is unusually wet, in which case it is dark with organic materials; $B$ horizon distinctly clay-enriched, as shown by clays in cracks and pores; abundant available cations indicated by unleached minerals and strong vegetation, which is typically forest or rank prairie growth; generally moist during growing season and thus a soil of humid climates.

**8. Spodosols** — moist sandy soils with a pale gray, loose $A_2$ horizon and a structureless $B$ horizon that is weakly cemented by iron and manganese-bearing organic complexes that give a reddish to black color; this colored layer may be underlain by a somewhat brittle clay-rich layer; and a thin ( <1 cm ) strongly cemented deep red to black layer may lie at or near the top of the $B$ horizon. In cool climates, these soils are common under coniferous forests.

**9. Ultisols** — $A$ horizon generally pale; $B$ horizon reddish or orange and clay-enriched; $C$ horizon reddish and commonly mottled. Kaolin clays are abundant to dominant and almost all primary silicates are altered to clay and iron oxides. Developed under hardwood-conifer forests in warm, moist climates south of glacial drift, and thus typically old, thick soils.

**10. Oxisols** — loamy to clayey soils with horizons developed weakly if at all; commonly red but also yellow or gray; so thoroughly leached as to consist only of kaolin, iron and aluminum oxides (thus lateritic), and quartz; typically meters thick and with all vestiges of rock structures destroyed by creep and bioturbation. These are soils of humid tropical to subtropical climates.

Classification of soils into further taxa (as the specific soil names used in most soils publications) has been described in detail by the Soil Survey Staff (1975), in condensed form by Buol and others (1980), and in brief by Birkeland (1984).

**Soil maps and soil data** are available for many areas within the United States in the Department of Agriculture series entitled *Soil Surveys*. These reports vary somewhat in content but typically include: (1) maps of soil boundaries plotted on an aerial photograph base, mainly at scales of 1:24,000 or 1:20,000; (2) a description of how the maps were made; (3) an overall soil map of the report area, typically at a scale of 1:250,000; (4) descriptions of all the mapped soils; (5) suggestions regarding agricultural and engineering practices on the soils; and (6) origin and classification of the soils. Partial analyses of sampled profiles (*pedons*) are included in some reports. Full analyses of certain pedons have been published (e.g., Gile and Grossman, 1979) or may be available from the Soil Conservation Service (contact a local office or inquire to Soil Conservation Service, U.S. Department of Agriculture, Washington, D.C.). The analyses are made for each horizon or sampled parts of a horizon, and include sediment size-data, bulk density, water content, organic carbon content, and certain measures of extractable bases and cation exchange capacities. Similar information and soil maps may also be available from state soil surveys.

The main cartographic units shown on soil maps are *soil series*, which are a taxon level given geographic names rather than soils names. Series are often subdivided into *phases* on the basis of slope, texture, or certain other variables, and these lesser units are those generally delineated during the actual mapping. Even at scales of 1:15,000 to 1:30,000, however, a mapped phase or series is likely to include small patches of other phases or series that can be recognized in the field but are too small to plot separately. The map units are thus somewhat generalized, and the maps record the *dominant* soil at and near a given locality.

A major geologic value of soil maps derives from the fact that soils tend to reflect the kind of material beneath them, whether they have developed from that material or not. Boundaries between soils series or phases may thus mark contacts between different rocks or deposits. Soil map units are commonly based on properties reflecting rate processes, such as leaching of carbonates or buildup of clay in the *B* horizon, and can thus be used to trace out older and younger units within arrays of drifts, alluvial sediments, eolian fields, or emergent lake or marine deposits. Areas blanketed by a specific loess or volcanic ash may be indicated by a separate soil series or phase; in fact, surficial layers so thin as to be incorporated entirely in the soil might otherwise go unnoticed by geologists. Soil maps may also provide clues to the correlation of landforms, such as terrace remnants and extensive erosion surfaces of low relief (Ruhe, 1975).

All of these uses require careful consideration of factors in addition to parent materials and age, which lead to different soil series or phases. The degree of slope, the geographic facing of the land surface, and past land uses are major influences on soil temperature and moisture regimes, and thus on formation of certain soil series or phases. These factors are discussed for local conditions in most soil survey reports and are described more broadly in texts on soil genesis (Buol and others, 1980; Birkeland, 1984).

**Paleosols and soil stratigraphy.** Paleosols ("ancient soils") are soils developed during a distinctly earlier climatic regime. Many paleosols are buried, but some lie at the surface either because they have been exhumed or because they are relict paleosols — soils that have never been buried, yet record long past conditions.

Specific kinds of buried paleosols can be recognized easily where eolian sand dunes, loess, ash falls, or alluvial accumulations on floodplains have buried soils without disturbing any horizon. Generally, however, *O* and *A* horizons are eroded prior to burial and much or all of the *B* and *C* horizons may be eroded. It is thus necessary to look for any vestige of a former profile: (1) reddish or yellowish relics of the *B* or *C* horizon; (2) zones of root tubes or bioturbation; (3) prismatic or other soil structure, (4) clay skins or carbonate, gypsum, opal, iron oxide, or manganese oxide cements in pores and cracks; (5) resistant layers of brittle clays or of sandy sediments cemented by iron and manganese oxides associated with organic materials; and (6) dark gray clays with root tubes.

An exhumed paleosol can be proven by finding places where it is still covered by younger deposits (Fig. 10-17*A*). Relict paleosols can often be recognized because they appear out of place in the present climatic regime and may lie under atypical vegetation, as Mollisols under forest cover or dried Spodosols under a grassland. Relict paleosols may in some cases be traced to the point where they are overlapped by younger deposits (Fig 10-17*B*). Many are overprinted by horizons, perhaps subtle ones, expressing the present climatic and vegetative regime (Fig. 10-17*C*). They may also be recognized where their initial landforms are partly eroded and they thus lie next to younger soils (Fig. 10-17*D*).

The stratigraphic meaning of a paleosol is determined by comparing its thickness and the nature of its horizons to modern data on soil genesis, as suggested briefly in the soil descriptions above, more fully by Buol and others (1980), and specifically in the soils literature. Generally, a well-developed extensive paleosol indicates a period of time when little erosion or deposition took place over a large area, and can thus be used to separate strata that represent periods of deposition (Morrison, 1967). Such a soil can be dated geologically by noting the youngest stratigraphic unit in which it is developed and the oldest unit that lies above it. Paleosols of particular value are those that are as well developed where they have been buried as

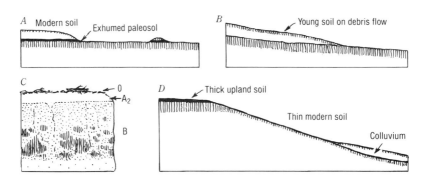

**Fig. 10-17.** *A*. Partially exhumed paleosol (*A* horizon black, *B* horizon lined). *B*. Paleosol locally covered by younger deposit, as by a debris flow on an alluvial fan. *C*. Developing Spodosol with relics of a paleo-Mollisol (vertical lines). *D*. Thick upland paleosol, passing into Recent soil on slope, which becomes intercalated with colluvium.

where they have been exposed continuously (Fig. 10-17*B*). These soils appear to have been formed almost entirely during periods of optimum conditions, thus representing warmer or wetter paleoclimatic periods within a Quaternary cycle. They may thus prove valuable for interregional correlation.

## References Cited

Birkeland, P. W., 1984, *Soils and geomorphology*: New York, Oxford University Press, 372 p.

Black, R. F., Goldthwait, R. P. and Willman, H. B., editors, 1973, *The Wisconsinan Stage*: Geological Society of America Memoir 136, 334 p.

Boulton, G. S., 1971, Till genesis and fabric in Svalbard, Spitsbergen, p. 41-72 *in* Goldthwait, R. P., editor, *Till; a symposium*: Columbus, Ohio State University Press.

Buol, S. W., Hole, F. D., and McCracken, R. J., 1980: *Soil genesis and classification*, 2nd edition: Ames, Iowa State University Press, 404 p.

Campbell, R. H., 1975, *Soil slips, debris flows, and rainstorms in the Santa Monica Mountains and vicinity, southern California*: U.S. Geological Survey Professional Paper 851, 51 p.

Carson, M. A., and Kirby, M. J., 1972, *Hillslope form and process*: Cambridge, England, University Press, 475 p.

Cooke, H.B.S., 1981, Age control of Quaternary sedimentary/climatic record from deep boreholes in the Great Hungarian Plain, p. 1-12 *in* Mahaney, W. C., 1981.

Dackombe, R. V., and Gardiner, V., 1983, *Geomorphological field manual*: London, George Allen & Unwin, 254 p.

Dreimanis, A., 1976, Tills: their origin and properties, p. 11-49 *in* Legget, R. F., editor, *Glacial till*: Royal Society of Canada Special Publication 12, 412 p.

Eyles, C. H., and Eyles, N., 1984, Glaciomarine sediments of the Isle of Man as a key to late Pleistocene stratigraphic investigations in the Irish Sea Basin: *Geology*, v. 12, p. 359-364.

Feth, J. H., 1964, *Review and annotated bibliography of ancient lake deposits (Precambrian to Pleistocene) in the western States*: U.S. Geological Survey Bulletin 1080, 119 p.

Fink, J., and Kukla, G. J., 1977, Pleistocene climates in central Europe: at least 17 interglacials after the Olduvai event: *Quaternary Research*, v. 7, p. 363-371.

French, H. M., 1976, *The periglacial environment*: London, Longman Group, 309 p.

Gile, L. H., and Grossman, R. B., 1979, *The desert project soil monograph; soils and landscapes of a desert region astride the Rio Grande Valley near Las Cruces, New Mexico*: Soil Conservation Service, U.S. Department of Agriculture, 984 p.

Goudie, A., editor, 1981, *Geomorphological techniques*: London, George Allen & Unwin, 395 p.

Hansen, W. R., 1965, *Effects of the earthquake of March 27, 1964, at Anchorage, Alaska*: U.S. Geological Survey Professional Paper 542-A, 68 p.

Hunter, R. E., 1977, Basic types of stratification in small eolian dunes: *Sedimentology*, v. 24, p. 361-387.

Hunter, R. E., Richmond, B. M., and Alpha, T. R., 1983, Storm-controlled oblique dunes of the Oregon coast: *Geological Society of America Bulletin*, v. 94, p. 1450-1465.

Johnson, B., 1978, Blackhawk landslide, California, U.S.A., p. 481-504 *in* Voight, B., editor, *Rockslides and avalanches*, 1: *Natural phenomena*: Amsterdam, Elsevier Scientific Publishing Co.

Keefer, D. K., and Johnson, A. M., 1983, *Earth flows: morphology, mobilization, and movement*: U.S. Geological Survey Professional Paper 1264, 56 p.

Larionov, A. K., 1982, *Methods of studying soil structure*: New Delhi, Amerind Publishing Co., 193 p.

Legget, R. F., Brown, R.J.E., and Johnston, G. H., 1966, Alluvial fan formation near Aklavik, Northwest Territories, Canada: *Geological Society of America Bulletin*, v. 77, p. 15-30.

McKee, E. D., 1979, Introduction to a study of global sand seas, p. 3-19, *and* Sedimentary structures in dunes, p. 83-113 *in* McKee, E. D., editor, *A study of global sand seas*: U.S. Geological Survey Professional Paper 1052, 429 p.

Mahaney, W. C., editor, 1976, *Quaternary stratigraphy of North America*: Stroudsburg, PA, Dowden, Hutchinson & Ross, 512 p.

Mahaney, W. C., editor, 1981, *Quaternary paleoclimate*: Norwich, England, Geo Abstracts Ltd., 464 p.

Mahaney, W. C., editor, 1984, *Quaternary dating methods*: Amsterdam, Elsevier Science Pub. Co., 431 p.

Moran, S. R., 1971, Glaciotectonic structures in drift, p. 127-148 *in* Goldthwaite, R. P., editor, *Till; a symposium*: Columbus, Ohio State University Press.

Morrison, R. B., 1967, Principles of Quaternary soil stratigraphy, p. 1-69 *in* Morrison, R. B., and Wright, H. E., Jr., editors, *Quaternary soils*: Reno, Center for Water Resources Research (Desert Research Institute), University of Nevada, 338 p.

Nilsen, T. H., 1982, Alluvial fan deposits, p. 49-86 *in* Scholle, P. A., and Spearing, D., *Sandstone depositional environments*: American Association of Petroleum Geologists Memoir 31.

Péwé, T. L., 1975, *Quaternary geology of Alaska*: U.S. Geological Survey Professional Paper 835, 145 p.

Péwé, T. L., 1983, The periglacial environment in North America during Wisconsin time, p. 157-189 *in* Wright, H. E., Jr., editor, 1983, v. 1.

Pharo, C. H., and Carmack, E. C., 1979, Sedimentation processes in a short residence-time intermontane lake, Kamloops Lake, British Columbia: *Sedimentology*, v. 26, p. 523-541.

Ray, L. L., 1974, *Geomorphology and Quaternary geology of the glaciated Ohio River valley — a reconnaissance study*: U.S. Geological Survey Professional Paper 826, 77 p.

Ruhe, R. V., 1975, *Geomorphology; geomorphic processes and surficial geology*: Boston, Houghton Mifflin Co., 246 p.

Self, S., and Sparks, R.S.J., editors, 1980, *Tephra studies as a tool in Quaternary research*: Dordrecht, Holland, D. Reidel Publishing Co., 481 p.

Shreve, R. L., 1968, *The Blackhawk landslide*: Geological Society of America Special Paper 108, 47 p.

Soil Survey Staff, 1951, *Soil survey manual*: Soil Conservation Service, U.S. Department of Agriculture Handbook 18, 503 p.

Soil Survey Staff, 1975, *Soil taxonomy: A basic system of soil classification for making and interpreting soil surveys*: Soil Conservation Service, U.S. Department of Agriculture Handbook 436, 754 p.

Washburn, A. L., 1980, *Geocryology; a survey of periglacial processes and environments*: New York, John Wiley & Sons, 406 p.

Williams, G. P., and Guy, H. P., 1973, *Erosional and depositional aspects of Hurricane Camille in Virginia, 1969*: U.S. Geological Survey Professional Paper 804, 80 p.

Woillard, G. M., 1978, Grande Pile peat bog: a continuous pollen record for the past 140,000 years: *Quaternary Research*, v. 9, p. 1-21.

Wright, H. E., Jr., 1962, Role of the Wadena lobe in the Wisconsin glaciation of Minnesota: *Geological Society of America Bulletin*, v. 73, p. 73-100.

Wright, H. E., Jr., editor, 1983, *Late-Quaternary environments of the United States; v. 1, The late Pleistocene; v. 2, The Holocene*: Minneapolis, University of Minnesota Press, 407 p., 277 p.

# 11

# Stratigraphic Sections ■

## 11-1. Preliminary Steps

Stratigraphic sequences are measured and described in many kinds of projects and are generally presented as columnar sections or detailed descriptive logs. These surveys have three basic purposes: (1) to obtain accurate thicknesses of mapped units; (2) to record full sequential descriptions of the rocks within the units; and (3) to obtain exact stratigraphic positions of fossils and rock samples.

Well-exposed sections are often surveyed during a mapping project in order to recognize stratigraphic position in places where the same rocks are poorly exposed or broken by faults. Stratigraphic measurements may also have specific purposes, such as determining areal variations in thickness or lithology; connecting surface mapping with subsurface logs of wells; and describing type sections of new formations (Section 5-3). In addition to their use in sedimentary sequences, stratigraphic sections may be compiled for volcanic sequences, for metamorphic rocks, and for stratified plutons. Additional suggestions for specific projects as well as for methods of measurement have been given by Kottlowski (1965).

Accurate stratigraphic measurements and descriptions require much time and effort and thus warrant thorough preparation. The preliminary steps that follow are suggested for field projects consisting chiefly of stratigraphic measurements. In cases where extensive geologic mapping precedes measurements, most of the steps will already have been taken.

1. *Researching rock units* will firm up the purpose of the project and indicate features that should be examined during the field work. Geologic mapping is reviewed in order to compile areal variations in thickness and broad structural relations, such as unconformities, gradations to other units, and positions of dated samples in nearby areas. The units are discussed with geologists who have studied them elsewhere.

2. *Selecting locations.* Ideally, each section should: (a) be well exposed; (b) be measurable within an area of small size (to be as nearly as possible a true stratigraphic column); and (c) be undisturbed by faulting or other deformation that cannot be resolved. These requirements are assessed mainly during field reconnaissance, but study of aerial photographs, geologic maps, topographic maps, and soil maps will help to narrow the selection. Road and railroad cuts, aqueducts, and stream courses often provide the most continuous exposures but may intersect bedding at such low angles as to extend

the study too far laterally. If dips are at low angles, steep valley walls scored by ephemeral streams may be ideal sites.

3. *Reconnaissance* is the most important step in planning (Section 1-4). Landowners or local residents may be questioned regarding access and permission and can often supply valuable information on locations of outcrops or on records from local wells and inaccessible mine workings. The best locations are examined and, if outcrops are scattered, possible courses for measurement are walked out to see if a complete section can be pieced together (Section 11-2). Landslides, faults, and folds that affect the sections are studied to determine if they can be resolved. Closely spaced exposures of the sequence are compared to test its lateral uniformity. Samples are collected to determine if materials are fresh enough for micropaleontologic or petrologic studies (Sections 3-7 and 3-8). The openness of the terrain is noted in cases where telescopic surveying instruments will be used.

4. *Precision* of stratigraphic measurements arises from two different aspects of a study. The first is the mechanical precision of the surveying methods that will be used (Sections 11-4 to 11-8). The second is how finely the rocks should be subdivided, measured, and described (Section 11-2). The time and funds available affect both aspects of precision. The scale of stratigraphic columns that will be prepared may also affect them (Section 11-9).

5. *Preparations for the field.* Besides accumulating equipment and supplies ordinarily used in field work (Section 2-1), stratigraphic measurements may require surveying equipment (Section 11-4 to 11-8), special equipment for sampling (mattock, shovel, sledge, rock drill, or crowbar), and equipment for scraping or washing rock faces. Field sheets or notebooks may be set up in advance to help systematize rock descriptions.

### 11-2. Subdividing and Describing a Section

The description of the section is generally undertaken in this order: (1) location and study of contacts between formations or members; (2) subdivision of these formal units into the informal units that will be described; (3) trial measurements and descriptions of several of these units; (4) systematic measurements and descriptions of the entire section; and (5) reexamination of parts of the section, as necessary.

Formations or members in the section provide a means of correlation with the same units elsewhere. Their contacts should be located exactly at the line of section and should be studied for at least several hundred meters on either side of the section. The contacts are based on prior mapping in the area or on published descriptions (Section 5-3). Any contact that cannot be located firmly is described as being within a certain measured interval.

The subdivisions used in the measurement are physically distinct groupings of rocks that will be of value in unifying parts of the description, in preparing a columnar section, and in interpreting the section. Other geologists may use these units to locate specific parts of the sequence in the field. Ideal stratigraphic units are based on lithologic characteristics that have genetic meaning, and the more practical are also based on obvious surficial characters, such as degree of exposure, steepness of slope, thicknesses of beds, kinds of vegetative cover, and color of rock and soil. Thickness is not an important criterion provided the unit is physically distinct. A thin tuff bed in a sedimentary sequence, for example, would be a valuable unit. Disconformities should be used as contacts between units even if they seem minor.

A trial description and measurement of part of the section serve to calibrate procedures to the time available and the purpose of the project. These trials may be unnecessary if the rocks have become well known during prior field work.

Systematic measurement and description of the full section is best made

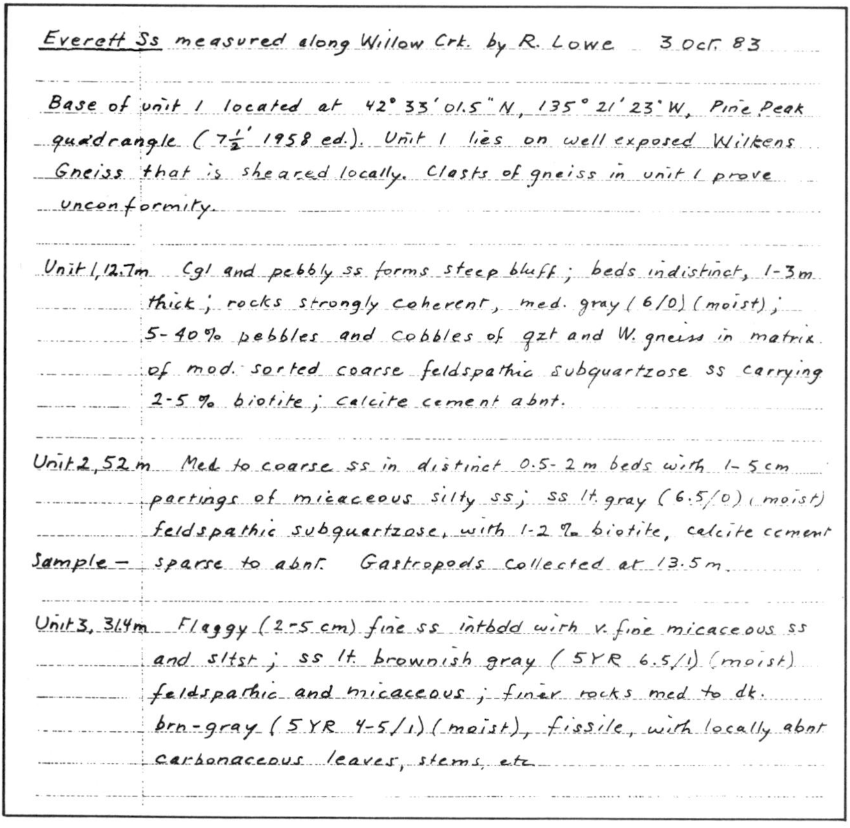

---

*Everett Ss measured along Willow Crk. by R. Lowe 3 Oct. 83*

*Base of unit 1 located at 42° 33' 01.5" N, 135° 21' 23" W, Pine Peak quadrangle ( 7½' 1958 ed.). Unit 1 lies on well exposed Wilkens Gneiss that is sheared locally. Clasts of gneiss in unit 1 prove unconformity.*

*Unit 1, 12.7m  Cgl and pebbly ss forms steep bluff; beds indistinct, 1-3 m thick; rocks strongly coherent, med. gray (6/0) (moist); 5-40% pebbles and cobbles of qzt and W. gneiss in matrix of mod. sorted coarse feldspathic subquartzose ss carrying 2-5% biotite; calcite cement abnt.*

*Unit 2, 52 m  Med to coarse ss in distinct 0.5-2 m beds with 1-5 cm partings of micaeous silty ss; ss lt. gray (6.5/0) (moist) feldspathic subquartzose, with 1-2% biotite, calcite cement Sample — sparse to abnt. Gastropods collected at 13.5 m.*

*Unit 3, 31.4m  Flaggy (2-5 cm) fine ss intbdd with v. fine micaceous ss and sltst; ss lt. brownish gray ( 5YR 6.5/1) (moist) feldspathic and micaceous; finer rocks med to dk. brn-gray ( 5YR 4-5/1) (moist), fissile, with locally abnt carbonaceous leaves, stems, etc.*

**Fig. 11-1.** Field notes from a detailed description of a measured formation.

by advancing up-section, in order to see the sequence in depositional order. This approach is especially important for sections with occasional scour or omission surfaces, or for sequences composed of depositional cycles or rhythms. Outcrops are generally easiest to find and examine when walking upslope, especially if the survey is somewhat pressed for time. In any case, all aspects of approach, study, and precision should be kept as consistent as possible during the survey.

Restudying or recollecting parts of the section may prove necessary if important features or relations are recognized after the measurement is well underway. The section may also have to be restudied after thin sections, fossil identifications, or analytical data have been examined. Reoccupying parts of the section should be no problem if a number of stations are marked permanently, or if the units are lithologically distinctive and they were described carefully and fully in the first place.

**Descriptions** are accumulated per measured unit as the survey proceeds. They should be as complete as time permits, but they are usually recorded in telegraphic style and with use of abbreviations. They are generally started with the name of the rock or rocks that make up the measured unit, followed by data in a systematic order suitable to the project and to the specific rocks being described. The order in Fig. 11-1 is appropriate for general study of a sedimentary sequence, and the characteristics listed in Section 3-3 and those described in Chapters 9, 10, 13, 14, and 15 give additional categories of data. Consistency of order is important because of the need to compare units later and to synthesize overall formation descriptions. It may be worthwhile to make up and duplicate check sheets if a large number of units must be measured in a short time or if data will be digitized (Fig. 11-2).

If a columnar section will be prepared from the data, a graphic log depicting shapes and relative thicknesses of beds and their primary structures is highly desirable (Fig. 11-3). Drawing the log forces one to look systematically at important features and to generalize them consistently. The log also affords a detailed basis for drawing the final columnar section.

Sample sites are recorded in the notes by specific measure and are plotted to scale on the graphic log. Sites of samples that will be analyzed or used biostratigraphically should be marked clearly on the outcrop or photographed with a hammer or other scale at the sample point.

### 11-3. Covered, Deformed, or Laterally Variable Strata

Offsets are commonly needed to piece together separated exposures of a stratigraphic sequence (Fig. 11-4). Ideally, an offset is made by walking along a specific bed or bedding surface that connects the stratigraphically upper part of one exposure to the lower part of a nearby exposure. In cases where the offset must be made across a covered area, a specific bed or a

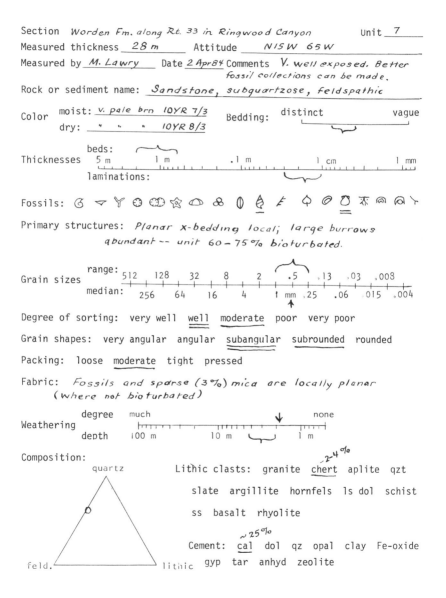

Section  *Worden Fm. along Rt. 33 in Ringwood Canyon*        Unit __7__

Measured thickness __28 m__    Attitude   *N15W  65W*

Measured by __M. Lawry__   Date __2 Apr 84__ Comments  *V. well exposed. Better*
                                                *fossil collections can be made.*

Rock or sediment name: *Sandstone, subquartzose, feldspathic*

Color  moist: *v. pale brn 10YR 7/3*    Bedding:   distinct         vague
       dry:  *" " " 10YR 8/3*

                    beds:
Thicknesses   5 m        1 m          .1 m         1 cm         1 mm
                    laminations:

Fossils:  (fossil symbols)

Primary structures:  *Planar x-bedding local; large burrows*
                     *abundant -- unit 60 - 75% bioturbated.*

              range: 512   128   32    8    2    .5   ,13   ,03   ,008
Grain sizes
              median:   256   64   16    4   1 mm  ,25  .06  ,015  ,004

Degree of sorting:  very well  <u>well</u>  moderate  poor  very poor

Grain shapes:  very angular  angular  <u>subangular</u>  subrounded  rounded

Packing:  loose  <u>moderate</u>  tight  pressed

Fabric:  *Fossils and sparse (3%) mica are locally planar*
         *(where not bioturbated)*

               degree   much                            ↓         none
Weathering
               depth   100 m           10 m                  1 m

Composition:                                           ~2-4%

            quartz          Lithic clasts:  granite  <u>chert</u>  aplite  qzt

                                      slate  argillite  hornfels  ls dol  schist

                                      ss  basalt  rhyolite

                                            ~25%
                            Cement:  <u>cal</u>  dol  qz  opal  clay  Fe-oxide

feld.                lithic   gyp  tar  anhyd  zeolite

**Fig. 11-2.**  Check sheet for describing measured stratigraphic units, with slant-lettering, brackets, arrows, and underlines showing data recorded for a specific case.

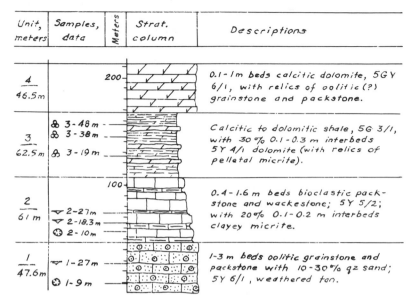

| Unit, meters | Samples, data | Meters | Strat. column | Descriptions |
|---|---|---|---|---|
| 4 46.5m | | 200 | | 0.1-1m beds calcitic dolomite, 5GY 6/1, with relics of oolitic(?) grainstone and packstone. |
| 3 62.5m | ✿ 3-48m ✿ 3-38m ✿ 3-19m | | | Calcitic to dolomitic shale, 5G 3/1, with 30 % 0.1-0.3 m interbeds 5Y 4/1 dolomite (with relics of pelletal micrite). |
| 2 61m | ▽ 2-27m ▽ 2-18.3m ◑ 2-10m | 100 | | 0.4-1.6 m beds bioclastic pack- stone and wackestone; 5Y 5/2; with 20% 0.1-0.2 m interbeds clayey micrite. |
| 1 47.6m | ▽ 1-27m ◑ 1-9 m | | | 1-3 m beds oolitic grainstone and packstone with 10-30% qz sand; 5Y 6/1, weathered tan. |

**Fig. 11-3.** Part of a graphic log and accompanying unit descriptions, constructed from the base of the formation upward. The fossil symbols are explained in Appendix 9, the rock symbols in Appendix 8, and the color notation in Appendix 6.

unique set of beds can sometimes be recognized in both exposures. Lacking these direct means of correlation, the offset must be made by walking along a visually projected line of bedding, as by the method described in Section 5-2 under *Using strike and dip to locate contacts.*

**Cover.** Despite careful selection, a section line may cross intervals that are unexposed and cannot be resolved by lateral offsets. If an interval is hidden only by soil, its lithology may be approximated from soil texture, from fragments brought up by burrowing animals, by augering through the upper part of the soil, or by digging shallow pits. Intervals covered by

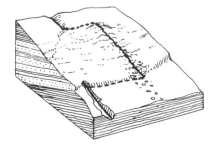

**Fig. 11-4.** Offsets (*heavy dashed lines*) between measured courses (*heavy solid lines*) in gullies and along a craggy outcrop.

surficial deposits may have to be recorded as unknown parts of the section; however, some can be resolved by unusually long offsets or by study of the same general part of the sequence in neighboring areas.

The approximate thickness of strata in covered intervals must be determined in order to obtain the thickness of the unit overall. Covered intervals are measured by using strikes and dips like those stratigraphically above and below the covered interval and measuring as if the interval were exposed.

**Faults** typically are not exposed and therefore constitute a major problem in measuring stratigraphic sections. The suggestions in Section 5-4 may help in recognizing unexposed faults. In particular, any linear strip of unexposed ground should be checked by tracing specific beds or other structures (dikes, veins, etc.) up to the strip and seeing if they project across it to the other side.

The section being measured is generally the best means of correlating stratigraphic units from one side of a fault to another. Where faults and units have been mapped, the relations described in Section 12-5 may help in correlating fragments of a faulted section and thus in determining the stratigraphic interval that is missing or duplicated along a specific fault.

**Landslides** can generally be recognized by their morphology, by variability of bedding attitudes in the slide mass, or by backward tilt of beds (Section 10-7). In measuring sequences of poorly exposed rocks or surficial deposits, large exposures are often formed at the main scarps of slides or in the gullies that tend to form along the lateral shear surfaces (Figs. 10-9 and 10-10).

**Folding** commonly changes thicknesses of folded units, most typically by thinning strata in the limbs of folds and seemingly thickening them in the hinge areas (Fig. 12-12). These effects are absent only in parallel (concentric) folds, which are probably far less common than generally supposed. The effects may be tested by measuring specific rock units on limbs and in the hinge area. If that is not possible, the shapes of outcrop-size folds may suggest the sort of changes likely in large folds. The only firm resolution, however, requires mapping folds and constructing a cross section at right angles to hinge lines, a so-called *profile view* or right section of the fold (Fig. 12-10).

To use this view in determining the approximate original thickness of a unit, it is first necessary to determine the mechanism of apparent thinning and thickening, which may have been (1) compactional closing of pore space during folding, or (2) plastic deformation of the solid grains in the rock. If the deformation was only compactional, the approximate original thickness is most nearly represented by the thickness at the hinge. If the solid grains of the rock were deformed plastically, the approximate original thickness can be determined as shown in Fig. 11-5. The result is generally a minimum thickness, because rocks are typically also condensed by compaction.

Compaction is indicated by low porosity in rocks normally with high porosity, as mudstone, and by undeformed grains and weak grain fabrics, if any, in limestone and sandstone. Plastic deformation is indicated by cleavage oblique to bedding in the hinge areas of folds, and by strong fabrics and deformed grains in limestone and sandstone. If the deformed grains are also elongated parallel to the hinge line, the amount of extension in that direction must be determined in order to further correct the thickness of the deformed rock unit. Sections 12-1, 12-2, and 12-4 include suggestions for recognizing compactive deformation and for measuring grain fabrics and rock strains due to plastic deformation.

Small folds in an otherwise unfolded sequence may or may not cause problems in stratigraphic measurements. Single beds containing soft-sediment folds (Section 9-3) are measured according to their present thickness. A group of beds affected by soft-sediment folding may be part of a major slump and should, if possible, be traced laterally to determine the original thickness of that interval. Finally, small overturned folds associated with shear zones along specific stratigraphic intervals may indicate detachment faulting (Section 12-5). These intervals may have to be mapped over a large area to determine if part of the section is missing.

*Lateral variations* in original thickness or lithology are not a problem if a section is measured on a fairly straight course at right angles to bedding and the results are intended to show thicknesses along that line. If the purpose of a section is to illustrate average or typical thickness and lithology over some larger area, additional sections should be measured and averaged to form the section. If exposures are abundant, lateral variations can be walked out and a typical section selected directly.

### 11-4. Measurement with the Jacob Staff

In this method, strata are measured in true thickness as the section is traversed and described. The method requires only one person and is especially suited to areas of at least moderate exposure where outcrops are fairly closely spaced near the section course. It may be the only usable method in

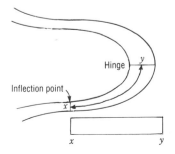

**Fig. 11-5.** To obtain an average thickness from a profile view: (1) mark the hinge and inflection points on both folded surfaces and connect them by lines perpendicular to the surfaces; (2) measure the average segment length $xy$; (3) place a sheet of transparent grid paper over the profile and count the squares included in the segment $xy$, thus determining its area; (4) obtain the thickness by dividing the area by the segment length. The block below shows the unfolded segment.

rough or brushy country where long taped measurements and long sights are difficult or impossible.

A Jacob staff is a board or pole graduated in suitable units (as decimeters and centimeters) and somewhat longer than the eye height of the person using it. A staff can be made from a piece of planed lumber measuring approximately 2.5 × 5 × 200 cm (1 × 2 × 75 in.) and of light but strong wood, as Douglas fir. The graduations are ruled across the broader face and must be exactly perpendicular to the staff's length. A Brunton or Silva compass, used as a clinometer, is held firmly against this face parallel to the graduations (Fig. 11-6A). Suppliers (see Section 2-1) generally stock a telescoping Jacob staff and attachments for mounting a Brunton compass to it.

In principle, a measurement is made by aligning the staff at right angles to bedding and sighting downdip to the point subtended by the measure (Fig. 11-6B). The distance from the base of the staff to the sighting axis of the clinometer is thus equal to the thickness of strata from the base of the staff to the point sighted. The complete procedure is as follows:

1. Measure the strike and dip of bedding at the place of the measurement; record the attitude and set the clinometer at the angle of dip.

2. Open the compass lid 60° and hold the compass firmly against the staff as shown in Fig. 11-6A, with its base parallel to the graduations on the staff and at a comfortable eye-height. Record this height and, if convenient, use it for all other full measures.

3. Place the staff at the base of the unit to be measured and tilt it downdip (exactly perpendicular to strike) until the clinometer bubble is centered.

4. Study the point sighted on the ground to determine if the staff can be placed on it; if so, note the point carefully by eye and advance the staff to it.

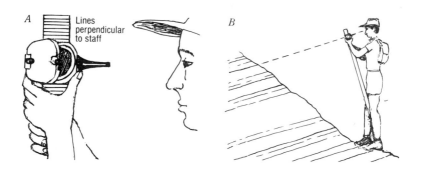

**Fig. 11-6.** *A.* Holding a Brunton compass against the Jacob staff in sighting position. *B.* Measuring the stratigraphic thickness between two bedding surfaces with a Jacob staff and clinometer (here, a Brunton compass).

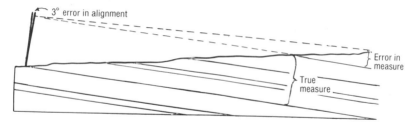

**Fig. 11-7.** Error in measurement resulting from a small error in alignment.

5. If the point sighted cannot be used, move the base of the staff along the same bedding surface until a suitable point can be sighted.

6. Before taking the next measure, tally the first.

7. Proceed similarly to the top of the unit, and for the last fractional measure hold the compass at whatever height is appropriate to sight the top, and record that partial measure.

The thickness of the unit is equal to the tallied number of measures multiplied by the staff-height used, plus the final partial measure.

Because the method requires sighting with a small instrument, it may be tempting to save time by aligning the staff by estimation. Moderate errors in alignment, however, can cause large errors in measurements (Fig. 11-7). In addition, when sighting up or down a slope, one tends to tilt the staff so that it is perpendicular to the ground surface. Errors thus tend to be systematic and to accumulate through a series of measures. When the staff is correctly oriented with the clinometer, the error should be no more than a few centimeters per measure and will tend to be compensated in successive measures. Other methods should nonetheless be considered when both the slope and dip are at low angles.

Sighting with a staff as just described becomes increasingly awkward as

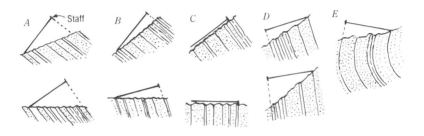

**Fig. 11-8.** Positions of a Jacob staff for specific combinations of slope and dip (see text for explanation).

dips become steeper. Lines of sight typically become shorter, however, and thus accuracy is maintained (Fig. 11-8*A*). For dips greater than 70°, the operator can kneel and look along strike, viewing the clinometer face-on and making the projection to the ground by estimation (Fig. 11-8*B*). Beds meeting the surface of the ground at 90° can be scaled by direct measurement (Fig. 11-8*C*), and the staff can be reversed to measure steeply dipping beds where the average slope is steep to very steep (Fig. 11-8*D*) or beds are overturned (Fig. 11-8*E*).

Steep walls of valleys eroded across strike often have the best exposures, and they must be measured by views parallel to strike. In such cases, the clinometer is set to the dip as usual but the lid of the compass is opened so as to make an angle of 90° with the compass face (Fig. 11-9*A*). The observer holds the compass against the staff and stands facing the compass and outcrop, looking exactly along strike (Fig. 11-9*B*). The outcrop is sighted along the upper edge of the lid, and the staff is held so that the side facing the observer is vertical. An exact vertical orientation can be obtained by holding a small carpenter's level against the staff.

An efficient procedure is to measure a full unit (or some 10 measures or so if the unit is thick) and then return to the base of this measured interval to start describing the rocks bed by bed or measure by measure. The top of the unit should first be marked so that it can be found again easily. Beds and intervals thicker than a staff-measure are measured in the same way as the section is measured. Thinner ones generally can be measured directly with the staff or with a roll-up tape. When the description of the unit or other measured interval is completed, the thicknesses are summed and compared with the overall measurement.

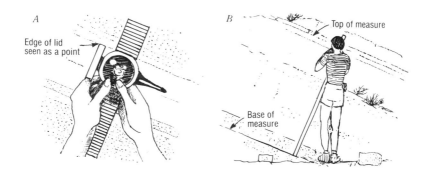

**Fig. 11-9.** Sighting along strike with a Jacob staff. *A.* Holding the Brunton compass against the staff and sighting along edge of lid. *B.* Position for taking a measure.

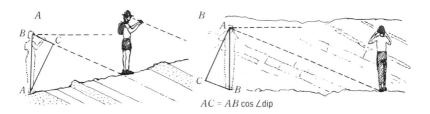

$AC = AB \cos \angle\text{dip}$

**Fig. 11-10.** Successive positions of an observer and formula for measuring strata with a Brunton compass, by sighting perpendicular to strike (A), and parallel to strike (B). AB is the eye height of the geologist, and AC the thickness measured.

## 11-5. Measurement using Eye Height and a Brunton Compass

The Brunton compass can be used to measure sections rapidly when no other equipment is available. It is not suitable for measuring the details of a variable section, however, because it cannot be used for partial measures. Horizontal strata are measured by setting the clinometer at 0 and taking eye-height measures as in leveling (Section 2-7). For inclined strata, the clinometer is set at the local angle of dip and the compass is used to sight down dip much like it is with a Jacob staff (Fig. 11-10A). The number of measures through a unit or exposed interval is tallied and multiplied by the trigonometric relation shown in the figure. Strata that can be observed most easily when looking along strike are measured by setting the clinometer at 0 and sighting eye-height measures that are then converted to thickness trigonometrically (Fig. 11-10B). The point sighted is projected by eye down the trace of bedding to a convenient place to stand for the next measure.

Precision depends on knowing one's eye height in the footwear being used, standing straight at the exact spot sighted, measuring the strike and dip frequently, and sighting exactly along lines of dip or strike.

Tanner (1953) noted the ease and adaptability of the Brunton method and added a version for places where the measurement must be made oblique to the dip, such as where a stream bed or roadcut provides the only exposure.

**Fig. 11-11.** Stratigraphic measurement along a linear exposure (RT) that is oblique to strike and thus shows apparent dip of beds. RTV is the direction angle used in Appendix 13.

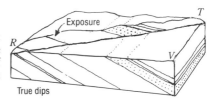

The clinometer is set to the apparent dip in the exposure and sights are made parallel to the exposure (line $RT$ in Fig. 11-11). The number of measures along the exposure is the same as it would be by using the true dip and sighting downdip (as along the line $RV$). The thickness is thus equal to the number of eye-height measures times eye height, multiplied by the cosine of the true dip (Fig. 11-10A). If true dip cannot be measured, it can be determined from the apparent dip by measuring the angle $RTV$ and using Appendix 13. If the line $RT$ is inclined more than 10°, a correction must be made for the fact that the angle $RTV$ is not a horizontal angle. Angle $RTV$ is reduced by 1° if $RT$ is inclined at 15°, by 2° if it is inclined at 20°, and by 3° if it is at 30°. The method also requires that the direction of the line $RT$ be measured with extra care (to the nearest degree) in cases where this line is within 15° of the line of strike.

## 11-6. Tape-Compass-Clinometer Method

A stratigraphic unit can be measured indirectly by taping its intercept on a topographic slope (the distances $AB$ in Fig. 11-12), measuring the dip of bedding and the inclination of the slope (angles $x$ and $y$ in Fig. 11-12), and calculating the thickness by the appropriate formula. This method is more precise than the Jacob staff method in cases where the ground surface crosses bedding at angles less than 20° (Fig. 11-7). It is ideally suited for

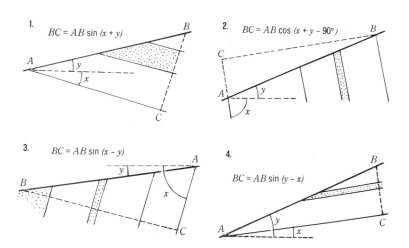

1. $BC = AB \sin (x + y)$

2. $BC = AB \cos (x + y - 90°)$

3. $BC = AB \sin (x - y)$

4. $BC = AB \sin (y - x)$

**Fig. 11-12.** Formulas used for determining thickness where the slope measurement *(AB)* is at right angles to strike: (1) slope and dip are opposed and slope angle plus dip angle is less than 90°; (2) slope and dip are opposed and slope angle plus dip angle is greater than 90°; (3) slope and dip are in the same direction and dip angle is greater than slope angle; and (4) slope and dip are in the same direction and dip angle is less than slope angle.

comparatively smooth slopes where outcrops lie along more or less straight courses for distances of 20 m or more. These courses need not be parallel, and offsets can be made as described in Section 11-3 or by a short compass-and-tape traverse (Section 5-1).

The calculation of thickness is simple in cases where the taped line is perpendicular to the strike of the beds (Fig. 11-12). Where the taped line is oblique to the strike, which is typical if the line is to follow the best array of exposures, the thickness is found by a formula derived by Palmer (1916) (Fig. 11-13). This calculation may be made fairly quickly with a pocket calculator or by using a nomograph (Palmer, 1916; Mertie, 1922).

If the difference between the angle of slope and the dip is less than 10°, and especially where the slope and dip are in the same direction, precision can become a problem because sines of angles around 10° change nearly 1 part in 10 for every degree. The following precautions are therefore important:

1. Check the accuracy of the clinometer before the survey (Section 2-4).

2. Use two persons for the survey if possible.

3. Be certain that the angle of slope is measured on a line exactly parallel to the tape. Thus: (a) in taping, hold the tape the same distance above the ground at both ends, and (b) in measuring the vertical angle, stand exactly at one point and sight to a point eye height above the ground at the far end of the taped line.

4. Measure strike and dip by sighting rather than by contact methods (Section 3-5).

A measure is made as follows:

1. Starting at the stratigraphic base of the section, hold (or secure) the 0

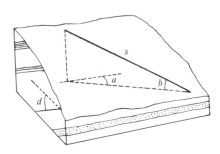

**Fig. 11-13.** Where the taped line *(s)* is oblique to the strike (angle *a*), the true dip (angle *d*) and vertical angle of the taped course (angle *b*) can be used to find the stratigraphic thickness by the formula: *thickness* = $s(\sin a \sin d \cos b + \cos d \sin b)$ for cases where dip and slope of the taped course are opposed. Where they are in the same direction, the + sign is changed to a − sign.

end of the tape at the base of the section and carry the reel to the first full length or the first break in slope. Pull the tape taut and mark the measure with a chaining pin, nail, or stake.

2. Record the distance, then measure the bearing of the taped line and the local strike and dip as exactly as possible, and record them.

3. Measure and record the slope angle as described in step 3 above.

4. Describe the sequence in stratigraphic order, positioning each lithic contact by projecting it (parallel to bedding) to the tape, which is held taut and on the same slope as that just measured (Fig. 11-14).

5. Measure strike and dip at several places along the line and determine an average as necessary. If the attitude changes abruptly, measure the distance to that point and treat the parts of the line separately.

6. Calculate the overall thickness by the appropriate formula (Figs. 11-12 and 11-13); then calculate the thicknesses of the units included, if any, and check their total thickness against the overall thickness. Or, if more convenient, calculate the thicknesses later and return as necessary to correct errors of measurement.

7. Advance the tape so that the 0 end is at the first point marked by the nail or stake, offset if necessary (Section 11-3), and continue by the steps just described.

### 11-7. Transit Method

The transit is ideally suited for measuring stratigraphic sections in areas of low relief where beds dip at angles less than 15°, and it should be considered for any section that must be measured with unusual precision. Advantages over the alidade and plane table are: (1) a transit can be set up more solidly and more quickly; (2) it can be used to measure vertical angles more quickly and more precisely; (3) it permits measurement of a greater range of vertical angles; (4) it can be used to sight down slopes too steep to be visible from a plane table; (5) it can be operated at stations where there is not enough footroom to work around a plane table; and (6) the mountaineering transit and most modern "micro" models are less cumbersome to carry. Descriptions of transits and instructions for checking, adjusting, and oper-

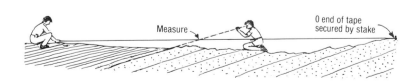

**Fig. 11-14.** Tape held taut on grade by the recorder while geologist determines measure to a unit contact by sighting down dip with a clinometer set to the local dip.

ating them are included in texts on surveying and, for recent models, are supplied with the instruments.

The transit method used most commonly is almost like the tape-compass-clinometer method (Section 11-6) except that vertical angles and angles between the line of strike and the taped course are measured with the transit. As with the ordinary clinometer, the transit must be aligned parallel to the tautly stretched tape, and this can be done by holding a Jacob staff (or any graduated rod) at one end of the line and sighting a point equal to the height of the transit axis above the station mark on the ground.

The stations where the transit is set up are generally marked permanently with a firmly set stake and a small nail. The strata in between may be measured and described as in Section 11-6, or by using a Jacob staff in cases where angles of dip and slope permit adequate precision. The staked stations are a basis for collecting samples or additional data at a later time. In some cases it may be most efficient for two or three persons to survey the line with a transit and tape, and then for one person to measure and describe the section on the basis of the staked stations.

### 11-8. Plane Table Methods

These methods are ideal where dips and slopes are too gentle for accurate measurements with either the Jacob staff or tape and clinometer and where outcrops are so scattered that a transit and tape survey is impractical. The alidade and plane table may also be needed where no suitable geologic map is available and the sequence is complicated by faults or folds. In such cases, a strip or skeletal map must be constructed along the section course, and the plane table and alidade permit doing this as the measurement proceeds.

An important initial consideration is the scale required to measure the section to some limiting degree of precision. If the section must be measured to the nearest meter or so, and plotting precision is approximately 0.2 mm on the plane table sheet, the map scale would have to be greater than 1:5000 and preferably 1:2500 or more. At a scale of 1:2500, a standard plane table sheet is just large enough to map a strip 1.5 km long.

The length of the strip to be mapped and the map scale also affect the selection of the kind of survey that will give adequate precision. An uncontrolled stadia traverse should be precise enough to map a strip less than 0.5 km long at scales of 1:2500 or more, providing stadia distances are kept under 100 m and checked by a backsight when the plane table is advanced from one station to the next (Section 8-7, last subsection). Stadia measurements may also be used to map outcrops off the section line, as needed to fill in the stratigraphic section or to determine positions of faults and geometry of folds. Detailed measurement and description of the section itself, however, are better made with a Jacob staff or tape, because otherwise one person will be standing at the plane table for long periods between stadia sightings. The stadia points would thus provide a geologic framework for the more detailed measurements.

If the section strip is between 0.5 and 1.5 km long, a traverse can still be adequately precise for scales of 1:2500 or larger, but the main traverse legs should be taped rather than measured by stadia. Strata exposed along the taped traverse legs can be measured directly by the tape, as in the transit method described above. Strata exposed on either side of the traverse line may be mapped by stadia.

Section strips longer than 1.5 km, and thus extending beyond one plane table sheet if mapped at scales of 1:2500 or larger, should probably be controlled by setting up and intersecting a chain of triangulation stations (Section 8-7). This control is especially advisable where structural relations must be mapped in order to measure the section or where there is no large-scale topographic map by which to check the overall dimensions of the plane table map. Mapping between the control stations is done by stadia methods (Section 8-9) and the details of the stratigraphic section can be accumulated by any of the direct methods described in this chapter.

In areas of gently dipping strata, errors in vertical position introduce almost equal errors in stratigraphic thickness. Vertical distances must thus be measured as precisely as possible, as by using the stepping method rather than measuring vertical angles (Section 8-5). An advantage of the plane table is that the three-point method can be used easily to determine accurate strikes and dips (Section 3-5).

## 11-9. Presenting Stratigraphic Sections

*Graphic columnar sections* are the traditional means of presenting measured sequences (Fig 11-15). Brief descriptions of the units may be lettered to the right of the column, as in the figure, or the column may be accompanied by an explanation consisting of a small box for each lithologic symbol and for the other symbols alongside the column. No explanation is included in Fig. 11-15, but symbol boxes of lithologic patterns are shown in Appendix 8 and other symbols in Appendix 9. The following elements of a stratigraphic column are essential and are keyed to Fig. 11-15 by numbers: (1) title, indicating topic, general location, and whether the section is single (measured in one coherent course), composite (pieced from two or more section segments), averaged, or generalized; (2) name(s) of geologist(s) and date of the survey; (3) method of measurement; (4) graphic scale; (5) map or description of locality; (6) major chronostratigraphic units, if known; (7) lesser chronostratigraphic units, if known; (8) names and boundaries of rock units (Section 5-3); (9) graphic column composed of standard lithologic patterns; (10) unconformities; (11) faults, with thickness of tectonic gaps, if known;

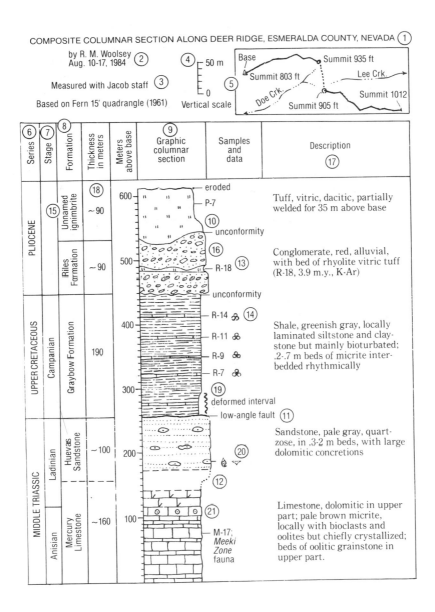

**Fig. 11-15.** Columnar section with title and accessory data. Numbered items are identified in the text.

(12) covered intervals, as measured; (13) positions of key beds; and (14) positions of important samples, with number and perhaps data.

Other kinds of information that may be included are: (15) designations of formal or informal measured units; (16) an irregular edge indicating relative resistance of the rocks; (17) summary descriptions of formations or other units (especially desirable if the section will not be accompanied by an explanatory text); (18) thicknesses of units; (19) intervals of deformed rocks; and (20) symbols or numbers indicating kinds of fossils, primary structures, porosities, cements, shows of petroleum, and so on. Some of the latter features may be added directly to the lithologic column, as at (21).

Columns are constructed from the stratigraphic base upward and should be plotted first in pencil in order to insure spaces for gaps at faults and unconformities. Sections that are thicker than the height of the plate can be broken into two or more segments, with the stratigraphic base at the lower left and the top at the upper right. Bedding and unit boundaries are drawn horizontal except in detailed sections or generalized sections of distinctly nontabular deposits, as some gravels and volcanic units (Fig. 11-15, and see Mullineaux, 1970, p. 37).

Uses of columnar sections in reports are described in Sections 16-2 and 16-3.

**Stratigraphic logs** are used to describe sections in the text of reports. Although telegraphic in style, they may describe each measured unit as fully as warranted, and thus present sections in greater detail than all but the most detailed columnar sections. They are not usually published unless they are the type section of a formation or member; however, they may be the main vehicle in an unpublished report.

Logs are arranged so that the youngest rocks appear first in the text. The smallest measured units are numbered to make the sequence clear. Order of presenting data should be kept as consistent as possible, as in the fragment that follows:

Smith formation

3. Shale, black, soft, locally leaf-bearing ..................... 5 m
2. Sandstone, dark gray, moderately resistant, carbonaceous, feldspathic, in beds 0.5-2 m thick, the thinner
   with carbonaceous 1-cm laminae; thick beds heavily
   burrowed .......................................... 23 m
1. Conglomerate, light gray, highly resistant, of
   rounded chert pebbles, fine sand matrix, imbricated ....... 2 m

  Total Smith Formation .................... 232 m

Base of formation is an unconformity on well-exposed Byron Shale.

An additional description at the base of the log is generally used to locate the base of the section exactly, and a description at the base or the top of the log gives methods of measurement, date, and personnel. Descriptions inserted at all major offsets give their locations and the bedding attitudes at the top of one exposure and the base of the other. The purpose of these descriptions is to guide other geologists to the base of the section and completely through it. Examples are given by Love (1973).

## References Cited

Kottlowski, F. E., 1965, *Measuring stratigraphic sections*: New York, Holt, Rinehart and Winston, 253 p.

Love, J. D., 1973, *Harebell Formation (Upper Cretaceous) and Pinyon Conglomerate (Uppermost Cretaceous and Paleocene), northwestern Wyoming*: U.S. Geological Survey Professional Paper 734-A, 54 p.

Mertie, J. B., Jr., 1922, *Graphic and mechanical computation of thickness of strata and distance to a stratum*: U.S. Geological Survey Professional Paper 129-C, p. 39-52.

Mullineaux, D. R., 1970, *Geology of the Renton, Auburn, and Black Diamond quadrangles, King County, Washington*: U.S. Geological Survey Professional Paper 672, 92 p.

Palmer, H. S., 1916, Nomographic solutions of certain stratigraphic measurements: *Economic Geology*, v. 11, p. 14-29.

Tanner, W. F., 1953, Use of apparent dip in measuring thickness: *American Association of Petroleum Geologists Bulletin*, v. 37, p. 566-567.

# 12

# Features of Deformed Rocks ■

## 12-1. Early Formed Deformational Features

The principal subjects of this chapter are features formed in solid rocks by tectonic processes—folding, faulting, and more evenly distributed strains. Mélanges are included because of their importance in tectonically active regions, even though many mélanges are of sedimentary origin (Section 12-7). Mélanges, in fact, point to the more general problem addressed in this first section: how to recognize tectonic deformation of unconsolidated sediments.

Unconsolidated and partly consolidated muds generally have porosities between 40% and 80% and may thus undergo strains by loss of pore water alone (Fig. 12-1A and B). At large strains (Fig. 12-1C), flaky grains in mudstone may be rotated sufficiently to produce a new planar fabric and thereby a slaty cleavage (Williams and others, 1969; Clark, 1970). The cleavage has most of the attributes described in Section 12-4, but its soft-sediment origin can be recognized where liquefied sand has intruded along the cleavage planes as well as across them (Fig. 12-2A) (Powell, 1972). The parallelism must be exact, however, because large postconsolidation strains will typically rotate sandstone dikes toward the plane of slaty cleavage in mudstone (Boulter, 1974). Additional indications of tectonic deformation of soft sediments are faults that break semiconsolidated layers but not unconsolidated ones (Fig. 12-2B), and beds of sand, silt, and clay that have become mixed to a sandy mudstone in shear zones that have indistinct or irregular margins (Fig. 12-2C).

Nontectonic deformational structures in soft sediments are described in

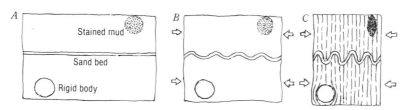

**Fig. 12-1.** Tectonic shortening of unconsolidated mud containing a rigid body (as an early formed concretion), a stained but otherwise identical mud body, and a bed of sand. A, initial state, porosity = 60%; B, after 25% shortening; and C, after 50% shortening and 10% vertical extension. Vertical lines represent cleavage.

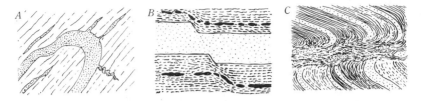

**Fig. 12-2.** Tectonic deformation of soft sediments. *A.* Sand intruded from folded bed along cleavage (parallel lines) in tectonically compacted mud. The folded dikelet on the right was presumably emplaced before or during deformation. *B.* Fault breaking sand layer but not soft mud. *C.* Right-lateral shear zone in interbedded sand and mud.

Section 9-3, and primary deformational structures of igneous rocks are described in Chapters 13 and 14. The forms and associations of these structures should be reviewed during mapping, because they can easily be mistaken for tectonic features.

### 12-2. Determining Directions and Amounts of Strain

The directions in which a rock has been deformed (strained) may be indicated by the shapes and orientations of grains or structures whose original shapes are known, such as oolites, sand grains, or fossils. The amounts of strain can be approximated by measuring the dimensions of the strained forms and comparing them with the original dimensions. The shapes of some deformed grains and structures indicate, further, the kind of strain that took place. *Flattening* is expressed by equidimensional grains and structures that have been pressed into symmetrical plates or disks (Fig. 12-3*A*). *Extension* is indicated by grains elongated and constricted so as to become prisms or rodlike ellipsoids with equidimensional cross sections (Fig. 12-3*B*). Measurements of the lengths and widths of these bodies give a

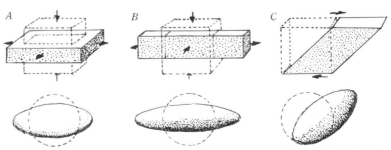

**Fig. 12-3.** Changes in the shapes of cubic and spherical bodies (dashed outlines) affected by flattening (or shortening) of 76% (*A*), by extension of 137% (*B*), and by simple shear rotation of 47% (*C*).

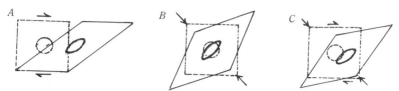

**Fig. 12-4.** Cubic and spherical bodies deformed by simple shear (*A*), by pure shear (*B*), and by a combination of simple and pure shear (*C*), giving new forms (the parallelograms and ellipses) of similar shape and orientation.

numerical ratio expressing the amounts of strain, or the changes can be expressed as percentages of flattening or of elongation, as noted in the figure.

The general kind of strain in both cases just described is called *pure shear*, a deformation that has taken place symmetrically around an axis of flattening or elongation, without rotation of the strained body. In *simple shear*, an axis of principal strain rotates progressively with increasing amounts of shear, and the grain or structure changes shape much as though it were a stack of very thin sheets displaced laterally (Fig. 12-3C). The height of the skewed cube (the axis perpendicular to the shear plane) remains unchanged, as does the axis lying parellel to the shear plane and perpendicular to the direction of shear. The sphere, for example, becomes a triaxial ellipsoid in which the axis roughly perpendicular to the page is the same as a diameter of the original sphere.

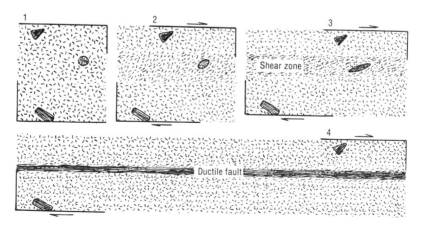

**Fig. 12-5.** Changes of shape and fabric in a granite body with inclusions, where simple shear and accessory flattening are concentrated in a tabular zone. The zone is considered a *ductile fault* when the displacement of the two walls is more than 5 to 10 times the thickness of the zone (Wise and others, 1984.)

Simple shear is an important kind of geological strain, such as in fault zones or near the contacts of intrusions. In deformed solid rocks, however, simple shear is not always distinguishable from pure shear, at least not on the basis of the shapes of grains or structures alone. The three ellipsoidal bodies in Fig. 12-4, for example, are similar in shape and orientation.

In the field, one can seek additional evidence for kinds of strain, and it may be helpful at the outset to assume that any distinct foliation includes at least one principal axis of the strain ellipsoid, and that a distinct lineation in this foliation is parallel to one of the axes. Additional features and measurements may then be referred to these axes to determine the kinds of strain that have contributed to the total strain of the rocks. Where simple shear has been dominant, rocks tend to show different amounts of strain from one part of a foliated body to another, giving rise to variation in the shapes and orientations of deformed features (Fig. 12-5). Also suggestive of simple shear are phacoidal bodies of rock separated by fine-grained or schistose sheets (Fig. 12-6A) or by faults. Veins, color bands, or cross-beds oriented at a large angle to the direction of simple shear may be cast into folded forms and, if so, the dimension parallel to the direction of shear remains unchanged from one part of a fold to another (Fig. 12-6B). In metamorphic rocks, porphyroblasts may show systematically rotated inclusions because the crystals rotated as they grew (Fig. 15-2B), and undeformed porphyroblasts in foliated rocks may have skewed deposits of quartz or carbonates (Fig. 12-6C). Patterns of cleavage and gash fractures may indicate simple shear (Fig. 12-6D).

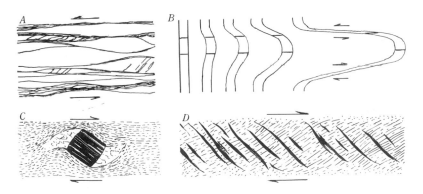

**Fig. 12-6.** Suggestions of simple shear. *A*. Section through granular rock cut into lenticular prisms by thin zones of schistose or slaty rock. *B*. Stages in the deformation of a vein by simple shear parallel to the short lines, which are all of the same length. *C*. Deposits of quartz in "strain shadows" next to crystal that apparently rotated due to simple shear during deformation. *D*. Orientations of cleavage (thin lines) and of veins in gash (extension) fractures in clayey limestone affected by simple shear in a tabular zone.

**Fig. 12-7.** *A.* Veins and folded bed subsequently flattened by vertical compression. *B.* Vertical flattening of conglomerates, indicated by symmetrically inclined intraclast faults (*left*); by pressure solution at contacts of limestone clasts (heavy lines on dotted clasts, *center*); and by slickensides on clast faces oriented at angles between about 30° and 60° to the axis of flattening (heavy-lined surfaces, *right*).

Cases where pure shear was dominant are indicated where shapes such as those of Fig. 12-3*A* and *B* lie parallel to rock fabric. Cases of flattening are indicated further by: (1) stylolites and other pressure solution surfaces oriented parallel to foliation or to flattened objects (Fig. 4-6*B* and *C*); (2) symmetrical folds associated with a plane of flattening (Fig. 12-1*C*); (3) veins, dikes, or folded beds that are thickest where they lie perpendicular to a plane of flattening (Fig. 12-7*A*); and (4) conglomerate or gravel with systematically oriented strain features (Fig. 12-7*B*).

Direction of extension in pure shear is indicated by: (1) symmetrical boudinage of layers less ductile than the surrounding rock (Fig. 12-8*A*); (2) grains or fossils that are broken and pulled apart (Fig. 12-8*B*); (3) symmetrically filled openings at opposite sides of pyrite or magnetite (etc.) in slates and other finely foliated rocks (Fig. 12-8*C*); and (4) veins in dilated fractures oriented approximately normal to planar or linear fabrics.

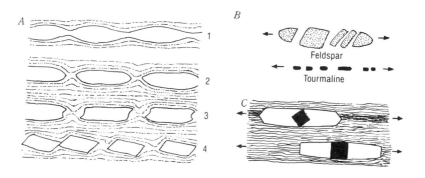

**Fig. 12-8.** *A.* Boudinage of ductile layers in more ductile matrix (*1* and *2*), of moderately brittle layer in ductile matrix (*3*), and of a brittle layer affected by both extension and rotation in a ductile matrix (*4*). *B.* Single grains that have been broken and extended in more ductile materials. *C.* Extension of phyllite matrix containing pyrite euhedra, the unpatterned material typically being quartz or a carbonate.

***Strain measurements*** can be made in the field from objects (markers) that were originally equidimensional, had about the same ductility as their matrix, and are large enough to measure with a scale. The most common possibilities are fossiliferous limestone; conglomerate with a matrix similar in composition to its clasts; mudstone or slate with pigmented spheres formed by diagenetic reduction or oxidation; pisolitic limestone or phosphorite; devitrified silicic lava containing spherulites; tuffs with accretional lapilli; basalts with varioles; and plutonic rocks with orbicules. Measurement and calculation may be made in the following steps:

1. Note the orientation of foliation and lineation, and find outcrop surfaces oriented perpendicular to the foliation, one set perpendicular to the lineation and one parallel to it (Fig. 12-9A).

2. Measuring marker by marker, record the dimensions across each one (a) parallel to the foliation, and (b) perpendicular to it (Fig. 12-9A).

3. After measuring approximately 50 markers on each surface, make a plot for each set on graph paper; draw lines through each array of points to obtain average ratios of the measured dimensions (Fig. 12-9B).

4. Combine the two ratios to give a full expression of the strain ellipsoid by setting the intermediate axis as 1. For example, if the two ratios obtained from the graphs are 1:7 and 1:20, the axial ratio of the ellipsoid is approximately 0.14:1:2.8 (if $w$ = 1 and $h/w$ = 7, $h$ = 1/7 or approximately 0.14; then, because $l$ = 20h, $l$ = 20 × 0.14 or approximately 2.8).

5. To obtain the strains, recalculate the triaxial ratio so that the resulting numbers will give an ellipsoid volume equal to a sphere of the same volume and with radius = 1. Thus, if 4/3 $abc$ (the volume of an ellipsoid) = 4/3 $r^3$ (the volume of a sphere), and $r$ = 1, $abc$ = 1. From the ratio already determined, $a$ = 0.14$b$ and $c$ = 2.8$b$. Substituting these values in $abc$ = 1, $b^3$ = 2.55, or $b$ = 1.37. Using this value and the ratio 0.14:1:2.8, $a$ = 0.20 and $c$ = 3.8, giving a full triaxial ratio of approximately 0.20:1.4:3.8. Strain may be calculated as

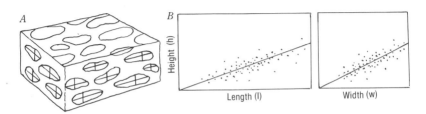

**Fig. 12-9.** *A.* Block with one side parallel to lineation and perpendicular to foliation; one side perpendicular to foliation and lineation, and the top parallel to foliation and lineation, showing the dimensions measured for each body. *B.* Graphs of measured dimensions, averaged by "best fit" lines.

the difference between each final and initial dimension, divided by the initial dimension (which is here 1). The shortest axis thus expresses a shortening of $(1 -0.20) \div 1$ or 80%; the intermediate axis an extension of $(1.4 -1) \div 1$ or 40%; and the longest axis an extension of $(3.8 -1) \div 1$ or 280%.

Probably the most reliable rocks to use for measuring strains are oolitic limestone and quartz-cemented quartz-rich sandstone. Their grains are too small to measure in the field, although approximate axial ratios can be estimated with a hand lens. Oriented samples should be collected for petrographic measurement (Section 3-8). Cloos (1947) has described measurement and calculation of strains in oolitic limestone. Ramsay and Huber (1983) have described a number of methods for measuring strains in thin section, and their center-to-center method might be studied before the field season, because it permits strain measurements in rocks consisting of materials with different ductilities.

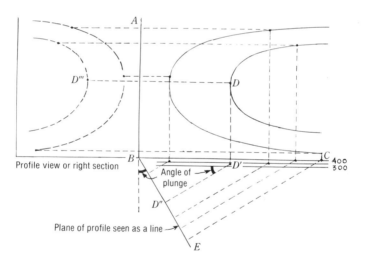

**Fig. 12-10.** Construction of a profile from a mapped fold, based on the precept that all points will project parallel to the hinge line. The steps in the construction are: (1) draw line $AB$ perpendicular to the hinge line of the map and $BC$ parallel to it; (2) give $BC$ a convenient elevation and lay off additional elevation lines below it (numbered on the right); (3) plot the line $BE$ by using the angle of plunge, as indicated; (4) starting with any point on the fold form (as $D$), draw a line perpendicular to $BC$ and plot point $D'$ according to the elevation of $D$, which is read from the map; (5) use the angle of plunge of the hinge to locate the point $D''$; (6) draw a line parallel to the hinge from $D$ through the profile line $AB$, and plot point $D'''$ by using the distance $BD''$ and making the elevation of $AB$ the same as $BC$; and (7) using the same procedure, transfer additional points from both surfaces of the mapped fold (see the other construction lines) until the form of the fold can be completed as two curved lines.

## 12-3. Folds

Large folds are generally detected by mapping rock units and small parasitic folds (Section 5-4). Even where no large folds have formed, small folds can indicate much about the orientation and distribution of deformation. In addition, maps of their distribution relative to major intrusions and faults may suggest causes of folding.

To describe and measure folds, it is necessary to observe them in profile—that is, on outcrop surfaces or constructed cross sections oriented perpendicular to the hinge lines or axial directions ("axes") of the folds (Fig. 12-10). Elements of fold shape and orientation can be measured readily if: (1) the folds are cylindroidal (their hinge lines are approximately parallel), and (2) their axial surfaces are at least roughly planar. Folds with distinctly curved hinge lines or axial surfaces must be measured in enough places to approximate their actual forms, or they must be averaged visually when measurements are made. In areas with numerous folds or complex folds, it generally will be helpful to carry a stereo net in the field in order to obtain averages of sets of data. One can then decide what degree of measurement and detail are likely to be suitable. Use of a stereo net in studying folds has been described by Turner and Weiss (1963) and is presented in some structural geology texts.

The following measurable elements will define most folds:

1. Plunge and bearing of the *hinge line*, which is an imaginary line connecting all points of maximum curvature at an anticlinal or synclinal hinge (Fig. 12-11*A*). If the curve is a circular arc, the hinge is at its center (Fleuty, 1964).

2. Strike and dip of the *axial surface*, which is approximated by an imaginary surface connecting the hinge lines along any one set of folded layers (Fig. 12-11*A*). The hinge areas should be examined for an axial-plane cleavage and for lineations (Section 12-4).

3. The *interlimb angle* ($\theta$ in Fig. 12-11*B*) (Fleuty, 1964).

4. The ratio between the length of the more or less straight limb segment of the fold (*s* in Fig. 12-11*B*) and the length of the arc between the limb and the hinge line (*c* in Fig. 12-11*B*). This ratio and the interlimb angle give numerical measures of the tightness of the fold. Different s/c ratios for the two limbs give a measure of the asymmetry of folds.

5. The strike and dip of an *enveloping surface*, which is approximately planar and touches the crests or the troughs of a train of folds in one layer (Fig. 12-11*C*).

6. The *wave length*, which is the distance between any two equivalent parts of adjacent folds (Fig. 12-11*C*).

7. One-half the distance between the two enveloping surfaces (Fig. 12-11*C*). This is a measure of the amplitude of the folds.

8. *Vergence*, which is the direction in which asymmetric folds are over-turned, and is an indicator of rotational displacement. In Fig. 12-11*B*, for example, all of the asymmetric folds verge toward the right and imply displacement of the upper part of the fold envelope over the lower part. Where the enveloping surfaces are not horizontal, vergence is inclined and is parallel to the enveloping surfaces; thus, in Fig. 12-11*C* it is upward and to the right. The amount and direction of vergence can be determined from the strike and the dip of the axial surfaces and the strike and dip of the enveloping surfaces.

If the folds are kink bands (Fig. 12-11*D*), the measured elements are: (1) bearing and plunge of either hinge line; (2) strike and dip of the lenticular body constituting the kink band; (3) the two interlimb angles; (4) thickness of the band; (5) if exposure is adequate, length of the band in profile view; and (6) typical distance between adjacent kink bands. Similar measurements are made of coupled folds (Fig. 12-11*E*), which may be equivalent to kink bands formed in more ductile materials. Note that the interlimb angles in the coupled folds vary from almost 180° at each end to a minimum at the most deformed part of the band; it is the minimum interlimb angle that should be measured.

Most of the elements described so far are measured on one folded surface. The layers themselves are also important, especially with respect to systematic variations in thickness within a fold. In Fig. 12-12, each folded

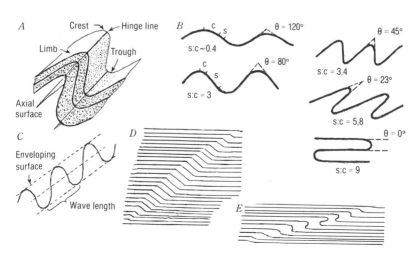

**Fig. 12-11.** *A.* Some descriptive elements of folds. *B.* Examples of axial angles (θ) and limb-to-hinge ratios (*s:c*). Where the limbs are unequal, ratios are determined for each. *C.* Elements of a fold train, based on one folded surface. *D.* Kink band. *E.* Coupled folds.

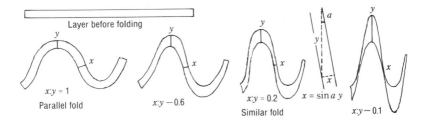

**Fig. 12-12.** Profiles of a folded layer thickened progressively at hinges and thinned in limbs. The numerical ratios give a measure of strain, and the insert illustrates the geometric relations unique to similar folds.

segment has the same cross-sectional area as the unfolded segment at the upper left. In only one case, the *parallel fold*, does the thickness remain equal throughout the fold. The one other unique case is the *similar fold*, which can be recognized by several criteria: (1) the curving surface that defines the top of the layer has the same form as the surface defining the base; (2) all lines transcribed across the layer, parallel to the axial surface, are equal in length to the thickness at the hinge, and (3) the layer thickness in any part of the fold is equal to the sine of the angle between the layer and the axial surface (angle *a* in Fig. 12-12) multiplied by the thickness at the hinge.

The parallel fold and the similar fold are only two specific possibilities in the broad spectrum suggested by the figure. A numerical measure of any fold in this spectrum is given by the ratio between the layer-thickness in the limb (or at the inflection point between the two curves expressing the antiform and synform) and the layer-thickness at the hinge ($x{:}y$ in Fig. 12-12). For parallel folds, the ratio is 1, for all folds between parallel and similar it is less than 1 and more than sin *a*; for similar folds it is equal to sin *a*, and for more "pressed" folds it is less than sin *a*. This sequence indicates an increasing degree of flattening toward the axial plane. For cases such as Fig. 12-7*A*, the ratio is greater than 1, indicating flattening toward a surface oriented at large angles to the axial plane.

**Fig. 12-13.** Names for specific fold shapes, shown here in more or less parallel folds.

       *Geology in the Field*

Some specific fold shapes have been named (Fig. 12-13). Folds should be drawn or photographed, in profile view if possible. Folds in which forms are not repeated from one layer to the next (*disharmonic folds*) must generally be photographed or drawn in detail in order to be described. Field notes must specify orientation as well as scale in these drawings.

Orientations and approximate amounts of strain should be measured or estimated as described in Section 12-2, with notation of the kinds of rocks or minerals involved and their geometric relations to the fold at points of measurement. Because folded rocks may have been extended parallel to hinge lines, strained grains should be examined and measured on surfaces that strike parallel to hinge lines. These measurements can often be made on the axial surface, because rocks tend to cleave along it.

**Interpretation of folds.** Mechanisms of folding are important aspects of geologic events, and they are sometimes suggested by geometric relations in folds. Classical theories of folding have been reviewed by Ramsay (1967), Hobbs, Means, and Williams (1976), and by Johnson (1977). Thorough analyses of folding have also been presented by Johnson (1977), and he includes many references on the subject. The analyses are generally based on simplified materials and ideal geometric situations, so that actual folds may differ because of preexisting foliation, residual stresses, special material properties, or perhaps because of two or more periods of deformation. Of great value are strain measurements of the folded materials (Chapple and Spang, 1974).

*Buckling* under compression more or less parallel to layers can be recog-

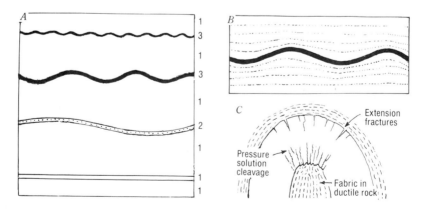

**Fig. 12-14.** *A.* Interbedded materials with viscosities increasing from 1 to 3, deformed by compression parallel to bedding. *B.* Buckled competent layer (black) and associated bend-folds in laminated ductile rock. *C.* Features associated with a sandstone bed buckled into a tight parallel fold.

nized by several relations (Ramberg, 1964, 1965). The tightness of folds in any one layer will depend on the layer's viscosity (ductility, competence) relative to surrounding layers (Fig. 12-14A). Note that the layer with the same viscosity as the matrix is not folded, and that the most viscous layer is folded most tightly. By unfolding the most tightly folded layer, one can obtain a minimum length of the block before folding began, and thus an approximate measure of the minimum shortening (flattening) strain parallel to the layer. A relation of great value in recognizing buckle folds is that layers of the same viscosity are folded into identical forms which vary in size with layer thickness (Fig. 12-14A). Another relation is that folds of layer markers in the ductile matrix (e.g., color lamination in mudstone) die out rapidly away from the more competent folded layers (Fig. 12-14B). Layers so competent as to buckle into tight parallel folds may show extensional features in the outer parts of arches or troughs and flattening features in the inner parts (Fig. 12-14C).

*Bending folds* (Ramberg, 1965) are those formed where relatively ductile materials are pressed against, or between, more rigid bodies, as in: (1) shales compacted around early-formed concretions (Fig. 9-11A); (2) ductile rocks pressed between boudins (Fig. 12-8A); and (3) laminated rocks next to more rigid layers folded by buckling (Fig. 12-14B). Anticlines and domes over diapirs are presumably also bending folds.

*Folding by extension* is illustrated by grasping a linen towel at two ends and stretching it. Natural folds of this kind would be expected to: (1) occur in rocks strained variably in extension, (2) have hinge lines parallel to the direction of extension, and (3) have axial surfaces mainly perpendicular to the enveloping surfaces of folded layers but locally dipping one way or the other.

**Deformed folds, age relations.** Folded rocks may later be folded again or may be modified by uniform strains of the entire rock body. Superimposed folding can be recognized easily on: (1) surfaces broken along bedding and thus exposing knobby or systematically cross-rippled forms (Fig. 12-15A); (2) paired surfaces that each show a profile view of a set of folds (Fig. 12-15B); (3) profile views of tightly folded layers that have been refolded more or less coaxially (Fig. 12-15C), and (4) surfaces eroded approximately parallel to the average direction of layering and thus showing quadrate, rhombic, or more bizarre patterns of layers (Fig. 12-15D). Thiessen and Means (1980) illustrated idealized outcrop sections through the principal types of superimposed folds, and Turner and Weiss (1963) and Ramsay (1967) have presented figures and stereographic solutions of several types.

Two periods of deformation are also indicated by deformed small-scale features commonly produced by folding: (1) lineations that wrap obliquely around fold hinges (Fig. 12-15E); (2) folded linear structures, such as stretched and flattened cobbles that have later been folded (Fig. 12-15F);

(3) crossing lineations; (4) folded cleavage or schistosity; and (5) folded veins that are known to be planar with respect to folds elsewhere in the area. In a broader context, steeply plunging small-scale folds generally imply two periods of deformation, because the layers in which they occur must have been rotated to steep dips before or after the smaller features formed. Similarly, small folds with steeply inclined axial surfaces in gently dipping inverted strata (which are generally on the lower limb of a recumbent fold) imply two periods of folding.

Geologic mapping should develop additional relations and may also locate subareas where only one age-set of folds is developed and can thus be studied free of an earlier or later set.

Once two deformations have been recognized, an important field procedure is determining their relative ages. In most cases, the axial surfaces and cleavages of the older set of folds are folded, whereas those of the younger set are more or less planar. Of two sets of crossing lineations, the younger is generally dominant at hinges, or on the shorter limbs of asymmetric folds. Folded triaxial forms such as the cobble of Fig. 12-15*F* imply an age relation. If both sets of folds have associated foliations or schistosity, the age criteria described in Section 12-4 can be used to determine the relative ages of the folds. Lacking field evidence, oriented samples can be collected for petrographic study of age relations; for example, of minerals that formed during one deformation and were kinked or pulled apart during a second.

Later strains that do not produce folds may deform earlier folds into shapes suggestive of two deformations. The relations in Fig. 12-7*A* suggest folds that were flattened by forces acting approximately parallel to their axial surfaces and at a time when the folded layers were more ductile than

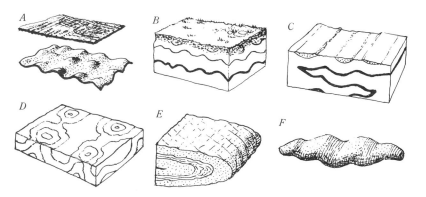

**Fig. 12-15.** Forms indicating superimposed folding (see text for explanation).

when they were first folded. Preexisting folds affected by strong extensional strains or by strong simple shear are likely to become rotated toward the axis of extension or the direction of simple shear (Fig. 12-16). These cases can be recognized if the strains are measured, for the amount of rotation must be proportional to the amount of strain at that place. Conversely, if the original shapes and orientations of the folds are known from studies outside the area of later strain, the kinds and amounts of strain can be estimated.

Folding can be bracketed within numerical age ranges by dating the youngest rocks affected by deformation and the oldest that are unaffected. Deformation can be dated directly by using igneous bodies intruded during a deformation, as shown by igneous structures and fabrics that are coaxial with folds and associated lineations in the surrounding country rocks (Todd, 1980). A minimum age of strain in metamorphic rocks can be determined by dating minerals that grew during deformation. Determining relative ages of deformation and metamorphism, and of two or more periods of metamorphic deformation, are described in Section 15-8.

## 12-4. Foliations, Cleavages, and Related Lineations

These structures and fabrics are developed in many deformed rocks, and are of great value in mapping the distribution of folds and shear zones. They can often be used in conjunction with relict features to determine orientations of strains (Section 12-2) and fold histories (Section 12-3). *Foliation* is a general term for any texture or structure that gives a rock a leaflike or platy character. Examples described in other sections are schistosity (Section 4-6), gneissose structure (Section 15-5), and compositional or textural layering in igneous rocks (Sections 13-4 and 14-4). *Cleavage* is a foliation due to rocks plitting readily into sheets, lens-shaped bodies, or linear bodies. It is a term used especially for rocks that are not macroscopically schistose or gneissose.

Cleavages may be classified nongenetically into *penetrative cleavage* and

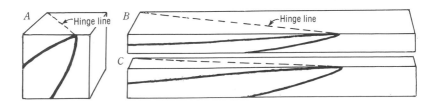

**Fig. 12-16.** Hinge line of fold (*A*) rotated by 340% extension in plane strain (*B*) and constrictional strain (*C*).

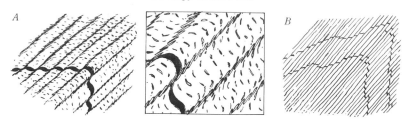

**Fig. 12-17.** *A*. Crenulation cleavage, with detail (*right*) showing that cleavage takes place along zones of rotated, appressed mica flakes, and that other minerals, typically quartz or calcite, were differentiated from the limbs to other parts of the rock. *B*. Axial-plane cleavage along which nonuniform slip has taken place.

*spaced cleavage.* Penetrative cleavage is pervasive at hand-specimen scale, so that the rock may be broken on any plane parallel to the cleavage and thus into very thin plates. Schistose rocks have distinct penetrative cleavage and semischistose rocks crude penetrative cleavage (Section 4-6). *Slaty cleavage* is a penetrative cleavage in aphanitic rocks and results from parallel orientation of platy mineral grains or of microscopic aggregates of mineral grains.

Spaced cleavages include any set of cleavage surfaces spaced at fairly regular, measurable distances apart, typically at 0.2 to 2mm in phyllite, slate, and similarly fine-grained rocks, and at 1 to 4 cm in sandstone, quartzite, and similarly coarse-grained rocks. Genetic varieties are:

1. *Crenulation cleavage*, which lies along the appressed limbs of small folds that crenulate a pre-existing schistosity or slaty cleavage (Fig. 12-17*A*).

2. *Pressure-solution cleavage*, which consists of a series of parallel pressure-solution surfaces along which material has been removed (Figs. 4-6*B* and 12-14*C*).

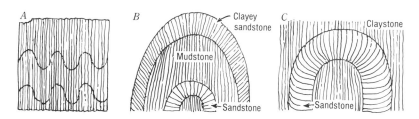

**Fig. 12-18.** Axial-plane cleavage in slate (*A*), in interbedded sandstone and slaty mudstone (*B*), and in a sandstone bed that grades upward to slaty claystone (*C*). Typically, the cleavages in the sandstones are spaced cleavage and those in the mudstone and claystone penetrative or crenulation cleavage.

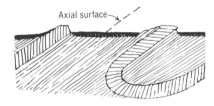

**Fig. 12-19.** Bedding-cleavage relations in isolated outcrops of upright and overturned limbs of an antiform, showing position of interpreted axial surface. Black is surficial cover.

3. *Slip cleavage*, which consists of closely spaced minor faults, such as in the matrix of some mélanges (Section 12-7) and in some fault zones (Fig. 12-6*A*). Commonly, slip occurs along preexisting surfaces, such as crenulation cleavage (Fig. 12-17*B*).

*Axial-plane cleavage* is a general term used for any cleavage that is parallel to the axial plane in the hinge areas of folds but may be oblique to the axial plane in the limbs (Fig. 12-18). Commonly, axial-plane cleavages in folds compose systematic fanning arrays with orientations that correlate closely with lithology (Fig. 12-18*B* and *C*). These relations imply that the cleavages formed during folding. In poorly exposed folds, the angular relations between cleavage and bedding can be used to recognize stratigraphically upright and overturned parts of limbs (Fig. 12-19). The angular relations can also be used to estimate position in a fold, even where bedding is not discernible.

*Transpositional layering* may develop where isoclinally folded rocks have a strong axial-plane cleavage. If the cleavage obscures hinge areas, and further deformation causes slip along the cleavage surfaces, the rock body may look like a simple layered sequence (Fig. 12-20*A*). Such cases can be recognized by the twisted remnants of layers and can be worked out by determining tops of beds in opposed fold limbs (Section 9-10). In some cases, a strongly developed spaced cleavage leads to transpositional layering (Fig. 12-20*B*).

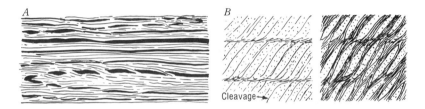

**Fig. 12-20.** Transpositional layering. *A*. Ductile foliated rocks with more competent beds (black) that were folded, separated along cleavage, and strewn out parallel to axial surfaces. *B*. Horizontal beds in metasandstone cut by spaced cleavage that becomes so dominant (*right*) as to form a transposed flaser layering.

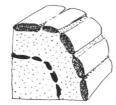
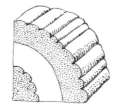

**Fig. 12-21.** Boudins (*left*), mullions (*center*), and rods, in their typical orientation with respect to folds formed at the same time.

*Lineations* in deformed rocks include a variety of grains and structures that are aligned in parallel arrays:

1. Relict primary grains or structures that have been strained systematically (Section 12-2).

2. Prismatic crystals or elongate aggregates of diagenetic or metamorphic minerals.

3. Hinges of small folds and crenulations.

4. Intersections of cleavage with the surfaces of beds or other layers.

5. Intersections of two cleavages.

6. Grooves, striations, or mineral smears on bedding surfaces, on slip cleavage, or on other minor faults.

7. Axes of boudins, mullions, and rods, the latter being relics of folded and sheared-out beds and veins (Fig. 12-21).

Lineations that trend parallel to fold hinges include intersections of cleavages and bedding, hinges of minor folds and crenulations, rods, and mullion. Boudins, prismatic minerals, and extended grains and structures typically are parallel to fold hinges but may be perpendicular or oblique to them (Fig. 12-22). Grooves, striations, and mineral smears are typically oriented perpendicular to fold hinges.

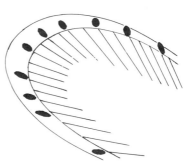

**Fig. 12-22.** Profile of a major fold, with ellipses indicating average amounts of strain measured from oolites that were originally spherical. Lines show orientations of cleavage. Locally, the oolites were also extended slightly parallel to the fold hinge. Data from Cloos (1947, p. 880, 885).

## 12-5. Faults

General suggestions for mapping faults are given in Section 5-4. Faults may consist of: (1) a single break with almost unstrained walls but typically with at least a thin sheet of gouge or cataclastic breccia; (2) a more or less tabular zone of two or more breaks (a *fault zone*); (3) one or more breaks in a zone of moderately strained rock (Fig. 12-23*A*); or (4) a tabular zone of intensively strained rock (a *ductile fault*) (Fig. 12-5). The strained zones of ductile faults are mylonitic or schistose rather than cataclastic, because ductile faults form without loss of intergranular cohesion (Wise and others, 1984). Brittle faults (the other faults enumerated above) form with loss of cohesion but may later become healed by growth of mineral grains. Mylonitic ductile faults are thought to require rapid strain rates under large loads, conditions that generate heat during faulting. Schistose ductile faults probably require slow strain rates in rocks already at metamorphic temperatures. Brittle faults superimposed on ductile faults thus imply changes in conditions and probably in strain rates. In addition, major brittle faults with little or no cataclastic material along them suggest presence of pore fluids at high pressures, a suggestion that may be supported by clastic dikes in the fault walls.

On geologic maps, faults are traditionally classified according to their attitude and the sense of displacement along them; however, it may be difficult to obtain enough information to do this at the outset. Names and map symbols based on direction of displacement should not be used when all that is apparent is the separation of strata along the fault trace (Crowell, 1959). The map symbols of Appendix 7 are thus organized in a hierarchy based on how much is known about a fault. If poor exposure and low topographic relief make it impossible to judge the direction and amount of dip of a fault, the fault is mapped as a simple line. In cases where the dip of a fault can be measured or estimated, the fault can be classified as *high-angle* (dip greater than 45°) or *low-angle* (dip less than 45°).

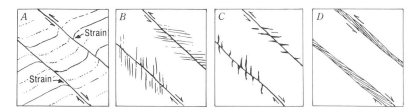

**Fig. 12-23.** Evidence of direction of fault movement: "drag" relations (strain in a zone along the fault) (*A*); feather joints (*B*); veins in gash fractures (*C*); and fibrous or acicular crystals that grew between the fault walls during slow displacements (*D*).

Sense of displacement can sometimes be determined, perhaps tentatively, by major grooves or striations on the fault surface; by the features in Fig. 12-23; or by the relations described for specific kinds of faults in the subsections that follow. High-angle faults may then be classified further as *normal faults* (the overlying rock body moved downward relative to the underlying rock body), as *reverse faults* (the sense of displacement is opposite that on normal faults), or *high-angle strike faults* (the displacement was dominantly horizontal). Low-angle faults can be classified as *thrusts* (the overlying rock body moved relatively upward); as *low-angle normal faults, detachment (décollement) faults,* or *low-angle gravity faults* (the overlying rock body moved relatively downward); or as *low-angle strike faults.*

The complete solution of fault movement generally requires finding a linear feature or a rock body of unique shape that can be located on both sides of the fault. The two most common linear features are (1) fold hinges along unit contacts, and (2) intersections of two planar features, such as a unit contact cut by a dike, an unconformity, or an older fault. Examples of uniquely shaped rock bodies are sedimentary units that have distinct lateral changes in thickness, and elongate or unusually shaped intrusions or veins. When suitable features have been mapped on both sides of the fault, their projected positions on the fault plane can be constructed by the methods of descriptive geometry (Billings, 1972, p. 564). The results will show the amount and direction of net displacement as precisely as mapping and projection methods will permit. Such faults may then be specified as *normal-slip faults, reverse-slip faults*, and so on, as proposed and discussed by Kupfer (1960) and Hill (1962). The plunge of the measured slip direction may be shown by a half-arrow (Appendix 7).

**Normal Faults.** Hubbert (1951) showed mathematically and experimentally that uniform materials under horizontal extension tend to collapse along faults dipping approximately 60° (Fig. 12-24A). He called these *high-angle gravity faults*, and the dips and slip directions of most normal faults are comparable. Normal faults formed in extension are likely to have porous breccia between the fault walls, especially at irregularities (Fig. 12-24B). High porosity due to extension may lead to alteration and mineral filling by

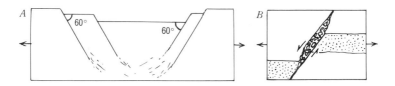

**Fig. 12-24.** *A.* Vertical section through normal faults produced by extension. *B.* Breccia body along an irregularity in a normal fault.

groundwater. Normal faults also tend to form in more or less parallel systems or in conjugate sets. Their mapped traces are generally moderately to highly irregular and may branch or turn angles of 90° or more. Major normal faults may also pass along strike into monoclines.

At the ground surface, dips of normal faults tend to change to vertical and even reverse, and in appropriate materials, such as vertically jointed clays or lavas, dips may be vertical for large exposed distances. Recently-active normal faults may be accompanied by gaping vertical fissures, and if their strike is oblique to the most recent direction of extension, the youngest features tend to be *en echelon* sets of faults (Duffield, 1975). Dips of normal faults may decrease with depth and in some cases pass into gently dipping

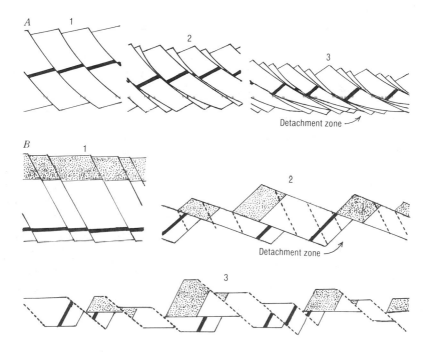

**Fig. 12-25.** *A.* Somewhat curved normal faults that dip approximately 60° and interact with a horizontal surface at depth (*1*), modified by splays that form with 60° dips (*2*), so that original faults rotate progressively to shallower dips, whereas bedding rotates to steeper dips (*3*). The underlying surface thus becomes a zone of detachment. After Gans and Miller (1983). *B.* Diagrammatic model showing how normal faults may be rotated by continued extension so as to dip at low angles. Faults initially dipping 60° (*1*) rotate during extension until they dip so gently that new 60°-faults form (dashed lines in *2*). Continued extension and rotation cause the original faults to dip at a low angle or even in the opposite direction (*3*). In spite of the strewing out of faulted fragments, note that nowhere do older rocks come to lie on younger ones. After Miller and others (1983).

bedding surfaces, becoming detachment faults. The curved surfaces may form at the outset (*listric faults*) and become accentuated by splaying (Fig. 12-25*A*). Normal faults may also terminate downward in an angular junction with a detachment surface and become broken and rotated on younger normal faults (Fig. 12-25*B*).

Mountain fronts initiated as normal faults tend to appear abrupt, to cut off rock units and structures of the mountain range, to lie next to a set of alluvial fans in the adjoining basin, and to have spurs ending in steep frontal slopes (*facets*) that strike parallel to the fault and have their bases at or near the trace of the fault. Wallace (1978) used the degree of degradation of these frontal slopes, together with the degree of weathering of faulted materials, to judge the age of young fault-block offsets in the Basin and Range Province. Relatively young faulted mountain fronts are also suggested by small alluvial fans and by playas lying against the mountain front.

**High-angle reverse faults** may contrast with normal faults in tending to have thicker zones of gouge or mylonite, or zones of interlaced fault strands that tend to subdivide the rocks into lenses. Reverse faults are typically associated with compressional folds, and apparent drag may thus be happenstance to where a fault cuts a somewhat older fold. Commonly, however, the sense of offset is appropriate to the fold, which may have localized the reverse fault (Fig. 12-26*A*).

**High-angle lateral or strike-slip faults** commonly differ from normal and reverse faults by their nearly straight or smoothly curving mapped traces. Gouge, mylonite, and phacoidal fault zones are typical. Major fault zones developed by recurrent offsets may be 0.1 to 1 km wide and include large lens-shaped bodies of unbroken rock. Fault scarps and fault-line scarps tend to face one way and then the other, depending on the lateral offset of topographic highs and lows, or of more and less resistant rock bodies. Major

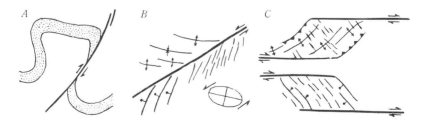

**Fig. 12-26.** *A*. Profile section through a fold with a reverse fault on an overturned limb. *B*. Map of a left-lateral strike-slip fault and associated folds (lines with opposed arrows), normal faults (bar and ball on downthrown side), and vertical dikes and open fractures (lines). Simplified from a map by Anderson (1973), with strain ellipse added. *C*. Maps of left-stepover (*above*) and right-stepover of strike-slip faults, showing typical associated structures in stepover area. Symbols are identified in *B*, above.

faults may be flanked by accessory structures that are oriented at about 45° to the faults and commonly curve to more nearly parallel orientations near the faults (Fig. 12-26*B*). Major fault segments may be distinctly *en echelon*, and Aydin and Page (1984) have described typical orientations of secondary structures that tend to form in the stepover areas between the segments (Fig. 12-26*C*). Where large rock bodies are deformed between two parallel strike-slip faults, fold hinges and thrust or reverse faults may be parallel to the main faults, which may themselves develop large components of dip-slip movement (Sylvester and Smith, 1976).

Indications of recent activity on major strike-slip faults are especially well documented because of interest in earthquake prediction. Streams and their terraces are offset consistently, and small alluvial fans or strips of alluvium are offset from the gullies that once fed them (Sieh, 1978). Sunken areas marked by sag ponds or by marshy ground, or pressure ridges, commonly mark stepovers between two *en echelon* breaks. Eroded furrows or scarplets along recent breaks may form low linear mounds (mole tracks). Breaks may also be indicated by split or felled trees or by offset roads and fences. Open fissures in and near an active trace may lead to collapse and to alignments of unusually luxuriant vegetation (Clark, 1972).

Average rates of recent displacement based on $^{14}$C dates can sometimes be determined where peat or wood have been deposited along a major fault trace. Sieh (1978) trenched a formerly marshy site on the San Andreas fault and used peat layers cut by the fault or cut by sand boil feeders (or overlain by sand boil deposits) to determine the recent history of displacements.

**Low-angle faults** often lie parallel to bedding or cut it at a low angle and are thus not nearly as obvious as high-angle faults. One or more of these features or relations may help in recognizing them:

1. Bedding or other primary structures end abruptly against a surface of low dip.

2. Older rocks lie on younger ones.

3. Inverted strata lie on a stratigraphically normal sequence.

4. Evidence of formerly high fluid pressures, as: (a) clastic dikes that pass into the hanging wall; (b) hydrofracturing of the hanging wall, espe-

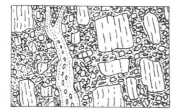

**Fig. 12-27.** Clastic dike of vertically bedded sand and gravel surrounded by brecciated country rock in which the lineation in large clasts (dashed lines) was rotated only slightly.

cially near clastic dikes (Fig. 12-27); (c) gypsum converted to anhydrite (Heard and Ruby, 1966); and (d) rocks altered hydrochemically just below a contact with relatively impermeable rock.

5. Channels filled with sorted detrital sediments under a more or less solid rock body.

6. Part of a stratigraphic sequence is missing.

7. Rocks of a distant facies lie on rocks of a local facies of the same age.

The presence of an underlying (unexposed) low-angle fault is suggested by subparallel high-angle strike-slip faults that divide the allochthon into segments and trend parallel to the direction in which it was displaced. These faults are sometimes called *tear faults*. The segments between the faults typically show different degrees of folding and different stratigraphic thicknesses of formations.

Displacement directions on low-angle faults are suggested by: (1) the strike direction of high-angle strike-slip faults that cut the displaced sheet but not the footwall; (2) the vergence of overturned folds above or below the fault; (3) the sense of drag of flexed beds that abut against the fault; (4) the updip direction in sets of imbrications; (5) displaced rocks that have a unique distribution in the autochthon; and (6) displaced structures or rock bodies of unique shape or character.

**Thrusts** are low-angle faults that have an upward as well as a lateral sense of displacement of the hanging wall. Thrusting is suggested by older rocks lying on younger; however, this arrangement can be produced locally by other kinds of low-angle faults, so that extensive mapping is generally necessary to prove a thrust relation. A critical relation is that the overthrust allochthon roots back into the ground. The following may also be helpful in recognizing thrusts:

1. Evidence that compression was more or less horizontal at the time of faulting, as: (a) folds with steeply inclined axial planes and with hinges parallel to the strike of the fault surface; (b) mineral grains or fossils flattened into a steeply inclined fabric; (c) small-scale faults dipping roughly 30° toward the fault trace or away from it; (d) faults deformed by folds that trend parallel to the strike of the fault; and (e) faults lying along attenuated limbs of overturned folds.

2. Surficial materials, notably conglomerate and sedimentary breccia consisting of detritus eroded from the overlying sheet, overridden by the sheet (an *erosion thrust*).

**Detachment (décollement) faults** typically form in unusually ductile units (salt, gypsum, or clay-rich rocks), at contacts between units with contrasting ductility, or where rocks with low permeability lie on rocks that contain large amounts of water or other fluids. Metamorphism may promote detachments by producing large amounts of pore fluids and by increasing ductility.

Detachment faults can be recognized by the absence of strata normally found at that horizon, or by the presence of those strata as discontinuous slices and lenses. No matter how thin these fragments may become, they retain their stratigraphic order, except where the overturned limb of a recumbent fold has been incorporated into the fault zone. In many cases, the overlying rocks are folded at least locally, whereas the underlying ones remain unfolded. Additional relations that suggest detachment faults are:

1. Simple shear and extensional strains above and below the fault, producing gently inclined fabrics, pressure-solution surfaces, and general attenuation of rock units.

2. Extensional high-angle fracture swarms, in some cases filled by dikes or associated with volcanic activity.

3. Minor detachment faults above and below the main fault.

4. Normal faults branching upward from the main detachment (Fig. 12-25).

5. In metamorphic rocks, schistosity, lineations, or axial-plane cleavages more or less parallel with the fault zone.

6. Less metamorphosed rocks displaced downward onto more metamorphosed rocks, thus attenuating metamorphic zonation.

Detachment allochthons possibly grade in size from those covering many tens of thousands of square kilometers to well-exposed gravity detachments of several thousand square kilometers (Pierce, 1973) and to ordinary landslides (Section 10-7). Sliding may emplace large coherent slabs of bedrock into basins, and these slabs can be distinguished from major allochthonous sheets by their association with breccias of the same kinds of bedrock, by their position relative to local bedrock sources, and by contemporaneous deformation of the basinal sediments onto which they have slid.

### 12-6. Joints

These fractures form by simple parting, with little or no shear displacement of their walls. They tend to be planar or nearly so, although they commonly curve near their edges. Many extend through a given rock layer and end at its top and base. Their spacing varies with layer thickness and lithology: They are closely spaced (a few centimeters) in thin-bedded coal and chert, and very widely spaced (tens of meters) in sandstone layers tens of meters thick. Study of extensive, widely spaced joints requires exceptionally large exposures. Aerial photographs are helpful because major joints are accentuated by vegetation and by erosion of joint walls, which tend to be weakened by groundwater alteration. Open joints may also be coated by iron oxide, clay, or carbonates, and in hydrothermally altered rock bodies joints are likely to become veins (Sections 14-7 and 14-8).

*Joint systems* are composed of parallel single joints or of *zoned joints*, each of which is a group of several closely spaced parallel joints (Dyer, 1983). In moderately folded layered rocks, joint systems commonly strike parallel to fold hinge lines and dip approximately perpendicular to layers. In strongly folded rocks, and in rocks that have undergone extension, joint systems are approximately perpendicular to the axis of folding or extension. Joint systems that are roughly parallel to the ground surface are thought to result from extension (expansion) due to erosional unloading. Polygonal joint sets form by contraction in lavas, welded tuffs, clay-rich soils, and periglacial patterned ground. Plutons may show a variety of joint systems (Section 14-7).

Although conjugate joint sets (two intersecting systems of joints) can theoretically form simultaneously as shear fractures, carefully studied natural sets have formed by two or more periods of extension. The principal evidence is that shear displacements on an earlier system are suitable in orientation and magnitude for extensional origin of a later system (Dyer, 1983) (Fig. 12-28). The most useful criterion for determining an age relation is that the younger joints commonly curve and terminate near older ones. Joints can thus be used to work out age relations and orientations among extensional events that might otherwise be difficult to detect. For example, Nickelsen and Hough (1967) described joints in coals that became lithified at an early stage and thus recorded stresses that produced no recognizable effects in associated rocks which were lithified later.

### 12-7. Mélanges

These units of fragmental rock have a chaotic aspect because of the variety of sizes, shapes, orientations, or kinds of rock fragments that compose them. Their fine, abundant matrix is typically clay-rich or serpentinous

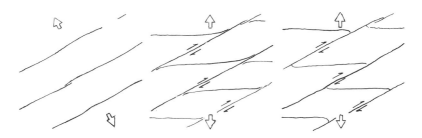

**Fig. 12-28.** A joint system formed by extension (*left*) and two conjugate sets resulting from a second period of extension at a different orientation. Small arrows show sense of displacement along the older joints. Note that the curvature of the younger joints need not correlate with the sense of displacement. After Dyer (1983).

and commonly breaks into small flakes or prisms along a myriad of minute faults. Many mélanges contain some fragments that appear to have been soft when incorporated, as well as fragments that were lithified and separated along joints (Fig. 12-29). Maximum sizes of fragments may differ greatly from one mélange to another, but most major mélanges contain fragments over 10 m across and some have slabs measuring up to 20 km (Saleeby, 1978). As noted by Hsu (1968), both matrix and fragments may consist of *native* materials (sediments or rocks formerly deposited together) and *exotic* materials (those of extraneous origin). The mixing required to introduce the exotic materials is inherent to the origin of mélanges. Therefore where only one rock unit is represented, the term *brecciated formation* is more suitable or, if the unit is broken along many small faults but is more or less intact, *faulted* or *broken formation.*

Mélanges are typically so heterogeneous, and some of the blocks are generally so large, that they are best studied and interpreted in the field. They have been classified genetically into *olistostromes*, which are deposits of subaqueous debris flows (Section 9-2), and *tectonic mélanges*, which result from tectonic fragmentation and mixing. Recognition of olistostromes is complicated by the fact that many form in tectonically active regions and are especially susceptible to later deformation because they are weaker than the rocks between which they are stratified. Olistostromes originating from oversteepened inner walls of active oceanic trenches are likely to be subducted, and olistostromes that flow from the front of an advancing allochthonous sheet are likely to be overridden by the sheet. Fragments of older mélanges in younger ones are evidence that mélange formation is a long continuing process at some plate junctions and in regions of extensive

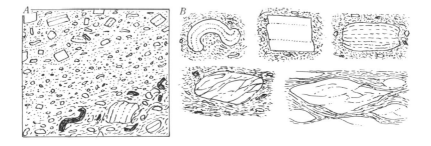

**Fig. 12-29.** *A.* Three melange units based on textural and compositional differences. *B.* Large mélange clasts showing, from upper left to lower right: a plastically deformed fragment of a sedimentary bed; a block bounded by joints; a moderately lenticular clast with fabric indicating plastic deformation; clast with internal phacoidal cleavage oblique to matrix fabric; and strongly phacoidal clasts conformable with a strong matrix fabric.

décollement, thrust faulting, or gravity sliding (Page, 1978; Elter and Trevisan, 1973).

Criteria for recognizing olistostromes are condensed here from a review by Page (1978):

1. Mapping shows the unit to be intercalated with nonchaotic aqueous deposits of nearly the same age.

2. The strata immediately under the mélange have soft sediment folds and pulled-apart beds and, at its top, the mélange grades into fine, nonchaotic sediments.

3. The unit consists of more than one flow, as shown by abrupt to moderately gradational boundaries marked by contrasting colors, textures, or compositions (Fig. 12-29A).

4. The clasts are of a greater variety than can be accounted for by tectonic transport and mixing alone.

5. Petrographic study shows the matrix to consist of sedimentary grains (sand, silt, and clay).

The following, on the other hand, are criteria for a tectonic mélange:

1. Mapping shows the unit to grade laterally through less deformed and disrupted rock or sediments into moderately or slightly deformed rock or sediments.

2. The upper contact of the unit is a fault or fault zone beneath a less strained hanging wall that extends laterally for many kilometers.

3. Some of the fragments in the mélange are of rocks in the hanging wall, or are rocks known to be younger than rocks of the hanging wall.

4. The fragments are dominantly lenticular (phacoidal) (Fig. 12-29B, lower right) or show textures indicative of solid-state necking and boudinage (Fig. 12-29B, upper right).

5. Petrographic study shows the matrix to be cataclastic and to be derived from rocks like those carried in the matrix.

6. The matrix has a persistent planar or linear fabric, and suitably shaped fragments lie parallel to this fabric (Fig. 12-29B, lower right).

Mapping mélanges consists of plotting the contacts of each mélange unit and any of these features within it: (1) unbroken rock bodies that can be mapped to scale, including their internal contacts and other structures (Fig. 12-29B); (2) strike and dip, or plunge, of the fabric of the matrix or the shapes of platy or elongate fragments; (3) subunits based on (a) the presence of specific kinds of rock fragments, (b) on specific associations of rocks, or (c) on the dominant kind of rock; (4) subunits based on textural characteristics, as maximum sizes, average sizes, or shapes of fragments; and (5) subunits based on color, composition, or texture of the matrix. One or more of these features can generally be used to distinguish between separate successive flows in olistostromes (Page and Suppe, 1981) or to trace separate mélanges across large areas (*collage terrains*) consisting largely or entirely of mélanges and broken formations brought together along faults (Decker, 1980).

Direction of emplacement of a mélange may be indicated where layers in the immediately underlying materials have been bent, extended, and pulled apart, or thrown into overturned folds. Direction may also be suggested by fragments of specific rocks which have a geographically limited source; however, these indicators need not have moved in a straight line nor in one event. Mapping of a large area may establish paleogeographic highs and lows, or related tectonic regions, which suggest directions of transport (Elter and Trevisan, 1973). Relative distance of movement of different mélanges may be suggested by size of fragments, proportion of matrix to fragments, degree of destruction of partially consolidated relics, and degree of abrasion of solid rock fragments.

The age of an olistostrome is bracketed by fossils in the underlying and overlying deposits. Lacking such evidence, a maximum age is indicated by the youngest contained fossils, and these will generally be embedded in soft sediment relics. Tectonic mélanges are older than the youngest rocks intruding them or lying unconformably on them. They are younger than the youngest strata they transect, or the youngest contained fossils or dated rock fragments. Saleeby (1978) found that early fault contacts between mélange units or between mélange and broken formations are generally shear zones gradational with regard to composition, clast sizes and shapes, or fabrics; and that younger faults typically form sharp contacts.

## References Cited

Anderson, R. E., 1973, *Large-magnitude late Tertiary strike-slip faulting north of Lake Meade, Nevada*: U.S. Geological Survey Professional Paper 794, 18 p.

Aydin, A., and Page, B. M., 1984, Diverse Pliocene-Quaternary tectonics in a transform environment, San Francisco Bay region, California: *Geological Society of America Bulletin*, v. 95, p. 1303-1317.

Billings, M. P., 1972, *Structural geology*, 3rd edition: Englewood Cliffs, N.J., Prentice-Hall, 606 p.

Boulter, C. A., 1974, Tectonic deformation of soft sedimentary clastic dikes from the Precambrian rocks of Tasmania, Australia, with particular reference to their relations with cleavages: *Geological Society of America Bulletin*, v. 85, p. 1413-1420.

Chapple, W. M., and Spang, J. H., 1974, Significance of layer-parallel slip during folding of layered sedimentary rocks: *Geological Society of America Bulletin*, v. 85, p. 1523-1534.

Clark, B. R., 1970, Origin of slaty cleavage in the Coeur d'Alene District, Idaho: *Geological Society of America Bulletin*, v. 81, p. 3061-3072.

Clark, M. M., 1972, Surface rupture along the Coyote Creek fault, p. 55-86 *in The Borrego Mountain earthquake of April 9, 1968*: U.S. Geological Survey Professional Paper 787.

Cloos, E., 1947, Oolite deformation in the South Mountain fold, Maryland; *Geological Society of America Bulletin*, v. 58, p. 843-918.

Crowell, J., 1959, Problems of fault nomenclature: *American Association of Petroleum Geologists Bulletin*, v. 43, p. 2653-2674.

Decker, J. E., Jr., 1980., *Geology of a Cretaceous subduction complex, western Chichagof Island, Southeastern Alaska*, Stanford University, CA, PhD Dissertation, 135 p.

Duffield, W. A., 1975, *Structure and origin of the Koae fault system, Kilauea Volcano, Hawaii*: U.S. Geological Survey Professional Paper 856, 12 p.

Dyer, J. R., 1983, *Jointing in sandstones, Arches National Park, Utah*: Stanford University, CA, PhD Dissertation, 202 p.

Elter, P., and Trevisan, L., 1973, Olistostromes in the tectonic evolution of the northern Apennines, p. 175-188 *in* DeJong, K. A., and Scholten, R., editors, *Gravity and tectonics*: New York, John Wiley & Sons.

Fleuty, M. G., 1964, The description of folds: *Geologists' Association Proceedings*, v. 75, p. 461-492.

Gans, P. B., and Miller, E. L., 1983, Style of mid-Tertiary extension in east-central Nevada, p. 107-160 *in* Gurgel, K. D., editor, *Geologic excursions in the overthrust belt and metamorphic core complexes of the intermountain region*: Utah Geological and Mineral Survey Special Studies 59.

Heard, H. C., and Rubey, W. W., 1966, Tectonic implications of gypsum dehydration: *Geological Society of America Bulletin*, v. 77, p. 741-760.

Hill, M. L., 1963, Role of classification in geology, p. 164-174 *in* Albritton, C. C., Jr., editor, *The fabric of geology*: Reading, MS, Addison-Wesley.

Hobbs, B. E., Means, W. D., and Williams, P. F., 1976, *An outline of structural geology*: New York, John Wiley & Sons, 571 p.

Hsu, K. J., 1968, Principles of mélanges and their bearing on the Franciscan-Knoxville paradox: *Geological Society of America Bulletin*, v. 79, p. 1063-1074.

Hubbert, M. K., 1951, Mechanical basis for certain familiar geologic structures: *Geological Society of America Bulletin*, v. 62, p. 355-372.

Johnson, A. R., 1977, *Styles of folding: Mechanics and mechanisms of folding of natural clastic materials*: Amsterdam, Elsevier Scientific Publishing Co., 406 p.

Kupfer, D. H., 1960, Problems of fault nomenclature: *American Association of Petroleum Geologists Bulletin*, v. 44, p. 501-505.

Miller, E. L., Gans, P. B., and Gaving, J., 1983, The Snake Range décollement: an exhumed mid-Tertiary ductile-brittle transition: *Tectonics*, v. 2, p. 239-263.

Nickelsen, R. P., and Hough, V.N.D., 1967, Jointing in the Appalachian Plateau of Pennsylvania: *Geological Society of America Bulletin*, v. 78, p. 609-630.

Page, B. M., 1978, Franciscan mélanges compared with olistostromes of Taiwan and Italy: *Tectonophysics*, v. 47, p. 223-246.

Page, B. M., and Suppe, J., 1981, The Pliocene Lichi mélange of Taiwan; its plate-tectonic and olistostromal origin: *American Journal of Science*; v. 281, p. 193-227.

Pierce, W. G., 1973, Principal features of the Heart Mountain fault and the mechanism problem: p. 457-471 *in* De Jong, K. A., and Scholten, R., editors, *Gravity and tectonics*: New York, John Wiley & Sons.

Powell, C. McA., 1972, Tectonic dewatering and strain in the Michigamme Slate, Michigan: *Geological Society of America Bulletin*, v. 83, p. 2149-2158.

Ramberg, H., 1964, Selective buckling of composite layers with contrasted rheological properties, a theory for simultaneous formation of several orders of folds: *Tectonophysics*, v. 1, p. 307-341.

Ramberg, H., 1965, Strain distribution and geometry of folds: *Geological Institutions of the University of Uppsala Bulletin*, v. 42, no. 4, p. 1-20.

Ramsay, J. G., 1967, *Folding and fracturing of rocks:* New York, McGraw Hill Book Co., 568 p.

Ramsay, J. G., and Huber, M. I., 1983, *The techniques of modern structural geology; v. 1: strain analysis:* New York, Academic Press, 307 p.

Saleeby, J., 1978, Kings River ophiolite, southwest Sierra Nevada foothills, California: *Geological Society of America Bulletin,* v. 89, p. 617-636.

Sieh, K. E., 1978, Prehistoric large earthquakes produced by slip on the San Andreas fault at Pallett Creek, California: *Journal of Geophysical Research,* v. 83, p. 3907-3939.

Sylvester, A. G., and Smith, R. R., 1976, Tectonic transpression and basement-controlled deformation in San Andreas fault zone, Salton Trough, California: *American Association of Petroleum Geologists Bulletin,* v. 60, p. 2081-2102.

Thiessen, R. L., and Means, W. D., 1980, Classification of fold interference patterns: a reexamination: *Journal of Structural Geology,* v. 2, p. 311-316.

Todd, V. R., 1980, Structure and petrology of a Tertiary gneiss complex in northwestern Utah, p. 349-383 *in* Crittenden, M. D., Jr., Coney, P. J., and Davis, G. H., editors, *Cordilleran metamorphic core complexes:* Geological Society of America Memoir 153.

Turner, F. J., and Weiss, L. E., 1963, *Structural analysis of metamorphic tectonites:* New York, McGraw-Hill Book Co., 545 p.

Wallace, R. E., 1978, Geometry and rates of change of fault-generated range fronts, north-central Nevada: U.S. Geological Survey, *Journal of Research,* v. 6, p. 637-649.

Williams, P. F., Collins, A. R., and Wiltshire, R. G., 1969, Cleavage and penecontemporaneous deformation structures in sedimentary rocks: *Journal of Geology,* v. 77, p. 415-425.

Wise, D. U., and seven others, 1984, Fault-related rocks: suggestions for terminology: *Geology,* v. 12, p. 391-394.

# 13

# Volcanic Structures and Field Relations ■

### 13-1. Map Units, Stratigraphy, and Ages

Extrusive volcanic rocks are treated as lithostratigraphic units, and thus subdivided into groups, formations, members, and beds (or flows) as are sedimentary rocks (Section 5-3). Many volcanic units, however, are much less tabular and extensive than sedimentary units. Individual lavas and pyroclastic flows thicken and thin rapidly where deposited on irregular topography. Intracanyon flows, for example, may form long, narrow ribbons that lie at a great range of elevations and on a variety of older units. Viscous lavas may have such steep ends and sides as to appear to be faulted against adjoining rocks, as may fluid lavas that are banked against steep slopes. Particularly complex stratigraphic relations result where lavas and tuffs with steep initial dips (as on a large volcanic cone) are partly eroded between each successive eruption.

Such complexities may require mapping individual lavas and fragmental beds in order to determine an eruptive history. Usually, however, sequences can be subdivided into groups of petrologically similar flows or fragmental beds, and poor exposure or time constraints may make it necessary to compose still larger units, perhaps including lavas, a variety of fragmental rocks, and minor intrusions. Composite units should be of genetically related rocks as much as possible, as suggested by petrologic similarity, eruption from one vent or group of vents, or sequences emplaced so rapidly as to lack intercalated soil or sedimentary deposits.

During observed eruptions, and in exceptionally exposed cases, separate lavas or pyroclastic flows may be seen to consist of *flow units*, which are subsidiary tongues or sheets emplaced over or alongside one another to compose single flows. In contrast to separate flows, flow units should have identical proportions of phenocrysts, show local mixing or branching, and totally lack intervening weathering profiles, erosional features, or deposits. In many cases, however, extensive flow units may be impossible to distinguish from separate flows erupted less than 100 years or so apart.

As described more fully in the sections that follow, individual lavas or pyroclastic beds may be used (with caution) in correlating stratigraphic position among exposed sections. Of the more extensive deposits, basaltic lavas are perhaps the most persistent in primary characteristics (Schmincke, 1967). Airfall tuffs may be unusually extensive but tend to be thin, to appear similar from one tuff to another, and to vary laterally in grain size and crystal content (Westgate and Gorton, 1981). Many ash-flow tuffs are extensive and have reasonably persistent primary characteristics; however, they may vary owing to pre-eruptive zonation in the magma chamber, to sorting during transport, and to geographic distribution of welding and later alterations. Hildreth and Mahood (1985) have listed criteria that seem most reliable for ash-flow correlations.

Most volcanic units and sequences pass both vertically and laterally into sedimentary strata. The sedimentary rocks and their fossils can be used to interpret environments of volcanism as well as geologic age, and volcanic rocks provide a means of dating sedimentary rocks by isotopic methods. Nonmarine lavas and tephra tend to be mixed or intercalated with a variety of deposits, and thus the procedures and references mentioned in Section 10-1 may be helpful.

Recognizing unconformities and faults in volcanic areas is made somewhat difficult by the local nature of many volcanic units, by the possibility of steep initial dip, by the irregular shapes of some bodies, and by the presence of intrusive rocks that look much like lava. A principal purpose of this chapter is to describe primary structures and characteristic stratigraphic relations within single volcanic units that should help resolve these difficulties. Geologic mapping or many isotopic dates are often needed to discover and resolve major unconformities. These unconformities may be suggested locally, however, by deep soil profiles that cross volcanic units, by surfaces truncating the deeper parts of volcanic intrusions, by abrupt superposition of greatly different volcanic rocks, by gravels containing clasts of volcanic rocks known to occur well-down in the local sequence, and by sedimentary deposits indicating major changes in environment.

K-Ar dating is a powerful tool in studies of volcanic rocks, even those younger than 100,000 years (Mahood and Drake, 1982). Fresh rocks containing two or more K-rich species are preferred because they provide a check on the determined ages. Rocks younger than 40,000 years or so can be dated isotopically by the radiocarbon method, as on carbonized wood that is preserved in an uncontaminated state where lava or densely welded pyroclastic flows form an air-tight cover on woody vegetation (often roots are the best preserved) (Lockwood and Lipman, 1980). The polarity of remanent magnetization can be used to distinguish and date flows in sequences that accumulated during long intervals which included polarity reversals (Hooper and others, 1979). For shorter intervals, direction of magnetization of rocks from one area can be used to date rocks if a local history of geomagnetic secular variation can be determined from rocks dated by other means (Holcomb, 1980).

The sections that follow are intended to help in field recognition of rocks and primary relations. They will have far more meaning when used in conjunction with literature describing volcanic activity (see the references cited). Broad coverage of interpretive volcanology has been presented by Williams and McBirney (1979) and Fisher and Schmincke (1984), and a variety of studies of a major recent eruption were edited by Lipman and Mullineaux (1981).

## 13-2. Subaerial Basalts and Other Fluid Lavas

The subaerial origin of basalts and similar lavas can be recognized by relict soil profiles on flows and by intercalation with alluvial, eolian, or glacial deposits (Chapter 10). These surficial deposits also provide evidence of the time lapse between successive flows, and may contain fossils or carbonized wood with which to date a sequence. Pillow structures and glassy breccias do not rule out subaerial origin of a lava sequence, because continental lavas may flow over swamps or into lakes and rivers.

In sequences of basalt flows, individual flows or flow sets may have certain discriminating characteristics. The most reliable characteristics are those intrinsic to the erupted magma, and the kinds and amounts of phenocrysts are generally the most useful. As examples, some basalts have abundant olivine phenocrysts (*olivine basalt*); some may contain large plagioclase phenocrysts (*plagiophyric basalt*); and still others may be free of phenocrysts (*aphyric basalt*). In sequences of flows with the same kinds of phenocrysts, individual flows may be distinguished by: (1) specific phenocryst percentages (sometimes measured as numbers of phenocrysts per square unit area); (2) sizes of phenocrysts; (3) shapes of phenocrysts; and (4) specific colors of phenocrysts. Aspects that may prove useful but may vary laterally or vertically are: rock color; grain size and glass content of groundmass; vesicularity; toughness or ring to hammer blows; joint patterns; kinds of mineral-fillings in amygdules; and deuteric alteration. Studies utilizing most of these criteria have been described by Schmincke (1967), Wright and others (1973), and Grolier and Bingham (1978).

The structures and primary features described below will help in recognizing specific flows, stratigraphic position and flow direction in poorly exposed flows, and eruptive attributes such as viscosity and rate of effusion.

**Thickness** of flows varies with viscosity, ground slope, and distance from the vent. Olivine basalt and similarly fluid lavas average 1.5 m thick on the steeper slopes of Hawaiian shield volcanoes, and 7 m thick on nearly flat ground near the shoreline (Wentworth and Macdonald, 1953). Olivine basalts are generally less than 6 m thick on Cascade Mountains shields (Waters, 1960), and 2 to 8 m thick on the flatter slopes of the Oregon plateaus (G. W. Walker, 1970). On the lava plateaus of Iceland, olivine basalts average 7 m thick and olivine-poor basalts 11 m thick (G.P.L. Walker, 1959). Most olivine-poor flows of Yakima Basalt on the Columbia River plateau are 10 to 70 m thick and average perhaps 30 m, thicknesses due mainly to ponding and partly to enormous rates of lava effusion during most eruptions (Swanson and others, 1975).

**Flow types** are initially *pahoehoe* in the more fluid lavas and generally *aa* in more viscous ones. Pahoehoe is recognized by its ropy, billowy, or complexly folded upper surface, which is often lustrous and lineated in detail, with festoons of surficial folds indicating direction of flow in specific streams (Fig. 13-1). The initial gas-rich phase of major eruptions on Kilauea Volcano has produced thin sheets of cavernous pahoehoe that tend to collapse into stacked plates or curving shells (Swanson, 1973). This *shelly pahoehoe* is thus an early near-vent facies which later may be covered by thicker flows of dense, hummocky pahoehoe. Pahoehoe commonly changes downstream to aa when its viscosity increases and it undergoes increased rates of shear, as on a steep bluff (Peterson and Tilling, 1980). Aa is rough-surfaced, with piles of clinker and occasional large lava balls lying among slabs and spines extruded from the main lava body beneath (Fig. 13-1). Sizes and shapes of clinker may characterize certain flows and thus have stratigraphic value. Aa flows usually have a layer of clinker at the top and bottom, with solid lava between; however, flows transitional from pahoehoe may have clinker only at the top.

Pahoehoe is further characterized by small tongue-shaped *lava toes* that are extruded one by one at the front of a flow, some cracking open and draining and thus becoming hollow (Fig. 13-1). Pahoehoe that is deeply weathered or overgrown by vegetation can be identified by *tumuli* (lava

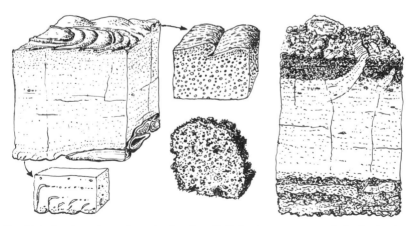

**Fig. 13-1.** Sections through a pahoehoe flow *(left)* that has overrun a set of pahoehoe toes, with detail below showing pipe vesicles that indicate flow toward the right, and detail at center showing filamented skin, flow wrinkles, and vesicles. Aa flow *(right)* has a large lava ball at its surface (in distance) and a protruding plate of the core. The detail at lower center is a lump of clinker, broken on the front to show vesicles. Chiefly from photographs by Wentworth and Macdonald (1953).

**Fig. 13-2.** Tumulus *(left)*, hornitos *(center)*, and arcuate pressure ridge with dark squeeze-out of lava along the crestal break.

domes with thick crusts cracked by extension) and *hornitos* (small spatter cones fed from the underlying flow) (Fig. 13-2). Elongate *pressure ridges* form where a congealed surface on a pahoehoe flow is under lateral compression, or where aa is forced to pile up periodically, usually as ridges that are perpendicular to the direction of flow and are convex downstream.

**Lava channels** mark a flow's principal streams and therefore its flow directions. In aa they are generally bounded by vertical walls and levees capped by overflow lava, and the channels may serve for later flows (Fig. 13-3A). Channels in pahoehoe have low walls of overflow lava that may

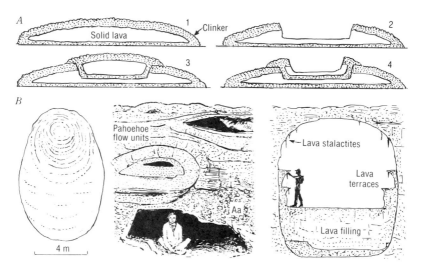

**Fig. 13-3.** *A.* Transverse sections of an aa lava stream, showing development of high levees: (1) initial stream; (2) channel formed by evacuation of lava downstream; (3) filling by resurgent flow; and (4) evacuation of that stream. From R. T. Holcomb, personal communication, 1984. *B.* Lava tubes in transverse sections: *left,* with concentric filling (from Wentworth and Macdonald, 1953); *middle,* partially filled small tubes in tube-fed pahoehoe of Mauna Ulu, Kilauea, exposed in a roadcut (after a photograph by L. R. Kanter, in Holcomb, 1980); and *right,* large tube with lava terraces marking high stages of filling flows.

accrete inward and close over the channel to form lava tubes. **Lava tubes** are common in most large pahoehoe flows and absent in aa flows. Filled tubes are marked by concentric flow banding, vesicle sheets, or concentric and radiating joints (Fig. 13-3*B*). In pahoehoe, tubes range from small openings in lava toes to caverns 10 m or more high.

**Vesicles** in pahoehoe are typically spherical or ellipsoidal and are most abundant at the tops of flows, where the rock may be pumiceous but typically contains about 30% of vesicles. Vesicles in aa are irregular and are concentrated near flow tops but typically make up less than 30% of the rock. Aa clinker has spiny, rough surfaces and only moderately vesicular cores (Fig. 13-1). Vesicles may be flattened and elongated on the upstream side of obstacles, providing an indication of flow direction (Waters, 1960). *Pipe vesicles* are elongate and typically turn downstream, thus indicating flow direction (Fig. 13-1). *Spiracles* are openings formed by injection of steam or other gas from wet ground or vegetation overrun by a flow. In dissected flows, spiracles can be seen to turn downstream and typically change into trains of vesicles (*vesicle cylinders*). *Pegmatite* carrying euhedral pyroxene and zeolites forms horizontal sheets and pipe amygdules in some thick olivine basalt flows in Iceland (Walker, 1959).

**Joints** in thin basalt flows typically compose crude systems, one parallel to the top of the flow and one or more perpendicular to it (Fig. 13-1). In Hawaii, columnar joints occur only in relatively thick flows emplaced over wet ground (Macdonald, 1967). Thick plateau lavas are typically jointed on some variant of a three-tier columnar system, although the columns in the upper colonnade are seldom well-formed (Fig. 13-4). The subhorizontal platy

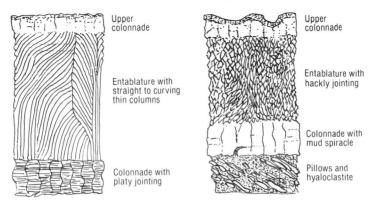

**Fig. 13-4.** Joint systems (schematic) typical of many thick Columbia Plateau flows. The hackly jointing tends to form where cooling is unusually rapid, as in water bodies or during heavy rains. Based chiefly on Swanson (1967 and personal communication, 1984) and Schmincke (1967), who described additional variations.

joints shown in the figure are along sheets of small vesicles that develop at a late stage in some flows (Waters, 1960).

### 13-3. Subaqueous Basaltic Lavas and Hyaloclastic Deposits

Basalt and other fluid lavas that flow into water or are erupted under water or ice develop one or more of these structural varieties: sheet flows (largely pahoehoe), pillow lava, and hyaloclastic (or *hydroclastic*) deposits. Fragmental rocks may also result from explosions, as described in Section 13-5. Subaqueous pahoehoe is probably far more common than once thought, and has been observed in abundance at some spreading rises (Ballard and others, 1979).

A predictable difference between subaerial and deep-sea lava is that the latter should be much less vesicular, because the size and abundance of vesicles decrease with water depth. Submarine basalts studied by Moore (1965) and Moore and Schilling (1973) indicate that vesicles decrease in average size from approximately 1 mm at 500 m depth to 0.5 mm at 1000 m to 0.2 mm at 3000 m to nil at 4000 m, and that vesicles comprise approximately 15% of lavas at 500 m, 5 to 10% at 1000 m, roughly 3% at 2000 m, and less than 2% at depths of 3000 m or more. These data were measured on pillow lavas but probably can be applied to the rinds of subaqueous sheet flows.

*Pillow lava* is a name given because of the shapes of lava flow units seen in cross sections perpendicular to the direction of flow (Fig. 13-5A). The bodies are actually tongues, generally five to ten times longer than they are wide, and commonly branching in the direction of flow (Moore and others, 1973; Vuagnat, 1975). Separate oval pillows may form, however, where lava has advanced to the top of a steep slope, and tongues emplaced onto the slope separate and roll down it (Fig. 13-4, lower right). Some pillows have cavities produced when lava drains downstream, and most have details similar to

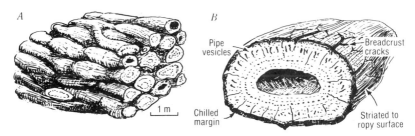

**Fig. 13-5.** *A.* Lava tongues, showing branching down-flow and pillow-like sections perpendicular to flow direction, with some centers drained and hollow. *B.* Part of a hollow tongue (a "pillow"), with radial fractures, concentric bands of vesicles, and pipe vesicles concentrated in upper part.

those shown in Fig. 13-5*B*, except that vesicles would be scarce or missing in deep-water varieties. Several of these features help in distinguishing pillows from pahoehoe lava toes (Fig. 13-1). Lava toes lack surficial striations and their transverse sections tend to be flat ellipses, whereas pillows are more nearly cylindrical and striated. In addition, only pillows have a matrix of hyaloclastic breccia, or are associated with hyaloclastic rocks.

**Hyaloclastic deposits** are thought to form when lava is quenched to glass in water and then granulates because of rapid change in volume during cooling (Honnorez and Kirst, 1975). The resulting breccias and fine-grained clastic rocks have been called *hyaloclastites* to set them off from exploded pyroclastic rocks (as tuff) and autoclastic volcanic rocks (as flow breccia). Hyaloclastic fragments are mainly in the size range of 1 to 30 mm and are angular blocks and chips which often have smoothly curving faces (Fig. 13-6). The commonest deposits are poorly sorted breccias that may have cements or matrixes of calcite, zeolite, or clay. Separate pillows and angular fragments of disintegrated pillows are a common constituent, and if these clasts are dominant, the deposit is a *pillow breccia*. Pillow-free hyaloclastite forms massive beds that may be so dark, compact, and vitreous that separate glass fragments are difficult to see. These rocks may grade upward or laterally into sorted, stratified hyaloclastite and, further, into beds of reworked hyaloclastic grains associated with other sedimentary materials (Silvestri, 1963). Basaltic glass (sideromelane) is typically clear and pale yellow when first formed, but tends to be altered to *palagonite*, which is opaque in hand specimens and may be nearly black, dark olive green, orange, or yellow brown. The resulting palagonite breccia or tuff is generally so mottled that original textures are obscure.

Several possible structural relations may be expected in hyaloclastite sequences. Carlisle (1963) described regular sequences of: (1) limestone, chert, or wacke overlain by (2) pillow lava, (3) hyaloclastite with whole pillows, (4) hyaloclastite with pillow fragments, and (5) either massive lava or more sedimentary strata. Silvestri (1963) suggested that hyaloclastite may be produced directly by subaqueous fissure eruption if water enters the fissure as it begins to open; thus all parts of a deposit may be hyaloclastite. In contrast, basalt erupted subaerially and then flowing into lakes or the sea has been seen to produce distinctive deltaic deposits of pillows and

**Fig. 13-6.** Hyaloclastic granules of pale yellow to black glass, formed at depths too great to permit explosion or much vesiculation. In contrast, fragments formed by explosion at shallow depth (Surtseyan activity) are irregular, vesicular, and much smaller (see Wohletz, 1983).

hyaloclastite that are overrun in due course by subaerial lava (Fig. 13-4, lower right).

### 13-4. Flows and Domes of Viscous Lava

A number of features can be used to determine: (1) the original extent of eroded viscous lava flows, (2) the positions of the vents from which they emerged, (3) the approximate stratigraphic position of a given outcrop within a flow, and (4) something of the eruption itself. Viscous lavas range in composition from andesite and trachyte to rhyolite; however, the viscosity of any of them may vary greatly due to factors other than rock composition, so that each case must be treated without assumptions. For example, although most siliceous flows are thick and stubby, Bailey and others (1976) mapped a flow 16 m thick that extended for 6 km.

Next to mapping the distribution of the flow itself, the most valuable features are its *flow structures*, which may consist of any or all of these components: layering (banding); platy joints with surface lineation (but not all platy jointing is parallel to flow layers); flattened and elongated vesicles; folds; faults oriented at low angles to layering; oriented crystals and inclusions; and extension cracks. The layers (bands) may be as thin as a fraction of a millimeter and as thick as several meters. They are generally discernible by color or tonal differences caused by differences in crystallinity or vesicularity, by sheets of spherulites or lithophysae, or by alteration along platy partings. Where the rock parts easily along the flow layers, lineation and extension cracks may suggest the direction of flow (Fig. 13-7A). Folds commonly form with hinges at right angles to the direction of flow; however, their hinges may be parallel to the direction of transport where a flow was constricted laterally (as in folding by extension, Section 12-3). Folds may also become rotated or superimposed where flow orientation varied with time. Folds with consistent vergence may be associated with local thrust faults (Fig. 13-7B).

The bases and tops of flows are typically vesicular to pumiceous and commonly brecciated. Intermediate lava flows, as those of andesite, may consist largely of angular blocks, many larger than 1 m in diameter. Solid spines and craggy ridges on the surface of silicic flows typically stand many meters above crevasses or alongside piles of angular fragments (Fig. 13-7C). Basal pumice breccias tend to mix with pyroclastic pumice erupted just before the flow, so that the basal contact may be gradational, and the underlying pumice may be compacted and fused into dense obsidian. Vesicular fragments from the top or base are often caught up in the main flow, where they are compacted, fused, and drawn out parallel to other flow structures. Where oxidized pumice is incorporated, the fused rock may be streaked with red, yellow, or brown. Williams (1942) described low caves formed where the base of a flow was arrested by obstacles and was arched upward (Fig. 13-7C).

Some flows have inclined layers that reveal a sense of flow direction and provide a means of charting the sequential spread of the flow as well as its source (Fig. 13-7C). Steeply inclined flow structures that formed in an entirely different way, however, were described by Fink (1983), who made detailed studies of a number of flows in which a layer of light pumiceous lava lies beneath dense obsidian. The resulting gravity instability led to rhythmically spaced diapirs that rose and expanded so as to rotate originally horizontal flow structures to vertical orientations.

The use of aerial photographs in interpreting patterns of flow structures and vent positions is well illustrated by Koringa (1973, p. 3859) and by Fink (1983). Vents in eroded flows may be indicated in the field by greater concentrations of phenocrysts or inclusions, by a greater degree of crystallinity of the groundmass, by abundant tridymite, cristobalite, or hematite in cavities or along fractures, and by alteration to white, red, or varicolored rocks consisting largely of clays, tridymite, opal, chalcedony, or zeolites. On uneroded flows, the vent is generally under the highest part of the flow and typically marked by a low dome.

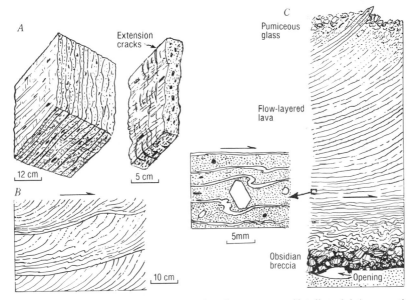

**Fig. 13-7.** *A.* Somewhat irregular lineated platy flow structure (detail on right) suggesting extension perpendicular to the sets of late cracks. *B.* Vertical section through flow layering with two thrust faults indicating movement upward to right. *C.* Vertical section through flow that moved to right, with hand-lens view of folded flow structures in the zone of horizontal layering. Based chiefly on the Cleetwood flow, Crater Lake National Park (Williams, 1942).

**Domes** result where lava is so viscous as to pile up over a vent, or to intrude and balloon upward and laterally into sediments or previously erupted volcanic deposits. The sides of many steep-sided domes are covered by an apron of talus formed by disintegration of lava during eruption (Fig. 13-8*A*). Lava spines and craggy ridges on top of the dome may be pressed upward or may result from subsidence of the surrounding parts of the dome. Graben or extension fractures on domes may indicate orientations and perhaps positions of underlying feeding systems (Rose, 1972; Fink, 1983). A pumiceous carapace generally covers more dense lava, which may be nearly structureless or have systematically oriented flow structures. A common structural configuration is an outward fanning of flow surfaces (Fig. 13-8*A*). Another common kind of dome consists of a pile of short flows (Fig. 13-8*B*). Several other configurations were illustrated by Williams and McBirney (1979, p. 190).

### 13-5. Pyroclastic Deposits Produced by Explosions

Three genetic kinds of pyroclastic deposits may be formed by explosions: (1) *airfall* (or *fallout*) *deposits*, consisting of materials fallen from high eruption clouds, or projected outward by explosions; (2) *base surge deposits*, accumulated from clouds that are initiated by strong explosions and sweep outward along the ground at hurricane velocities; and (3) *pyroclastic flow deposits*, formed from hot debris streams and the turbulent clouds that rise above them. These three kinds of deposits may occur as isolated layers or as genetically related sequences as will be described below. Components of pyroclastic deposits are illustrated in Fig. 4-17, and a textural classification is given in Fig. 4-16.

**Airfall deposits** blanket the topography quite evenly. They are typically poorly to moderately sorted on the volcano and well to very well sorted at increasing distances from it. They also become thinner and finer grained away from the volcano. Repeated explosions of different strengths lead to distinct stratification; however, it may be modified by creep and bioturbation. The deposits are also easily eroded. Blocks and bombs produce com-

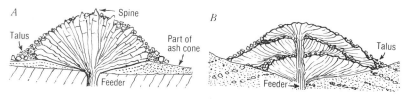

**Fig. 13-8.** *A.* Vertical medial section through a dome that has grown chiefly by injection of viscous lava within the mass (an *endogenous dome*). *B.* Dome formed chiefly by accumulation of viscous flows erupted to the surface (an *exogenous dome*).

pactive dimples ("sags") in underlying strata near the volcano. Uncommonly, the deposits are welded or agglutinated and can be distinguished from welded pyroclastic flow deposits by distinct stratification, by rapid vertical variation in degree of welding, by mantling the topography, and by the fact that the degree of welding falls off rapidly away from the vent.

Walker (1973) noted that magnitude of an explosive eruption is indicated by the volume of detritus produced, and eruptive violence by (1) the size of the area covered by ejecta and (2) the abundance of fine tephra in it. He developed a classification of explosive eruptions based on the percentage of fine tephra and the thicknesses of deposits at a fixed distance from the vent, and noted characteristics of airfall deposits that help in recognizing these eruptive types. Briefly, deposits consisting largely of glassy fragments shaped plastically during eruption (Fig. 4-17) suggest mild explosions or fountaining of liquid basalt (Hawaiian-type activity). Abundant ragged scoria and shaped scoria bombs accompanied by less than 25% of clasts finer than 1 mm suggest somewhat more violent (Strombolian-type) activity. In contrast, exceptionally widespread deposits consisting of irregular vesicular fragments of basaltic glass mostly finer than 1 mm indicate violently explosive (Surtseyan-type) activity, caused by mixing of water and basalt melt. Tuffs produced by these subaqueous explosions are commonly vesicular, and base surge deposits often accompany this type of airfall deposit.

Deposits consisting mainly of pumice lumps and colorless glass shards are generally felsic or silicic in composition. Those produced by unusually powerful but continuous gas blasts (Plinian-type activity) can be recognized by a near-vent facies consisting of a thick unstratified pumice layer that contains only moderate amounts of clasts finer than 1 mm, and in some cases is graded coarser-upward (Lirer and others, 1973). Violent explosions caused by mixing of water and abundant felsic magma (phreatoplinian-type activity) have been interpreted by Self and Sparks (1978) as the cause of deposits of blocky, sparsely vesicular glass fragments that: (1) are much finer grained than Plinian-type tephra (median diameter of 0.2-0.1 mm); (2) show little decrease in grain size away from the vent; (3) occur with base surge deposits; and (4) commonly include accretionary lapilli.

***Base surge deposits*** are characterized by laminae that are typically cast into dune, antidune, and smaller sand-wave forms near the volcano, and are planar at greater distances from the volcano (Fig. 13-9*A*, *B*, and *C*). In addition to this spatial relation, Wohletz and Sheridan (1979) noted that unlaminated deposits of poorly sorted clasts are abundant throughout the deposits, commonly forming elongate lenses in the lee of antidunes (Fig. 13-9*D*). The antidunes indicate the direction of the blast and therefore of the vent, as do deposits plastered against large blocks, tree trunks, or other steeply inclined surfaces (Moore, 1967). Plastered-on deposits and locally abundant accretionary lapilli indicate moist ash, and the moisture implies mixing of water with magma. Some surge blasts, however, are so hot and dry that they sear vegetation. Dry or moist, the blasts generally are so powerful as to sweep away all vegetation near vents and to sandblast and topple trees throughout large areas.

Typically, a base surge deposit extends about as far from its vent as the diameter of the vent (Wohletz and Sheridan, 1979); however, exceptional surges, as the one from Mount St. Helens on May 18, 1980, are far more extensive (Moore and Sisson, 1981).

*Pyroclastic flow deposits produced by explosions (ignimbrites)* include several varieties classified by primary textures, and additional secondary variants classified by degree of welding and alteration. Of the primary kinds of deposits, *ash flows* are by far the most common and voluminous, and have, by definition, a median grain size less than 2 mm. Pumice fragments larger than 2 mm are common in ash flows and predominate in *pumice flows*. *Block flows* are typically localized to the vicinity of the volcano. Although mainly subaerial, pyroclastic flow deposits may occur in sequences of marine rocks, some being erupted underwater (Fiske and Matsuda, 1964) and some evidently flowing into the sea and along the seafloor for many kilometers (Francis and Howells, 1973).

All pyroclastic flow deposits are characterized by: (1) very poor to moderate sorting; (2) lack of bedding within a single flow unit; (3) thicknesses of meters to tens of meters (in some cases hundreds of meters); (4) presence of accessory or accidental lithic fragments; and (5) level or smoothly sloping upper surfaces and uneven (topographic) lower surfaces. Many ash-flow and pumice-flow deposits can also be recognized by their secondary features: gradational layers or zones due to differing degrees of compaction and welding, patterns of columnar joints, and gradational zones due to vapor-induced

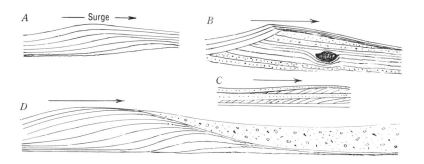

**Fig. 13-9.** Depositional structures formed by a base surge current near a vent. *A.* Simple antidune. *B.* Antidune cast onto eroded strata with a bomb sag (impact structure made by a falling block or bomb). *C.* Cross-bedding dipping toward the volcano. *D.* Antidune with unsorted coarse accumulation to its lee.

alteration and crystallization. These secondary features are likely to vary from one deposit to the next and thus are helpful in correlation, although lateral variations are considerable (Smith, 1960). Zones of compaction and welding may be distinctive enough to suggest approximate stratigraphic position in a given flow and have been used to measure fault displacement (Peterson, 1979). Differences in jointing and resistance of the secondary layers give rise to consistent topographic steps that are useful in reconnoitering large areas of pyroclastic flow deposits. These variations are described more fully in the list that follows:

1. *Depositional variations* are inconspicuous in most flows but distinct in some (Fig. 13-10*A* and *B*). Lithic fragments may accumulate at the base of the flow near the volcano and decrease in amount upward, whereas pumice increases upward (especially if the pumice is very light). In distal parts of a flow, pumice commonly forms a coarse concentration on top of each flow or flow pulse (Sparks, 1976). An upper ash layer, formed by settling of particles from a turbulent cloud above the original flow, is greatly depleted in crystals relative to the flow itself (Walker, 1972); this subunit, however, is commonly eroded away.

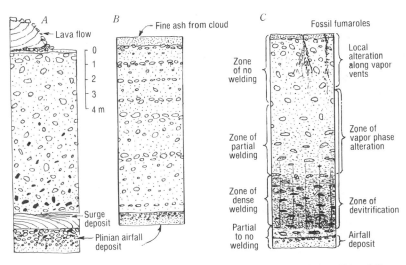

**Fig. 13-10.** Vertical sections through thick pyroclastic flows near volcano *(A)* and distant from volcano *(B)*, showing idealized textural variations and depositional subunits. Pumice lumps are unpatterned and lithic fragments are black. The layering in *B* may represent separate flows or differentiated flow units within one flow (Sparks, 1976; Sheridan, 1979). *C.* Zones due to welding are labeled on the left and zones due to crystallization and vapor effects on the right. Flattening of pumice lapilli indicates degree of compaction. Smith (1960, plate 20) has illustrated additional vertical arrangements as well as typical lateral variations.

2. *Compaction and postdepositional flow* are best shown by the systematic strains of pumice fragments (Fig. 13-10C). Density of the deposit increases and porosity decreases with depth, but these changes may be nonsystematic because of the additional effects of crystallization and transport of materials by vapor (see 4 below). Degree of compaction is customarily estimated from the average width-to-height ratio of many pumice fragments measured at a given locality. Ratios of about 6:1 have been found to represent completely compacted pumice that was originally highly vesicular (i.e., had a bulk density of 0.4-0.5 gr/cc) (Sheridan and Ragan, 1976). Rocks in which flattened pumice lapilli (*fiamme*) have ratios greater than 7:1 must have spread laterally under load, which indicates exceptionally hot or thick deposits (Peterson, 1979). Compacted glass shards and globules develop lenticular curving forms that drape markedly over crystals and lithic fragments, a fabric called *eutaxitic* (Fig. 13-11).

Pyroclastic flows are compacted approximately parallel to underlying topography, and thus their compaction fabrics dip moderately to steeply close to steep buried slopes. The lower parts of unusually hot or thick deposits may flow by extension and simple shear (Fig. 13-12). Magmas rich in alkalies and halogens have such low viscosities as to weld densely and flow secondarily even if thin.

3. *Zones due to welding* (Fig. 13-10C) are intergradational and may vary considerably in thickness from flow to flow and laterally within one flow (Smith, 1960). The densely welded zone is often marked by distinct columnar jointing and thus by a ledge or cliff. Densely welded rocks that are not devitrified are visibly glassy throughout and have distinctly conchoidal fracture. Pumice fragments have been compressed into obsidian fiamme with maximum flattening ratios, and rock porosities are less than 10%.

Rocks in the lower third of the partially welded zone have a silky (somewhat porous) groundmass, break with a hackly fracture, and may contain

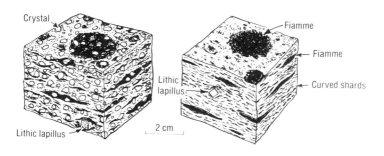

**Fig. 13-11.** Axial eutaxitic fabric in crystal-rich *(left)* and crystal-poor welded vitric tuffs, both showing compacted pumice lapilli (fiamme) and a few undeformed lithic lapilli.

some pumice fragments that are still slightly vesicular. Above them is a complete gradation into more porous but still coherent material with diminishing flattening ratios of pumice fragments. The average flattening ratio is approximately 2:1 at the boundary between the partially welded zone and the porous, friable rocks of the nonwelded zone.

4. *Zones due to crystallization and vapor effects* may be superimposed on either welded or unwelded flow deposits (Fig. 13-10C). Densely welded rocks in the devitrified zone become stony in appearance, pale colored, and may have macroscopic spherulites and lithophysae. Their flattened glass pyroclasts consist of radiating (axiolitic) fibrous crystals that may be visible through a hand lens. Rocks in the zone of vapor-phase alteration are pale and variably colored from place to place, generally in tints of red or orange. Most are noticeably porous, often with pumice fragments partially dissolved. Openings are typically coated with minute crystals of tridymite, sanidine, cristobalite, hematite, or iron-rich silicates. Tuff above the zone of vapor-phase alteration appears unaffected except where gases escaped through what are now fossil fumaroles and altered the surrounding rocks (Sheridan, 1970).

5. *Cooling units* are pyroclastic sequences that accumulated so rapidly as to compact and cool as a single unit, as shown by a complete gradation of compaction, welding, devitrification, and vapor-phase alteration. Cooling units are generally the most obvious mapping units in areas of extensive pyroclastic flow deposits. Many cooling units consist of more than one pyroclastic flow, as indicated by repetitions of some or all of the depositional subunits of Fig. 13-10A and B, or by a parting surface or an abrupt change in crystal content, rock color, or weathering characteristics. Flow contacts in welded or altered parts of cooling units are generally obscure and difficult to map except locally. *Compound cooling units* do not have a regular gradation of compaction forms, welding, or alteration, and probably result when

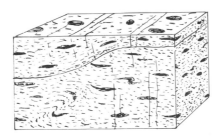

**Fig. 13-12.** Evidence of lateral flow in a welded ash-flow deposit: linear shard-fabric, linear fiamme (commonly pulled apart on minute extension fractures), folds, a minor thrust fault (ramp), and extension fractures (here, approximately vertical).

some flows were deposited after preceding flows had begun to cool but had not cooled completely.

**Reworked pyroclastic materials** are generally classified as pyroclastic even though they were transported and deposited by streams, winds, or waves. They are common as rocks because many pyroclastic deposits are eroded easily. They can be recognized by the presence of nonpyroclastic detritus or by sedimentary structures and associated rocks. It may be helpful to name these rocks *water-laid tuff*, *lacustrine tuff*, and so on.

### 13-6. Fragmental Rocks Formed Without Explosion

Several important kinds of fragmental volcanic rocks are formed by processes other than volcanic explosions or continuous gas-blast eruptions. One genetic group is composed of the hyaloclastites, described in Section 13-3, and other groups are described in this section.

**Lahars** are wet volcanic debris flows that form mainly where blocky lava or pyroclastic flows enter streams or flow onto snow and ice; where eruptions take place through bodies of water; or where pyroclastic deposits are eroded catastrophically during heavy rains (Williams and McBirney, 1979). Lahars formed from recently emplaced pyroclastic flows typically travel much farther than the hot flows. In some extensive accumulations, lahars extend as far as hundreds of kilometers from their sources (Curtis, 1954, Lydon, 1968).

Most lahar deposits are very poorly sorted breccias that are intercalated with volcaniclastic conglomerate, sandstone, and mudstone, and that grade into these rocks laterally as well as vertically. Individual lahars are typically a few meters to a few tens of meters thick. The layers in a lahar sequence may be obscure except when viewed from a moderate distance. A sand-mud matrix is the principal constituent of most lahars, distinguishing them readily from blocky lava flows. Most of the larger clasts are within pebble and cobble size-classes, although some are likely to be more than 1 m in diameter. The larger clasts are typically angular near the source of a flow and may become subangular tens of kilometers downstream. Sparse rounded stream cobbles and wood fragments generally become included in the flows; otherwise, most coarse lahars consist of several textural varieties of a given volcanic rock, typically andesite.

Deposits of unusually hot lahars may be reddish and contain charred wood; however, similar characteristics can result where loose pyroclastic flow deposits have been reworked as cool lahars. Lahar deposits can be distinguished from pyroclastic flow deposits by several characteristics already mentioned, and also by their lesser proportion of pumice fragments and glass shards, and their lack of welding, compaction, vapor alteration, or ash-flow stratigraphy (Fig. 13-10).

**Autobrecciated lava flows** can be recognized by local unbrecciated lava at their core or base. They are also composed of fragments that are texturally and compositionally alike, except for glassy, reddened, or more vesicular clasts that may be mixed in from the top or base during movement of the flow. The upper parts of brecciated flows tend to consist of large blocks between which are abundant open spaces that may eventually become filled with tephra, soil, or other sediment. Finely-broken lava tends to be abundant in the lower parts of blocky flows, where it may be pressed into a coherent breccia with low porosity. Blocky flows may change laterally to lahars when they enter streams or flow over snow and ice.

**Volcanic block flows and avalanches** are chiefly dry debris flows resulting from the collapse of oversteepened volcanic slopes, as the sides of domes, or recently formed walls of calderas or craters. Avalanches from hot domes knock down and sear vegetation, and individual blocks develop radial sets of columnar joints when they come to rest and cool, often disintegrating along these fractures (Fig. 13-13). Hot avalanche deposits differ from exploded pyroclastic flows in that they contain no freshly erupted pumice or ash (Section 13-5). Blocky deposits around domes become interlayered with talus deposits, which can be recognized by their monolithologic composition, their steeply inclined crude bedding, and the talus-sorting of large fragments toward their base (Fig. 10-11).

**Caldera-collapse breccias,** described by Lipman (1976), form by rockfalls and avalanches from caldera walls during and shortly after collapse of a volcanic center. The deposits can be recognized by their intercalation with pyroclastic flow deposits erupted into the caldera during its collapse, and by containing fragments derived from nearby caldera walls. The deposits are most abundant near the caldera walls, where they are associated with talus deposits. They also tend to be most voluminous in the upper parts of the sequence accumulated within the caldera. These deposits can thus be used in locating caldera walls that have been eroded away.

Lipman also described *megabreccias* that contain blocks measuring many meters across. The blocks occur in the lowest part of intracaldera sequences, and thus appear to have formed by collapse of the volcanic center rather

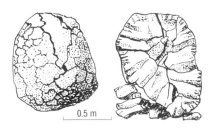

0.5 m

**Fig. 13-13.** Lava blocks that were hot when ejected, showing typical cooling fractures. The block on the left has remained whole, that on the right has broken in half since cooling.

than secondarily from the caldera walls. Megabreccias may help in locating and interpreting the lower parts of calderas in deeply eroded areas. They may also lead to recognition of intracaldera pyroclastic flow deposits. Other features that may help in recognizing calderas have been reviewed by Christiansen (1979).

### 13-7. Volcanic Feeders and Related Intrusions

Where a volcano is largely eroded away, the intrusions that fed it can be used to determine much of the volcano's history. Age relations are particularly important, and are determined by mapping and examining: (1) cross-cutting intrusive bodies; (2) inclusions of country rocks and of intrusive rocks; (3) chilling of one intrusive body against another; (4) explosive shattering of one rock and not another; and (5) metamorphism or alteration of certain rocks and not others. Exceptional examples of mapping, outcrop study, and petrography used to unravel histories of deeply eroded volcanic centers have been described by Richey (1961), Fiske and others (1963), and Verwoerd (1967). The descriptions in this section pertain mainly to the upper parts of these systems, and some in Chapter 14 to deeper parts. As noted in Section 5-3, volcanic intrusions should be treated as lithodemic units.

**Dikes** are generally the most abundant minor intrusions in and beneath volcanoes. Those at shallow depth typically have glassy margins, contain fairly numerous vesicles, have distinct columnar joints, and have little or no effects on their wall rocks. Exceptionally shallow intrusions mix with breccias of volcanic cones or intrude weakly consolidated sediments, in which they tend to splay out, to become invaded by mobilized clastic detritus, and perhaps to form intimate mixtures (*peperite*) with fragmental rocks. Dikes that fed pyroclastic eruptions may themselves be pyroclastic and become welded with eutaxitic fabrics parallel to their walls. Some tuff-breccias in lahar sequences are erupted from andesite-breccia feeders (Lydon, 1968). Hyaloclastite dikes are mentioned in Section 13-3.

Several dikes are commonly emplaced along one original fracture, with the later dikes intruding either along a contact or along the medial plane of an earlier dike. In *multiple dikes* the intrusions are of one kind of magma, and the internal intrusive contacts may be marked by a chilled margin or a thin layer with strongly developed flow fabric, vesicles, inclusions, or flow-differentiated phenocrysts. Columnar joints commonly develop at right angles to dike walls, providing an additional means of recognizing multiple dikes or dikes cutting sills or lava flows (Fig. 13-14A). In *composite dikes* the separate intrusions have different compositions and are generally apparent. If intruded in rapid succession, they may show: (1) broadly mixed gradations; (2) mafic magma chilled against silicic magma, and (3) each rock locally intruded and included by the other.

Dikes are most commonly radial to a volcanic pipe, or they may pass through it or near it in more or less parallel sets. In some cases they are tangential or describe concentric arcs or complete rings that dip toward the volcanic center (*cone sheets*) or stand more or less vertically (*ring dikes*). Cone sheets indicate episodes of forceful elevation caused by a magma body whose depth can be estimated by the dip of the sheets. Ring dikes are typically thicker than cone sheets, have mylonitic margins produced during subsidence of the central block, and are more likely to include explosion pipes.

**Sills** in volcanic sequences may be distinguished from lava flows by locally cutting across adjoining layers, by lack of oxidized scoria or breccia at their base and top, and by coarser grain, less glass, fewer vesicles, and simpler columnar jointing than in lavas (Fig. 13-14*A*). Very shallow sills tend to mix with fragmental deposits. Thick sills are commonly multiple or composite, and their internal contacts can be recognized as described above for dikes. Early-formed mafic grains may accumulate in the lower parts of sills. Because most sills are emplaced by magma lifting overlying strata, they may be expected in or under sequences of relatively light rocks, especially under light rocks that lie on heavier rocks along an unconformity or an extensive low-angle fault. Fiske and others (1963, p. 48) described widespread multiple and composite sills intruded preferably at and near a major unconformity.

**Volcanic necks and pipes** range from a few meters to several kilometers in diameter and are typically circular or elliptical in horizontal section, but may consist of an irregular plexus of dikes. The rocks composing them tend to be coarser grained, less glassy, less vesicular, and to carry more inclusions than flows once fed by the neck. Orientation of columnar joints may indicate approximate depth of exposure (Fig. 13-14*B*). Pipes and associated dikes of viscous magma generally have flow layering that is parallel to their contacts

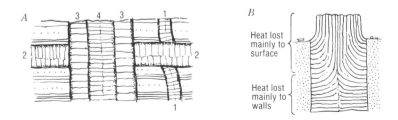

**Fig. 13-14.** *A*. Intrusive sequence, as numbered, for two dikes and a sill, as shown by cutting relations, chilled margins (closely dotted), and columnar joints. *B*. Butte developed on a cylindrical volcanic neck, showing changes in columnar joints with depth.

and continuous with the outward-fanning flow layers of the flows or domes they fed (Fig. 13-8).

The upper parts of volcanic necks are commonly mixed with fragmental rocks that slid or fell from the crater walls, and these materials may become emplaced to depths of hundreds of meters after powerful continuous gas discharges have cleared the upper part of the feeding column. Breccia pipes extending to still greater depths are *diatremes*, formed at depth by more or less continuous gas discharges, explosions, and rock bursts caused by release of confining pressure on the walls of a feeding pipe (Gates, 1959; McCallum and others, 1975). Diatremes may be recognized by: (1) dominance of country rock fragments; (2) attrition and rounding of the farthest traveled clasts; (3) rounded and polished crystals; and (4) accretional pellets consisting of crystal fragments mantled by fine volcanic matrix (Rust, 1937). The proportions of nonpyroclastic igneous materials in diatremes increase with depth, and some pipes extend to pluton cores (Gates, 1959; Richey, 1961).

*Alteration* in and near volcanic feeders is common and in some cases intense. Green rocks containing secondary epidote and chlorite and pale-toned rocks converted to clays and other fine, hydrated silicates are especially common. These alterations are typically produced by geothermal systems related to underlying plutons, as described in Sections 14-8 and 15-7. Except for the effects of vapor, however, volcanic feeders have limited effects on wall rocks. Basalt magma may fuse a thin skin of silicic walls and partially melt inclusions, but most magmas discolor or alter wall rocks for only a meter or so from intrusive contacts.

## References Cited

Bailey, R. A., Dalrymple, G. B., and Lanphere, M. A., 1976, Volcanism, structure, and geochronology of Long Valley caldera, Mono County, California: *Journal of Geophysical Research*, v. 81, p. 725-744.

Ballard, R. D., Holcomb, R. T., and van Andel, T. H., 1979, The Galapagos Rift at 86°W: 3. Sheet flows, collapse pits, and lava lakes of the rift valley: *Journal of Geophysical Research*, v. 84, p. 5407-5422.

Carlisle, D., 1963, Pillow breccias and their aquagene tuffs, Quadra Island, British Columbia: *Journal of Geology*, v. 71, p. 48-71.

Christiansen, R. L., 1979, Cooling units and composite sheets in relation to caldera structure, p. 29-42 *in* Chapin, C. E., and Elston, W. E., editors, *Ash-flow tuffs*: Geological Society of America Special Paper 180.

Curtis, G. H., 1954, *Mode of origin of pyroclastic debris in the Mehrten Formation of the Sierra Nevada*: University of California Publications in Geological Sciences, v. 29, no. 9, p. 453-502.

Fink, J. H., 1983, Structure and emplacement of a rhyolitic obsidian flow: Little Glass Mountain, Medicine Lake Highland, northern California: *Geological Society of America Bulletin*, v. 94, p. 362-380.

Fisher, R. V., and Schmincke, H.-U., 1984, *Pyroclastic rocks*: New York, Springer-Verlag, 472 p.

Fiske, R. S., Hopson, C. A., and Waters, A. C., 1963, *Geology of Mount Ranier National Park, Washington*: U.S. Geological Survey Professional Paper 444, 93 p.

Fiske, R. S., and Matsuda, T., 1964, Submarine equivalents of ash flows in the Tokiwa Formation, Japan: *American Journal of Science*, v. 262, p. 76-106.

Francis, E. H., and Howells, M. F., 1973, Transgressive welded ash-flow tuffs among the Ordovician sediments of NE Snowdonia, N. Wales: *Journal of the Geological Society of London*, v. 129, p. 621-641.

Gates, O., 1959, Breccia pipes in the Shoshone Range, Nevada: *Economic Geology*, v. 54, p. 790-815.

Grolier, M. J., and Bingham, J. W., 1978, *Geology of parts of Grant, Adams, and Franklin Counties, east-central Washington*: State of Washington Division of Geology and Earth Resources Bulletin 71, 91 p.

Hildreth, W., and Mahood, G., 1985, Correlation of ash-flow tuffs: *Geological Society of America Bulletin, in press*.

Holcomb, R. T., 1980, *Kilauea Volcano, Hawaii: chronology and morphology of the surficial lava flows*: Stanford University, CA, PhD Dissertation, 321 p.

Honnorez, J., and Kirst, P., 1975, Submarine basaltic volcanism: morphometric parameters for discriminating hyaloclastites from hyalotuffs: *Bulletin Volcanologique*, v. 39, p. 441-465.

Hooper, P. R., Knowles, C. R., and Watkins, N. D., 1979, Magnetostratigraphy of the Imnaha and Grande Ronde Basalts in the southeast part of the Columbia Plateau: *American Journal of Science*, v. 279, p. 737-754.

Korringa, M. K., 1973, Linear vent area of the Soldier Meadow Tuff, an ash-flow sheet in northwestern Nevada: *Geological Society of America Bulletin*, v. 84, p. 3849-3866.

Lipman, P. W., 1976, Caldera-collapse breccias in the western San Juan Mountains, Colorado: *Geological Society of America Bulletin*, v. 87, p. 1397-1410.

Lipman, P. W., and Mullineaux, D. R., editors, 1981, *The 1980 eruptions of Mount St. Helens, Washington*: U.S. Geological Survey Professional Paper 1250, 844 p.

Lirer, L., Pescatore, T., Booth, B., and Walker, G.P.L., 1973, Two Plinian pumice-fall deposits from Somma-Vesuvius, Italy: *Geological Society of America Bulletin*, v. 84, p. 759-772.

Lockwood, J. P. and Lipman, P. W., 1980, Recovery of datable charcoal beneath young lavas: lessons from Hawaii: *Bulletin Volcanologique*, v. 43, p. 609-615.

Lydon, P. A., 1968, Geology and lahars of the Tuscan Formation, northern California, p. 441-475 *in* Coats, R. R., Hay, R. L., and Anderson, C. A., editors, *Studies in volcanology: a memoir in honor of Howel Williams*: Geological Society of America Memoir 116.

McCallum, M. E., Woolsey, T. S., and Schumm, S. A., 1975, A fluidization mechanism for subsidence of bedded tuffs in diatremes and related volcanic vents: *Bulletin Volcanologique*, v. 39, p. 512-527.

Macdonald, G. A., 1967, Forms and structures of extrusive basaltic rocks, p. 1-61 *in* Hess, H. H., and Poldervaart, A., editors, *Basalts, the Poldervaart treatise on rocks of basaltic composition*, v. 1: New York, Interscience.

Mahood, G. A., and Drake, R. E., 1982, K-Ar dating young rhyolitic rocks: a case study of the Sierra La Primavera, Jalisco, Mexico: *Geological Society of America Bulletin*, v. 93, p. 1232-1241.

Moore, J. G., 1965, Petrology of deep-sea basalt near Hawaii: *American Journal of Science*, v. 263, p. 40-52.

Moore, J. G., 1967, Base surge in recent volcanic eruptions: *Bulletin Volcanologique*, v. 30, p. 337-363.

Moore, J. G., Phillips, R. L., Grigg, R. W., Peterson, D. W., and Swanson, D. A., 1973, Flow of lava into the sea, 1969-1971, Kilauea Volcano, Hawaii: *Geological Society of America Bulletin*, v. 84, p. 537-546.

Moore, J. G., and Schilling, J.-G., 1973, Vesicles, water, and sulfur in Reykjanes Ridge basalts: *Contributions to Mineralogy and Petrology*, v. 41, p. 105-118.

Moore, J. G., and Sisson, T. W., 1981, Deposits and effects of the May 18 pyroclastic surge, p. 421-438 *in The 1980 eruptions of Mount St. Helens, Washington*: U.S. Geological Survey Professional Paper 1250.

Peterson, D. W., 1979, Significance of the flattening of pumice fragments in ash-flow tuffs, p. 195-204 *in* Chapin, C. E., and Elston, W. E., editors, *Ash-flow tuffs*: Geological Society of America Special Paper 180.

Peterson, D. W., and Tilling, R. I., 1980, Transition of basaltic lava from pahoehoe to aa, Kilauea Volcano, Hawaii: field observations and key factors: *Journal of Volcanology and Geothermal Research*, v. 7, p. 271-293.

Richey, J. E., 1961, *British regional geology: Scotland, the Tertiary volcanic districts*, 3rd edition: Edinburgh, H. M. Stationery Office, 105 p.

Rose, W. I., Jr., 1972, Santiaguito volcanic dome, Guatemala: *Geological Society of America Bulletin*, v. 83, p. 1413-1434.

Rust, G. W., 1937, Preliminary notes on explosive volcanism in southeastern Missouri: *Journal of Geology*, v. 45, p. 48-75.

Schmincke, H.-U., 1967, Stratigraphy and petrology of four upper Yakima Basalt flows in south-central Washington: *Geological Society of America Bulletin*, v. 78, p. 1385-1422.

Self, S., and Sparks, R.S.J., 1978, Characteristics of widespread pyroclastic deposits formed by the interaction of silicic magma and water: *Bulletin Volcanologique*, v. 41, p. 196-212.

Sheridan, M. F., 1970, Fumarolic mounds and ridges of the Bishop Tuff, California: *Geological Society of America Bulletin*, v. 81, p. 851-868.

Sheridan, M. F., 1979, Emplacement of pyroclastic flows, a review, p. 125-136 *in* Chapin, C. E., and Elston, W. E., editors, *Ash-flow tuffs*: Geological Society of America Special Paper 180.

Sheridan, M. F., and Ragan, D. M., 1976, Compaction of ash-flow tuffs, p. 677-713 *in* Chilingarian, G. V., and Wolf, K. H., editors, *Compaction of coarse-grained sediments, II*; Developments in Sedimentology 18B:Amsterdam, Elsevier Scientific Publishing Co.

Silvestri, S. C., 1963, Proposal for a genetic classification of hyaloclastites: *Bulletin Volcanologique*, v. 25, p. 315-322.

Smith, R. L., 1960, *Zones and zonal variations in welded ash flows*: U.S. Geological Survey Professional Paper 354-F, p. 148-159.

Sparks, R.S.J., 1976, Grain size variations in ignimbrites and implications for the transport of pyroclasitc flows: *Sedimentology*, v. 23, p. 147-188.

Swanson, D. A., 1967, Yakima Basalt of the Tieton River area, south-central Washington: *Geological Society of America Bulletin*, v. 78, p. 1077-1110

Swanson, D. A., Wright, T. L., and Helz, R. T., 1975, Linear vent systems and estimated rates of magma production and eruption for the Yakima Basalt of the Columbia Plateau: *American Journal of Science*, v. 275, p. 877-905.

Verwoerd, W. J., 1967, *The carbonatites of South Africa and south west Africa:* Republic of South Africa Geological Survey Handbook 6, 452 p.

Vuagnat, M., 1975, Pillow lava flows: isolated sacks or connected tubes?: *Bulletin Volcanologique*, v. 39, p. 581-589.

Walker, G.P.L., 1959, Geology of the Reydarfjordur area, eastern Iceland: *Quarterly Journal of the Geological Society of London*, v. 114, p. 367-393.

Walker, G.P.L., 1972, Crystal concentration in ignimbrites: *Contributions to Mineralogy and Petrology*, v. 36, p. 135-146.

Walker, G.P.L., 1973, Explosive volcanic eruptions — a new classification scheme: *Geologische Rundschau*, v. 62, p. 431-446.

Walker, G. W., 1970, Some comparisons of basalts of southeast Oregon with those of the Columbia River Group, p. 223-237 *in* Gilmour, E. H., and Stradling, D., editors, *Proceedings of the Second Columbia River Basalt Symposium*: Cheney, WA, East Washington State College Press.

Waters, A. C., 1960, Determining direction of flow in basalts: *American Journal of Science*, v. 258-A (Bradley Volume), p. 350-366.

Wentworth, C. K., and Macdonald, G. A., 1953, *Structures and forms of basaltic rocks in Hawaii*: U.S. Geological Survey Bulletin 994, 98 p.

Westgate, J. A., and Gorton, M. P., 1981, Correlation techniques in tephra studies, p. 73-94 *in* Self, S., and Sparks, R.S.J., *Tephra studies*: Boston, D. Reidel Publishing Co.

Williams, H., 1942, *The geology of Crater Lake National Park, Oregon, with a reconnaissance of the Cascade Range southward to Mount Shasta*: Carnegie Institution of Washington Publication 540, 162 p.

Williams, H., and McBirney, A. R., 1979, *Volcanology*: San Francisco, Freeman, Cooper & Co., 397 p.

Wohletz, K. H. and Sheridan, M. F., 1979, A model of pyroclastic surge, p. 177-194 *in* Chapin, C. E., and Elston, W. E., editors, *Ash-flow tuffs*: Geological Society of America Special Paper 180.

Wohletz, K. H., 1983, Mechanisms of hydrovolcanic pyroclast formation: grain-size, scanning electron microscopy, and experimental studies: *Journal of Volcanology and Geothermal Research*, v. 17, p. 31-63.

Wright, T. L., Grolier, M. J., and Swanson, D. A., 1973, Chemical variation related to the stratigraphy of the Columbia River Basalt: *Geological Society of America Bulletin*, v. 84, p. 371-386.

# 14

# Field Studies of Plutons ■

## 14-1. Rock Units, Ages, and Depth Relations

The term *pluton* is used here in a general sense, implying only a subjacent body large enough to map to scale, typically intrusive but not necessarily so. Plutons may occur singly but typically are in groups, commonly forming linear chains of separate plutons or belts of overlapping plutons. The great batholiths that evidently formed beneath volcanic arcs, such as the Coastal Batholith of the Peruvian Andes, are composed of hundreds of plutons (Pitcher, 1978).

A pluton may consist of one rock unit or more than one, and nearby plutons may consist of the same unit(s) or of different ones. The most crucial step in mapping plutons is coming to know these rock units as exactly as possible. Ideally, one should be able to recognize a specific unit whether it recurs in another pluton tens of kilometers away or as a few inclusions in a dike nearby. Rock units must therefore be based on all primary features possible: (1) proportions among specific varieties of minerals; (2) all aspects of texture (Sections 4-4 and 14-2); and (3) kinds, shapes, and sizes of inclusions, layers, and schlieren (Sections 14-3, 14-4, and 14-5). Associated dikes, pipes, veins, and alterations may be helpful if they reflect the unit's original magmatic constitution; for example, its content of volatile substances or its tendency to segregate late-stage melts (Sections 14-6 and 14-7).

These fundamental mapping units have commonly been called *plutonic units* and are equivalent to formations in sedimentary rocks or to the lithodemes of the stratigraphic code (Section 5-3). Lesser units might be local textural or altered variants that are modifications of a given plutonic unit. A plutonic unit might thus have a *foliated facies* near certain contacts or a local *sericitic zone* due to superimposed alteration.

Plutonic units may be assembled into *plutonic suites* on the basis of features or relations indicating a close genetic relation. Chief among these indicators are geographic association, similar age, and minerals and mineral reactions indicating chemical affinity. A close genetic relation can be checked in the laboratory by determinations of numerical ages, chemical compositions, and isotopic compositions.

Plutonic suites are important because they give a basis for judging the source and history of magmatic sequences. For example, Shaw and Flood (1981) were able to classify a series of plutonic suites into *S-type suites* (thought to be derived mainly by melting of metasedimentary rocks) and *I-type suites* (derived from igneous or metaigneous rocks). This particular distinction was based largely on isotopic compositions of Sr, O, and S, but S-type suites were also indicated by minerals reflecting a high Al content (presence of cordierite, Al-rich garnet, or an $Al_2SiO_5$ polymorph), and I-type suites by minerals indicating relatively low Al content (hornblende with or without augite). Pitcher (1984) has summarized the characteristics and geologic environments of these two suites and two others, and has discussed pertinent literature.

*Contacts between plutons and country rocks* are usually distinct and readily mapped. Sharp contacts may be irregular due to blocky reentrants, cuspate forms, or folds, and these features should be mapped to scale, if possible, or recorded by suitable notes or by some design on the map. Where the contact is a broad gradation, the line is generally placed at the center of the gradation or where the plutonic rock forms a more or less continuous matrix around inclusions of country rock. In some cases two adjoining zones of mixed rocks can be mapped (Fig. 14-1). Pluton margins are likely to be broadly schistose or mylonitic in plutons emplaced diapirically when the body was more than 70% crystallized, and the contact may lie within this broad zone of ductile shear (Soula, 1982).

*Dikes near the contact* generally are of great value in interpreting a pluton's evolution. They may belong to one of these age groups: (1) dikes associated with the pluton but cut by it (Fig. 14-2A); these dikes may record the composition of the first magma emplaced at the observed level; (2) apophyses connected to the pluton, which, if porphyroaphanitic, will indicate the proportion of crystals to melt at this margin; (3) dikes emplaced when the pluton was still mobile and probably contained some melt (Fig. 14-2B); (4) dikes without chilled margins, emplaced in the pluton when it was still hot; and (5) dikes with chilled margins, emplaced in the pluton after it had cooled.

*Contacts between rock units within plutons* may be obscure where two rocks are almost identical or where they grade to one another. Such contacts may be marked by: (1) small differences in color and texture; (2) inclusions, schlieren, or layers in the younger rock, commonly forming a zone parallel

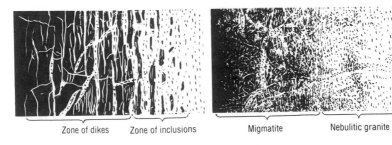

| Zone of dikes | Zone of inclusions | Migmatite | Nebulitic granite |

**Fig. 14-1.** Diked (*left*) and migmatitic gradational margins of plutons.

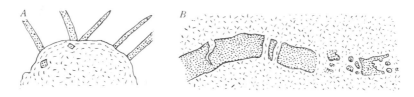

**Fig. 14-2.** *A.* Pluton emplaced into a radial array of somewhat older dikes. *B.* Dike broken, intruded, and partly granitized within the rock it intruded.

to the contact; (3) features in the older rock (inclusions, fabric, schlieren, layers, dikes) cut off by the younger rock; and (4) mild to moderate deformation of the grains in the older rock, which generally makes the rock darker than usual. The latter criterion must be used with caution, because magmas that are largely crystallized may develop a dark schistose or mylonitic contact facies due to ductile shear against an older rock (V. R. Todd, personal communication, 1984).

**Relative ages** can be determined from most of the features just listed. Where none of these features can be found, the younger unit commonly shows a broad gradation away from the contact, generally involving one or more of color, grain size, mineral content, numbers of inclusions, and abundance of schlieren (Moore, 1963, p. 43). Emplacement of the younger unit will typically remove the original marginal facies of the older, which will therefore be more uniform than the younger. Age relations are more difficult to determine where a septum of country rock lies between two intrusions; however, dikes of the younger may intrude the older, or dikes associated with the older may be cut off, deformed, or metamorphosed by the younger (Fig. 14-3).

Relative ages of emplacement are not necessarily resolved by cross-cutting relations, especially in areas where the country rocks are broadly metamorphosed. In a case described by Soula (1982), magmatic diapirs cut-

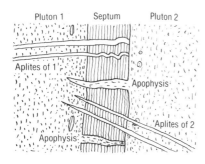

**Fig. 14-3.** Dikes indicating age relation between two plutons with a septum (intervening sheet of country rock).

ting upward into nonmagmatic gneiss domes have the same emplacement age as the domes but are discordant because they were less viscous than the domes. Viscosity contrasts may affect intrusive relations in other situations. Viscosity of magma can be predicted to decrease with decreasing $SiO_2$ content, with increasing content of water and halogens, and with decreasing proportions of suspended crystals and inclusions.

**Dating plutons** is done most conveniently and least expensively by K-Ar methods (Dalrymple and Lanphere, 1969). The data obtained give cooling rather than emplacement ages, however, which may be a major problem in areas heated broadly at a later time, or in cases where a pluton is heated by a younger intrusion. Even for a single pluton, one or two K-Ar dates may suggest an erroneously simple history. U-Pb dating of zircons from the Tatoosh pluton in Washington, for example, indicated an emplacement history lasting for approximately 12 m.y. (Mattinson, 1977).

U-Pb dating of zircons, however, requires large to very large whole-rock samples and even they may not yield zircons that will give a useful age. The Rb-Sr method will give a dependable emplacement age as long as the rocks used are fresh, have all developed from one starting magma, and represent a large range of rubidium concentrations, which generally increase between the initial melt and the late residual melts of a pluton. Late potassium-rich dike rocks, such as aplites, are typically rubidium-rich; however, they need not necessarily have formed from the pluton in which they occur.

**Mapping the country rocks** for considerable distances around a pluton is likely to be at least as informative as studying the pluton itself. This mapping may provide the only firm evidence of mechanisms of emplacement (Fig. 14-4) (Nelson and Sylvester, 1971; Pitcher and Berger, 1972). Studies of contact metamorphism will always be of unique value (Chapter 15). The stratigraphy and detrital content of sedimentary and volcanic rocks deposited during and after a pluton's emplacement may provide the only clear evidence of uplift or subsidence of the rocks over the pluton, of connected volcanic activity, and of the date at which the pluton was unroofed by erosion (Fiske and others, 1963, pp. 59, 63).

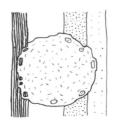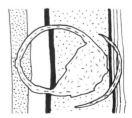

**Fig. 14-4.** Diagrammatic mapped relations of plutons and country rocks, indicating (from *left* to *right*) diapirism, piecemeal stoping, and cauldron (block) subsidence.

**Depth of emplacement** may be indicated approximately by stratigraphic relations and by metamorphic minerals and zones (Section 15-3). Additional kinds of evidence for a shallow level of intrusion are: (1) porphyroaphanitic marginal facies; (2) porphyroaphanitic dikes and sills associated with the pluton; (3) inclusions of porphyroaphanitic rocks; (4) breccia dikes or pipes containing porphyroaphanitic rocks; (5) nearby volcanic rocks of the same composition as the pluton; (6) miarolytic cavities; (7) granophyric rocks or patches of granophyre between larger grains; (8) breccia with open cavities (Tabor, 1963); (9) quartz veins, sulfides, and hydrothermal alteration widespread in the pluton or in the surrounding rocks (Section 14-8); and (10) evidence of rapid heating and cooling, such as patchy contact metamorphism and fine-grained alteration products (Pitcher, 1978, p. 164). Fiske and others (1963, p. 39-63) described most of these features in their study of the Tatoosh pluton.

Emplacement at unusually great depth is suggested by the following relations and features: (1) occurrence in extensive terrains of high-grade metamorphic rocks; (2) lack of a distinct contact metamorphic aureole, although migmatites may be more abundant near the pluton; (3) plutons more or less concordant and commonly phacolithic or broadly folded; (4) textures commonly allotriomorphic granular; (5) rocks commonly gneissose; (6) internal foliations and lineations parallel with those of surrounding country rocks, and (7) quartzofeldspathic rocks with moderately to distinctly dark feldspar and anhydrous mafic minerals, especially pyroxene and garnet. Many of these features, however, may be formed at moderate depths when country-rock temperatures are unusually high and plutons are synorogenic.

Plutons emplaced at intermediate levels tend to lack the two groups of features just noted, and otherwise have no unique characters.

## 14-2. Fabrics of Plutonic Rocks

Preferred orientation of mineral grains, inclusions, and miarolytic cavities can be used to judge flow directions in many plutons. Fabrics may be foliate, lineate, or both (Fig. 14-5). They are measured and plotted as described in Sections 3-5, 3-6, and 5-4. Lineations typically are parallel to ("lie in") the plane of foliation, but they may be oblique to it and may have more than one preferred orientation. Degrees of preferred orientation (fabric strength) have not been standardized; however, strongly developed fabrics are easily visible in hand specimens, moderately developed ones are clear only on outcrop surfaces of 1 sq m or more, and weakly developed ones require study of much larger outcrops. Oriented inclusions are often the most obvious indicators of otherwise weak fabrics; however, inclusions are not always parallel to grain fabrics. Where fabrics are obscure they can be measured on oriented samples sawed and etched with HF, which accentuates the feldspar forms (Duffield, 1968, p. 1363).

Mineral fabrics can form at any stage of crystallization of a magma. At an early stage, crystals are so few that they can rotate freely in the melt, thus producing a fabric that is sometimes preserved by oriented phenocrysts in the finer groundmass of chilled apophyses or marginal facies (Drever and Johnston, 1967; Compton, 1960, p. 1398). In nonporphyritic rocks, fabrics formed at an early stage are characterized by unstrained crystals and by orientation of only those grains that were distinctly shaped at the time: elongate prisms of amphibole and tablets or flakes of biotite, plagioclase (especially in gabbros), and potassium feldspar (especially in syenites). Also oriented but not nearly so obviously are stubby prisms (as pyroxenes) and somewhat elongate plates (as both feldspars in quartzofeldspathic rocks, and olivine). Quartz grains are equidimensional and thus not oriented. Feldspar orientations are often the main component of the fabric and can be recognized in hand specimens or large broken surfaces by numerous reflections of light from (010) cleavages. Such surfaces are common because the aligned cleavages determine the *freeway* — the direction in which the rock splits most readily.

Early-stage fabrics may be restricted to pluton margins, perhaps due to a steep gradient of simple shear there, or to flattening of the marginal facies by diapiric expansion of the pluton's core. Analyses and experiments by Willis (1977) indicate that inclusions or crystals free to rotate in a medium undergoing simple shear should develop degrees of preferred orientation dependent mainly on the ratio of their greatest to their least dimensions (Fig. 14-6A). Early-stage fabrics may also form by accumulation of platy or lineate grains on a substrate, and the grains may be aligned linearly if deposited from currents moving against the substrate (Morse, 1969, p. 63). Morse also noted accentuation of these fabrics near inclusions. Elongate

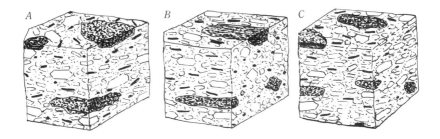

**Fig. 14-5.** Strongly developed fabrics in seriate rocks with the same mineral composition. *A.* Fabric planar with respect to all elongate and tabular components. *B.* Fabric linear with respect to all elongate components. *C.* Fabric linear with respect to elongate components (here, hornblende prisms and some inclusions) and planar with respect to tabular components (as tabular feldspar).

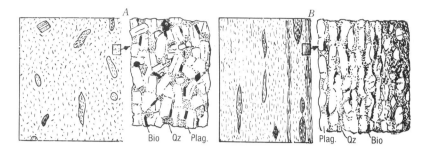

**Fig. 14-6.** *A.* Fabric typical of stage when inclusions (*left*) and crystals (*right*) can rotate more or less freely. Note that fabrics of some inclusions are not parallel to pluton fabrics. *B.* Strong fabric typical of late magmatic stage, with schistose zones parallel to inclusions (*left*) and sample showing gradation from moderately to highly deformed (mylonitic) rock. Mylonitic fabrics are shown at hand-lens magnification in Fig. 4-19*G* and *H.*

minerals oriented perpendicular to an interface suggest crystallization in exceptionally fluid magma (see *comb layers*, Section 14-4).

At a late magmatic stage, crystals typically bear on one another and flow of the magma will result in deformation of mineral grains. Ductile minerals such as quartz and olivine become flattened or elongated, often recrystallizing into fine-grained aggregates. Relatively strong minerals, such as feldspars, hornblende, and pyroxene, may rotate against ductile grains or may bend, break, or be pulled apart (Fig. 4-15*A*). Biotite and orthopyroxene may be kinked and recrystallized. Crucial for recognizing late-magmatic as opposed to postmagmatic flow are fillings of late-stage magmatic minerals between pulled-apart fragments of grains or in fractures opened during extension (Fig. 4-15*B*). Where deformation is unusually strong, the texture may become mylonitic (Fig. 14-6*B*). Late-stage fabrics tend to pass with little if any deflection into inclusions and wall rocks that were as ductile as the magma at that stage (Fig. 14-7).

Fabrics formed when the magma is entirely crystallized but still hot are similar to those just described except that the fabric will also pass through

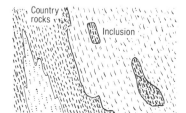

**Fig. 14-7.** Pluton fabric passing into inclusions and country rocks without deflection.

late-stage dikes, veins, and segregations. It will invariably pass through inclusions and pluton contacts and, when mapped, may appear as an overprint of parallel planes or lines over the entire pluton and its surroundings. Such fabrics can be distinguished from later regional fabrics by being restricted to the pluton and its aureole.

It must be emphasized that one pluton may have fabrics formed during all of these stages, so that criteria must be sought repeatedly and consistently from one place to the next. Age relations with respect to dikes emplaced at various times are especially valuable. Pitcher and Berger (1972) described studies of granitic rocks having late-stage to postmagmatic fabrics.

### 14-3. Inclusions in Plutons

The term *inclusion* is used here for any kind of rock body enclosed in plutonic rock, regardless of the body's origin. Most inclusions have the aspect of fragments or lumps of rock, but they may grade to less distinct clots or streaks called schlieren (Section 14-5). Typical sizes or abundances of inclusions may characterize certain plutonic units (Moore, 1963, p. 118). Specific kinds of inclusions of preexisting rocks (*xenoliths*) often provide clues to an intrusion's history; for example, metasedimentary xenoliths identifiable with formations in the surrounding country rocks can be used to measure net displacements since they were incorporated.

Igneous and meta-igneous xenoliths can sometimes be matched to older plutons, dikes, or volcanic sequences. Others are *cognate xenoliths*, which are composed of rocks formed during the intrusive history of the pluton that contains them. Cognate xenoliths can sometimes be correlated with earlier marginal or dike facies, and these correlations generally require study of the entire pluton and its surroundings (Fiske and others, 1963). Cognate xenoliths may also form if magma related to a pluton intrudes the pluton when the latter is viscous enough to fracture yet is hot or fluid enough to modify and incorporate the new magma (Fig. 14-2*B*). Such mixtures sometimes result in inclusion swarms that can be walked to places where a dike is still more or less intact. Cognate inclusion swarms may also form by accumulation of inclusions at certain levels or against solidified faces within a pluton.

Dark, igneous-appearing inclusions that cannot be matched to marginal or internal sources may be so widespread and evenly distributed as to suggest that they came up with the pluton magma from deeper sources (Moore, 1963, p. 120). Such inclusions may be *restites*, bodies residual to the partial melting of rocks at depth. They might also be cognate xenoliths of mafic igneous rocks emplaced at depth ahead of the ascending pluton. The term *autolith* (an inclusion formed from the magma that contained it) has also been applied to dark, granular inclusions in the belief that they have somehow crystallized as mafic segregations. A case that seems reasonably valid is that of *orbicules*, which are spherical or ellipsoidal bodies built up of one or more shells that commonly have a core or a radial fabric, suggesting growth outward from a core.

Finally, *skialiths* are inclusions remaining after the surrounding rock has been converted metasomatically to plutonic rock. They may have either sharp or gradational margins, and are suggested where country rock units can be projected through the plutonic body (Fig. 15-3*B*) (however, see Pitcher, 1970).

Inclusions are typically altered during and after their incorporation, in many cases greatly so. Quartzofeldspathic xenoliths in mafic and intermediate rocks may be partially melted and thereby associated with patches or schlieren of late-stage granophyre (Wager and Brown, 1967, p. 124). Silicic and intermediate magmas commonly react with rocks included in them, and Fig. 14-8 illustrates some common protoliths and reaction products. Thoroughly altered inclusions imply a long period of immersion or reaction with volatile-rich facies of the magma. Contact zones sometimes show all stages in the alteration of included rocks.

Characteristics of inclusions that are useful in systematic studies are: kinds of rock and their abundances; typical size and size-range; kinds and degrees of alteration; sharpness or fuzziness of borders; angularity or roundness, and degree to which borders are smooth or irregular. Additional characteristics that can be used in conjunction with host-rock fabric to

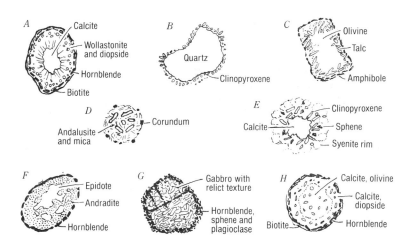

**Fig. 14-8.** Mineral and texture zones in and around inclusions that have reacted with silicic and intermediate magmas. All are about fist size. *A.* Limestone in granodiorite. *B.* Quartzite in granodiorite. *C.* Peridotite in granodiorite. *D.* Claystone in granite. *E.* Limestone in potassic granite. *F.* Epidote nodule (from metabasaltic or meta-andesitic tuff-breccia) in granodiorite. *G.* Gabbro in granodiorite. *H.* Dolomite in granodiorite.

determine the degree of late-stage deformation are: (1) degree of flattening or elongation per kind of rock, which should vary with ductility; (2) orientation of inclusion fabric and its relation to inclusion form; and (3) degree of preferred orientation of inclusions. Pitcher and Berger (1972) described a study that utilized data from inclusions, and Didier (1973) reviewed the nature of inclusions and their relations in a number of plutons.

### 14-4. Layering (Banding) in Plutons

Compositional and textural layers (bands) in silicic plutons have generally been called schlieren, whereas all planar features in mafic plutons, no matter how streaked or indistinct, have commonly been called layering (or banding). A consistent geometrical nomenclature should probably be used in the field, irrespective of how or where the features may have formed, and it is recommended that the term *layers* (or *bands*) be used for distinct bodies that are typically tabular or lenticular, but may have other shapes (Fig. 9-1), and that the term *schlieren* be used for discontinuous, streaky, or vague bodies (Section 14-5).

A terminology for layered plutonic rocks was proposed by Jackson (1967, p. 22) for supposedly cumulate rocks; however, it is essentially descriptive and thus can be used for layered rocks that are not necessarily cumulates. A *layer* is a sheetlike body of more or less uniform character, distinguishable by its proportions of minerals or by mineral or grain-size proportions that change gradually (as in graded layers). *Laminae* are the thinnest recognizable layers (generally less than 2 cm thick). A *horizon* (a single surface) is a *phase contact* when set at an abrupt appearance or disappearance of a (cumulus) mineral; it is a *ratio contact* when marked by the proportion of two (cumulus) minerals, or a *form contact* when marked by a sharp change in typical size or shape of a (cumulus) mineral. Layers forming cyclic sets, such as the harzburgite-orthopyroxenite-wehrlite sequence described by Irvine and Smith (1967), may be particularly informative.

The most-studied layers are those with sedimentary aspects such as cross-bedding, grading by grain size or mineral proportions, cut-and-fill structures, channels filled with lenticular layers, angular unconformities, and a variety of soft-sediment folds, flame structures (Parsons, 1979), faults, and slump-breccias, the latter reworked into size-graded conglomerate layers in some intrusions (Irvine, 1974, p. 63). These structures are not illustrated here because of their similarity to those shown in Chapters 9 and 10; many photographs of them have been published (Wager and Brown, 1967; Morse, 1969, Irvine, 1974; McBirney and Noyes, 1979).

Although many layer structures suggest sinking, transport, and accumulation of crystals in melts, they have also been interpreted in other ways. McBirney and Noyes (1979) suggested that the Skaergaard intrusion became layered due to gravitational segregation of melt combined with oscillatory crystal nucleation and growth. Komar (1972) described grain-sorting and concentration caused by mechanical interactions among grains being carried in melt, or by forces resulting from a rapid lateral variation in the rate of simple shear, as near a contact. Jahns and Tuttle (1963) interpreted layered pegmatite-aplite bodies as due to multiple intrusion followed by segregation during crystallization (Section 14-6). Nonuniform magma produced by local contamination or by partial mixing of two magmas may become swirled and drawn out into layered patterns during subsequent flow. Thayer (1963) described flow layering interpreted to have formed during emplacement of largely crystallized gabbro-peridotite bodies, the evidence being: (1) disrupted chromitite layers thought to be of cumulate origin; (2) lenticularity of the gabbro and peridotite layers; (3) interlayering, in any order, of widely different monomineralic and polymineralic rocks; (4) xenomorphic (allotriomorphic) textures; (5) foliation and lineation discordant to layers; (6) thickening of layers at fold hinges, and (7) widespread undeformed dikes of gabbro, including pegmatitic gabbro.

*Comb layers* consist of elongate crystals of plagioclase, pyroxene, or hornblende that appear to have grown outward into a body of magma, either from a pluton wall or from an inclusion (Fig. 14-9A). These structures may be of great value in interpreting processes in plutons, for Moore and Lockwood (1973) found that the crystals tend to branch inward and to bend upward, as though affected by fluids flowing upward along the contact (Fig. 14-9B). They interpreted the fluids to be volatiles that were channeled along inverted troughs in overhanging walls of plutons, and thus comb layers may help in identifying volatile-rich parts of plutons. Distinct comb layering has been described only from intermediate to moderately mafic plutons. Some mafic gabbroic intrusions, however, have thick layers or bodies of border-zone rocks with elongate plagioclase or olivine crystals oriented approximately perpendicular to contacts (Wager and Brown, 1967).

Layering may also form when plutons that are largely crystallized are

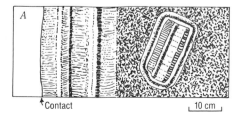
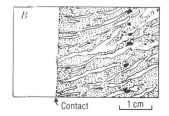

**Fig. 14-9.** *A.* Comb layers at contact of a pluton and forming an inclusion (orbicule) of comb-layered rock. *B.* Part of a single comb layer, showing elongate plagioclase crystals branching and curving upward and away from the contact.

injected by another magma that enters, sill-like, along foliation planes, or along parallel extension fractures. The resulting layering can be identified where the sills break across septa of the older rock or where the younger rock includes fragments of the older. In addition, the younger rock will commonly be finer grained than the older and often will lack the fabric of the older rock. Pitcher and Berger (1972, p. 215) described banding (layering) of this kind associated with extensive planar swarms of xenoliths in the main granitic pluton of Donegal, Ireland.

Late-stage layering may also form where residual melt is segregated into subparallel fractures, or where transient components, as $K^+$ ion, are carried by fluids to fractures, to foliation partings, or to ductile faults. This kind of layering may be recognized where fractures or faults locally cross, or by the distribution of specific minerals, such as potassium feldspar (Pitcher and Berger, 1972, p. 215). Segregated layers may also show locally diffuse contacts, partially feldspathized inclusions, or other early-stage relics. Finally, when all the melt has crystallized but the rock is still hot, it is susceptible to all the processes of differentiation that produce layering in metamorphic rocks, as described in Section 15-5.

Because all late-stage processes may be guided by earlier mineral fabrics or layer structures, several kinds of layers may be parallel, and evidence of their ages must be sorted out with care.

### 14-5. Schlieren and Related Structures

Schlieren (literally, *streaks*) are compositional or textural domains that are typically lenticular, planar, or elongate but may be more or less equidimensional. They differ from inclusions in being less distinctly circumscribed, especially at their ends, and from compositional layers in being less tabular, and less likely to compose sets. We know little for certain about the origin of schlieren, but they are useful physical characteristics of certain plutons or pluton facies, and can be classified in a descriptive way. Some schlieren grade into layered rocks or inclusion swarms, and in the latter case appear to have formed by partial assimilation of the inclusions (Fig. 14-10*A*). In other cases, schlieren are deflected around inclusions as sinuous flow streaks (Fig. 14-10*B*).

Schlieren may be classified into those darker and those lighter than their host rock. The two may occur separately or together. Light schlieren, which are far less obvious than dark ones, are commonly aplitic (Didier, 1973), but they may be as coarse or coarser than their host, in some cases being pegmatitic. Light schlieren that form sheets and patches in miarolytic facies may be late-stage segregations, akin to pegmatite and aplite. Dark schlieren in granitic rocks are commonly thin sheets and wisps that cut across one another, in some cases so systematically as to suggest regular late-stage displacements or cross-bedding (Fig. 14-10*C*). Others, however, have bizarre shapes, appearing like parachutes, medusas, or segmented funnels (Fig. 14-10D). Coarsely dioritic (pegmatitic) or granophyric schlieren in the border zones or upper parts of gabbro intrusions may result from partial assimilation of quartzofeldspathic or perhaps micaceous xenoliths. Widespread streakiness and lenticular layers in otherwise layered gabbros suggest late-stage flow or solid-state deformation (Davies, 1971, p. 19).

Didier (1973) pointed out that schlieren occur throughout migmatitic plutons, whereas they are usually restricted to the marginal zones of uniform plutons. Schlieren zones deep within uniform plutons may thus mark internal intrusive contacts between succeeding plutonic units. Regardless of their origin, schlieren are valuable indicators of late deformation; for example, by showing folds or faults. It is therefore desirable to fix schlieren in time sequence as much as possible. Their relations to inclusions and to the various kinds of fabrics described in Section 14-2 may be helpful in doing this.

### 14-6. Pegmatite and Other Volatile-related Rocks

Concentration and action of volatiles in plutons are suggested by pegmatite, aplite, miarolytic rock, orbicular rock, plumose structure, intruded breccia, and lamprophyre. Additional effects of hydrothermal fluids are described in Section 14-8. A principal value in mapping these variants is that of discovering which parts of a pluton were relatively rich in volatiles. Mapping should also include a large area around the pluton in order to determine whether the volatile-rich materials are indigenous to the pluton

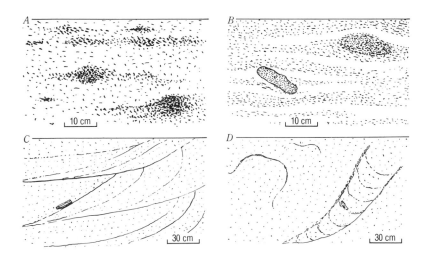

**Fig. 14-10.** Schlieren in plutons. See text for descriptions.

or whether they came from elsewhere, perhaps from another pluton. The suggestions for mapping zones of hydrothermal alteration (Section 15-7) and migmatites (Section 15-6) may be helpful.

**Pegmatite** is generally the most obvious of the volatile-related rocks. The influence of volatiles in its crystallization is indicated by large grain-size, euhedral feldspar, pronounced mineral segregation, alterations of early-formed minerals to minerals containing volatile components, dissolution of minerals, and filling of vugs by euhedral late-stage minerals. Pegmatite in gabbro typically forms dikes, lenses, or schlieren of coarse plagioclase and pyroxene, of plagioclase and hornblende, or of plagioclase, pyroxene, and fayalitic olivine, all with fine-grained quartz or granophyre. These rocks typically occur in the upper halves of differentiated sheets, or in their marginal facies, in the latter case commonly associated with partially assimilated quartzofeldspathic inclusions. Pegmatite bodies in peridotites described by Moores (1969) have margins of coarse pyroxene, with plagioclase locally in their cores. Pegmatites of intermediate and silicic plutons

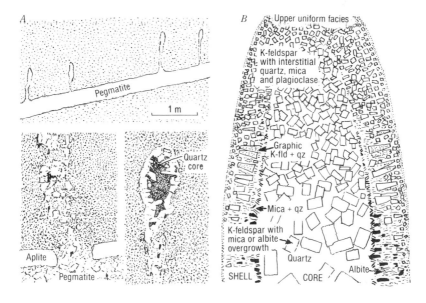

**Fig. 14-11.** *A.* Vertical face exposing gently inclining pegmatite-aplite dike with vertical spiracles that cut across the dike's aplitic margin (see detail, *lower left*) and are headed by zoned pods of pegmatite (detail, *lower right*). From photographs by R. A. Osiecki (and see Osiecki, 1981). *B.* Diagrammatic vertical section through granite pegmatite dike with zoned lower facies. The aligned white crystals of the shell are K-feldspar euhedra, graphically intergrown with quartz where they are stippled. The black elongate crystals in the lower shell are beryl, topaz, or minerals containing Li or P. After Uebel (1977, p. 92). © E. Schweizerbart'sche Verlagsbuchhandlung, copied with permission.

typically form tabular dikes or lens-shaped pods. Other structural varieties range in size and complexity from single large potassium-feldspar crystals laced with graphically intergrown quartz (Osiecki, 1981) to compositionally zoned dikes and pods many meters across. In exceptional cases, structural relations of zoned pods may indicate intrusion of pegmatite magma (Fig. 14-11*A*).

Mineral zones in most of the zoned granitic pegmatites described by Cameron and others (1949) are, from outside inward: (1) plagioclase-quartz-muscovite; (2) plagioclase-quartz; (3) quartz-perthite-plagioclase, with or without muscovite and (or) biotite; (4) perthite-quartz; and (5) quartz. Crystals tend to be oriented perpendicular to contacts. Grain size and degree of segregation tend to increase inward. The zones in some pegmatites appear symmetrical in most horizontal sections. Many are asymmetric in vertical sections, in some cases strongly so (Fig. 14-11*B*). Sodium, chiefly in albite, tends to be concentrated in the lower parts of pegmatites, whereas potassium, chiefly in potassium feldspar, tends to be concentrated in the upper parts and may form micas from aluminosilicates in the surrounding rocks. Vugs dissolved by volatiles cut across earlier-stage minerals or zone boundaries and have radial fabrics of volatile-deposited minerals.

**Aplite** is common in silicic and intermediate plutons and is occasionally associated with gabbro. It commonly grades texturally to fine pegmatite and may be porphyritic, carrying potassium-feldspar phenocrysts in a groundmass that consists mainly of albite and quartz. Some aplite forms separate bodies, but most occurs with pegmatite in composite dikes or pods. Aplite is usually marginal or central to pegmatite, and textural relations at the contacts may suggest continuity of crystallization (Fig. 14-12*A*). Layered bodies may form by multiple injection of sheets that are differentiated into albite-quartz aplite below and potassium-feldspar-quartz pegmatite above (Fig. 14-12*B*). Compositional layering in the aplitic parts of large pegmatite-aplite bodies is typically thin (1–10 mm) and marked by concentration of garnet, tourmaline, or muscovite (Jahns and Tuttle, 1963). Aplites in composite bodies may also show faint schlieren that are locally contorted near projections of wall rocks or large crystals and are probably primary flow structures (Jahns and Tuttle, 1963, p. 82).

According to the model of Jahns and Burnham (1969), aplite is typically produced by pressure-quenching of a residual melt, caused either by sudden opening of fractures or by rapid ascent of largely crystallized granitic magma. A volatile-rich phase released during the quenching then crystallizes to pegmatite, or the volatiles may form pegmatite where they stream along fractures. The total amount of aplite and pegmatite thus gives a measure of the amount of residual melt extracted from the pluton, and the ratio of aplite to pegmatite indicates the concentration of volatiles in the separated residual melt.

***Miarolytic structures*** are most common in silicic plutons. They may be vesicles formed by volatiles exsolved from melt or may be dissolution cavities formed by volatiles. Vesicles tend to be pressed around already crystallized grains, whereas grains are cut off at the margins of dissolution cavities. Euhedral grains may line or fill both kinds of cavities, and they are presumably deposited from the volatiles. The filling minerals are typically quartz, alkali feldspars, carbonates, epidote, or chlorite. Unfilled miarolytic cavities are not particularly obvious in hand specimens unless late-stage alteration forms colored halos around them.

***Orbicular rock*** and the comb-layered marginal facies often associated with it are suggestive of transport in a volatile medium, as already mentioned (Section 14-4).

***Plumose structure*** consists of groups of elongate crystals which radiate from centers that may be closely or widely spaced. The crystals are commonly micas or tourmaline in silicic rocks and epidote, chlorite, or amphibole in intermediate to mafic rocks. Because the structure is most common in pegmatite, it probably results generally from the presence of volatiles.

***Intrusions of breccia*** or conglomerate (pebble dikes) similarly suggest deposition from a fluid phase, for the fragments may be moved large distances from their sources, may be rounded, or may be faceted and pitted as though sand-blasted. Although these rocks are suggestive of explosive activity associated with volcanoes (Section 13-7) or with the uppermost parts of plutons (Section 14-8), Wright and Bowes (1968) and Pitcher and

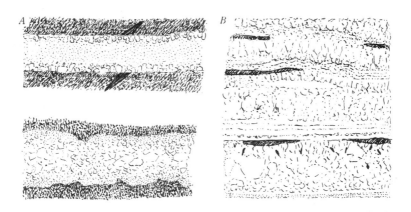

**Fig. 14-12.** *A.* Two aplite-pegmatite dikes, the marginal pegmatite of the upper one consisting largely of elongate K-feldspar crystals that suggest inward growth, and the lower with an irregular aplite margin (stippled) that grades inward to coarse pegmatite. *B.* Four parallel dikes of pegmatite and layered aplite (stippled) composing a multiple intrusion. The base of each successive dike is indicated by partial screens of country rock (dark).

Berger (1972, p. 156) documented and discussed cases that appear plutonic.

**Lamprophyre** is usually emplaced during or just after the late magmatic stage of a pluton and is thought to be formed when basaltic or basanitic magma mixes with the pluton's residual melts or volatiles (Rock, 1977). A point in evidence is the close age relation between lamprophyre and aplite or pegmatite (Fig. 14-13). Indications of volatiles in lamprophyres include: high degree of alteration; euhedral forms of most minerals; and, in alkaline varieties, nodules (*ocelli*) of feldspathoids or carbonates.

In addition, lamprophyre often occurs with small stocks and thick dikes of rocks characterized by long prisms of early, euhedral hornblende. These gabbroic and dioritic rocks are commonly associated further with granular diorite and may grade to hornblendite and hornblende-pyroxene-olivine rocks. Many of the dark rocks are cut by networks of feldspathic veins that intersect at pegmatitic nodes. Abundant volatiles are also suggested by biotitic facies and by hydrothermal alteration, especially to chlorite. This association of dark hornblendic rocks is called the *appinite suite* in Britain, and has been described and reviewed by Pitcher and Berger (1972, p. 143-168).

### 14-7. Fracture Systems in Plutons

Fractures form at a late magmatic or postmagmatic stage in most plutons and are commonly filled by fine-grained phaneritic rocks similar in composition to those of the pluton, or by one or more of the volatile-related rocks described in Section 14-6. A close age-relation to the pluton is indicated where these minor bodies: (1) form hybrids with the pluton; (2) occur as fragments in the pluton (Fig. 14-2B); (3) are locally intruded by the pluton; (4) are offset along faults that are healed within the pluton (Fig. 14-14A); (5) have a fabric continuous with that of the pluton (Fig. 14-14B); or (6) are overgrown by megacrysts that are also in the pluton (Fig. 14-14C). Late

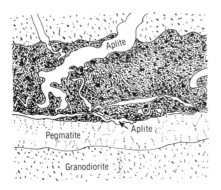

**Fig. 14-13.** Lamprophyre dike (dark) in granodiorite, with a pegmatite dike along the lower contact. Ptygmatic and pinch-and-swell shapes of aplite dikes suggest emplacement when the lamprophyre was still plastic.

magmatic fractures that are not filled with magma can be recognized by coatings or drusy fillings of late magmatic to postmagmatic minerals: quartz, potassium feldspar, albite, micas, epidote, actinolite, chlorite, magnetite, hematite, sulfides, carbonates, and so on (Section 14-8). These mineralized fractures may cut late dikes as well as the pluton, thus affording evidence of the late magmatic age of the dikes (Neff, 1973).

Fractures can often be classified genetically where they cut small-scale structures of a pluton. Most fractures are due to simple extension, as expressed by drusy fillings or by dilation of features cut by dikes (Fig. 14-14*D*). Faults and zones of simple shear are indicated by displaced crystals, inclusions, layers, dikes, schlieren, and so on, and locally by thin sheets of schistose or mylonitic rock that may include folds or other lineations suggesting a direction of displacement (Fig. 12-5). Dikes offset by faults soon after emplacement may show distinctly plastic strains (Fig. 14-14*E*), or may have sigmoidal fabrics in cross sections aligned parallel to the direction of simple shear (Fig. 14-14*F*).

The different kinds of fractures and fracture fillings should be mapped with different symbols as much as practicable; for example, by a small letter placed by a strike-and-dip symbol, or by a color. Except for large or otherwise unusual dikes, each symbol generally is used for localities where several dikes or fractures are parallel or nearly so.

Maps of fracture systems and rock fabric may be interpreted in light of the mapped shape of the pluton or one of its local phases. Steeply inclined

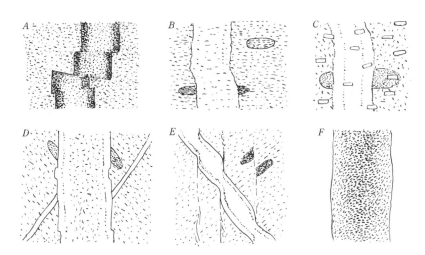

**Fig. 14-14.** Geometric relations between dikes and the plutons they intruded. See text for interpretations. *F* is after Moore (1963, p. 126) and is of a dike approximately 1 in. across.

dike swarms in elongate groups of plutons commonly strike parallel to the group (Pitcher, 1978, p. 171). Steeply inclined dilative fractures perpendicular to foliation are often radial in more or less circular plutons and suggest upward bulging, perhaps due to magma intruded at a late stage (Fig. 14-15*A*). Dilative fractures roughly perpendicular to linear fabrics suggest extensional strain in the solid or near-solid state. For example, radial fractures perpendicular to horizontal lineations in the outer parts of a pluton suggest diapiric extension when the marginal facies was nearly solid (Fig. 14-15*B*). The sense of displacement on faults may correlate with one of these patterns (Fig. 14-15*A*) or may be younger and express some other strain system. Balk (1937, p. 104) described systematic marginal reverse faults ("upthrusts") that indicate outward bulging of a pluton, and Bateman and others (1963, p. 25) used the gash-fracture shapes of dikes at pluton contacts to interpret the direction of movement along the contact (Fig. 14-15*C*). In an unusually complete analysis, upward dilation of the near-pluton edges of sills intruded from an early core of the Tatoosh pluton could be related to doming of the country rocks (Fiske and others, 1963).

*Fractures in volatile-rich cupolas.* Complete three-dimensional data on fracturing in the uppermost parts of volatile-rich stocks have come from detailed surface and underground studies of copper-bearing and molybdenum-bearing plutons, reviewed by Beane and Titley (1981) and White and others (1981). Myriads of quartz-sulfide veins in these bodies suggest hydrofracturing due to expansion of volatiles exsolved from melt at a late stage of crystallization (Burnham, 1979). These fractures dip steeply in a variety of sets, indicating outward expansion of the pluton under a vapor pressure close to load pressure. In regions broadly under stress, the frac-

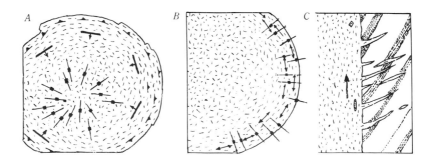

**Fig. 14-15.** *A.* Simplified map showing radial fractures, foliate fabric, and reverse faults suggesting upward bulging of an underlying core. *B.* Map of pluton with lineated marginal zone broken along fractures suggesting extension parallel to the margin. *C.* Vertical cross section through pluton margin and country rocks that have steeply inclined bedding. Orientation of the dikes and their systemmatic dilation indicate upward movement of the magma relative to the walls.

tures may have a preferred strike direction (Rehrig and Heidrick, 1972). The characteristics of the early veins are: (1) small size (generally less than 1 cm thick and 1 m or so long); (2) irregular shape and generally nonparallel walls; (3) laterally discontinuous or segmented forms; (4) lateral gradation to altered mineral grains; (5) lack of internal symmetry of mineral fillings; (6) lack of distinct external halos; and (7) association with high-temperature alteration minerals (typically potassium feldspar or biotite).

The same plutons tend to be broken by a somewhat younger set of near-horizontal fractures that carry simple tabular veins as thick as a meter or so, a relation indicating extension parallel to the vertical axis of the stock. Other late-formed sets may be radial with respect to some point within a pluton. At El Salvador, Chile, fractures of a radial array were intruded by latite porphyry, and pebble dikes formed when the latite magma intruded rock saturated with groundwater at depths of a few kilometers (Gustafson and Hunt, 1975, p. 875).

### 14-8. Autometamorphism of Plutons

Most plutons are altered to some degree by magmatic volatiles during late stages of crystallization. Many are also altered by geothermally circulated groundwater. Detailed studies of copper-bearing and molybdenum-bearing shallow plutons have provided a general model for these auto-metamorphic processes (Gustafson and Hunt, 1975; Burnham, 1979; White and others, 1981). Expansion of magmatic volatiles exsolved at an early stage result in either miarolytic cavities or, in more quenched facies, a myriad of small irregular fractures, as described above under *Fractures in volatile-rich cupolas*. Alterations at this stage produce minerals that are stable at high temperatures and in the presence of abundant volatiles: quartz, potassium feldspar, biotite, sodic plagioclase, and sulfides in silicic rocks, and quartz, biotite, sodic plagioclase, and sulfides in intermediate ones. Mafic rocks have not been studied in these associations but would probably be altered to hornblende (possibly actinolite), epidote, sodic plagioclase, quartz, biotite, and sulfides. In rocks of silicic composition, phenocrysts are easily recognizable and groundmass is completely reconstituted to a sugary (aplitic) mixture consisting mainly of quartz and alkali feldspar. Mapped as an alteration zone, these rocks lie completely within the pluton or extend for short distances beyond it. High volatile pressure at this stage keeps groundwater out of the pluton, but geothermal circulation may form a hydrothermal aureole outside it (Section 15-7).

As a pluton cools, the model predicts further fracturing, growth of larger and more tabular veins, and alteration of potassium feldspar and plagioclase to sericite, typically with quartz and sulfides and in some cases tourmaline, topaz, or fluorite. Igneous textures in these rocks are fuzzy to obscure due to the abundant sericite, which defines a broad zone of sericitic alteration.

Meteoric groundwater may form a major share of the altering fluid at this stage and may become dominant in still later, lower temperature alterations that produce clays and a variety of other fine-grained minerals (pyrophyllite, diaspore, alunite, etc.). The latter rocks are pale, locally vuggy, and generally without relict igneous texture. They are generally mapped as a zone of argillic alteration.

The close spatial relation between the concentric alteration zones and a given pluton, and the characteristics of the fractures and veins (Section 14-7), are the basic evidence for autometamorphism.

In plutons that are not so strongly nor so pervasively altered, or where alterations are of other kinds, one or more of the following relations may suggest autometamorphism by magmatic volatiles: (1) alteration is confined to the pluton, perhaps not even affecting its marginal facies; (2) if alteration is not confined to the pluton, it is most intense within it and is concentric around it; (3) in composite plutons or closely grouped intrusions, each intruded body has its associated alterations, a relation proven where younger intrusions cut across veins or alteration zones of an older intrusion (Gustafson and Hunt, 1975, p. 881); (4) alteration is widespread within any one intrusive unit, is selective of the minerals altered, and is unrelated to fracture-controlled alterations of regional extent; (5) alteration is confined to, or centered around, miarolytic or pegmatite-bearing units or facies; (6) alteration is concentrated in areas of small irregular veins that appear to have formed at a late magmatic stage (Section 14-7, last part); and (7) alteration minerals are those requiring temperatures higher than 550° to 600°C and the presence of abundant volatiles.

Relations suggesting alteration of a pluton by geothermal (largely meteoric) fluids are: (1) alteration is in wall rocks as well as in the pluton, and is especially intense in the outer and upper parts of the pluton; (2) alteration is concentrated near fracture systems; (3) minerals are dominantly OH-bearing species and are commonly OH-rich (chlorite, sericite, clays, etc.); and (4) successive alterations are overprinted on earlier ones (because geothermal circulation is likely to continue throughout the cooling history of the pluton).

### References Cited

Balk, R., 1937, *Structural behavior of igneous rocks*: Geological Society of America Memoir 5, 177 p.

Bateman, P. C., Clark, L. D., Huber, N. K., Moore, J. G., and Rinehart, C. D., 1963, *The Sierra Nevada batholith; a synthesis of recent work across the central part*: U.S. Geological Survey Professional Paper 414-D, 46 p.

Beane, R. E., and Titley, S. R., 1981, Porphyry copper deposits, Part II: hydrothermal alteration and mineralization, p. 235-269 *in* Skinner, B. J., editor, *Seventy-fifth anniversary volume*: El Paso, Economic Geology Publishing Co.

Burnham, C. W., 1979, Magmas and hydrothermal fluids, p. 71-136 *in* Barnes, H. L., editor, *Geochemistry of hydrothermal ore deposits,* 2nd edition: New York, John Wiley & Sons.

Cameron, E. N., Jahns, R. H., McNair, A. H., and Page, L. R., 1949, *Internal structure of granitic pegmatites:* Economic Geology Monograph 2, 115 p.

Compton, R. R., 1960, Contact metamorphism in Santa Rosa Range, Nevada: *Geological Society of America Bulletin,* v. 71, p. 1383-1416.

Dalrymple, B. G., and Lanphere, M. A., 1969, *Potassium-argon dating; principles, techniques, and applications to geochronology:* San Francisco, W. H. Freeman Co., 258 p.

Davies, H. L., 1971, *Peridotite-gabbro-basalt complex in eastern Papua: an overthrust plate of oceanic mantle and crust:* Bureau of Mineral Resources (Australia), Geology and Geophysics Bulletin 128, 48 p.

Didier, J., 1973, *Granites and their enclaves; the bearing of enclaves on the origin of granites:* Amsterdam, Elsevier Scientific Publishing Co., 393 p.

Drever, H. I., and Johnston, R., 1967, Picritic minor intrusions, p. 71-82 *in* Wyllie, P. J., editor, *Ultramafic and related rocks:* New York, John Wiley & Sons.

Duffield, W. A., 1968, The petrology and structure of the El Pinal tonalite, Baja California, Mexico: *Geological Society of America Bulletin,* v. 79, p. 1351-1374.

Fiske, R. S., Hopson, C. A., and Waters, A. C., 1963, *Geology of Mount Ranier National Park, Washington:* U. S. Geological Survey Professional Paper 444, 93 p.

Gustafson, L. B., and Hunt, J. P., 1975, The porphyry copper deposit at El Salvador, Chile: *Economic Geology,* v. 70, p. 857-912.

Irvine, T. N., 1974, *Petrology of the Duke Island ultramafic complex, southeastern Alaska:* Geological Society of America Memoir 138, 240 p.

Irvine, T. N., and Smith, C. H., 1967, The ultramafic rocks of the Muskox intrusion, Northwest Territories, Canada, p. 38-49 *in* Wyllie, P. J., editor, *Ultramafic and related rocks:* New York, John Wiley & Sons.

Jackson, E. D., 1967, Ultramafic cumulates in the Stillwater, Great Dyke, and Bushveld intrusions, p 20-38 *in* Wyllie, P. J., editor, *Ultramafic and related rocks:* New York, John Wiley & Sons.

Jahns, R. H., and Burnham, C. W., 1969, Experimental studies of pegmatite genesis: I. A model for the derivation and crystallization of granitic pegmatites: *Economic Geology,* v. 64, p. 843-864.

Jahns, R. H., and Tuttle, O. F., 1963, Layered pegmatite-aplite intrusives: *Mineralogical Society of America Special Paper 1,* p. 78-92.

Komar, P. D., 1972, Mechanical interactions of phenocrysts and flow differentiation of igneous dikes and sills: *Geological Society of America Bulletin,* v. 83, p. 973-988.

Mattinson, J. M., 1977, Emplacement history of the Tatoosh volcanic-plutonic complex, Washington: ages of zircons: *Geological Society of America Bulletin,* v. 88, p. 1509-1514.

McBirney, A. R., and Noyes, R. M., 1979, Crystallization and layering of the Skaergaard intrusion: *Journal of Petrology,* v. 20, p. 487-554.

Moore, J. G., 1963, *Geology of the Mount Pinchot quadrangle, southern Sierra Nevada, California:* U.S. Geological Survey Bulletin 1130, 152 p.

Moore, J. G., and Lockwood, J. P., 1973, Origin of comb layering and orbicular structure, Sierra Nevada batholith, California: *Geological Society of America Bulletin,* v. 84, p. 1-20.

Moores, E. M., 1969, *Petrology and structure of the Vourinos ophiolitic complex of northern Greece:* Geological Society of America Special Paper 118, 74 p.

Morse, S. A., 1969, *The Kiglapait layered intrusion, Labrador*. Geological Society of America Memoir 112, 204 p.

Neff, T. R., 1973, Emplacement of a dike swarm in the Buffalo Mountain pluton, Nevada: *Geological Society of America Bulletin*, v. 84, p. 3689-3696.

Nelson, C. A., and Sylvester, A. G., 1971, Wall rock decarbonation and forcible emplacement of Birch Creek pluton, southern White Mountains, California: *Geological Society of America Bulletin*, v. 82, p. 2891-2904.

Osiecki, R. A., 1981, *Textural development of pegmatite, aplite, and associated rock types in the Mason-Milford granite*: Stanford University, CA, PhD Dissertation, 158 p.

Parsons, I., 1979, The Klokken gabbro-syenite complex, South Greenland: cryptic variation and origin of inversely graded layering: *Journal of Petrology*, v. 20, p. 653-694.

Pitcher, W. S., 1970, Ghost stratigraphy in intrusive granites: a review, p. 123-140 *in* Newall, G., and Rast, N., editors, *Mechanism of igneous intrusion*: Geological Journal Special Issue No. 2, Liverpool, Gallery Press.

Pitcher, W. S., 1978, The anatomy of a batholith: *Journal of the Geological Society of London*, v. 135, p. 157-182.

Pitcher, W. S., 1984, Granite: typology, geological environment and melting relationships, p. 277-285 *in* Atherton, M. P., and Gribble, C. D., editors, *Migmatites, melting and metamorphism*: Nantwich, England, Shiva Publishing, Ltd.

Pitcher, W. S., and Berger, A. R., 1972, *The geology of Donegal: a study of granite emplacement and unroofing*: New York, Wiley-Interscience, 435 p.

Rehrig, W. A., and Heidrick, T. L., 1972, Regional fracturing in Laramide stocks of Arizona and its relationship to porphyry copper mineralization: *Economic Geology*, v. 67, p. 198-213.

Rock, N.M.S., 1977, The nature and origin of lamprophyres: some definitions, distinctions, and derivations: *Earth-Science Reviews*, v. 13, p. 123-169.

Shaw, S. E., and Flood, R. H., 1981, The New England Batholith, eastern Australia: Geochemical variations in time and space: *Journal of Geophysical Research*, v. 86, p. 10530-10544.

Soula, J.-C., 1982, Characteristics and mode of emplacement of gneiss domes and plutonic domes in central-eastern Pyrenees: *Journal of Structural Geology*, v. 4, p. 313-342.

Tabor, R. W., 1963, Large quartz diorite dike and associated explosion breccia, northern Cascade Mountains, Washington: *Geological Society of America Bulletin*, v. 74, p. 1203-1208.

Thayer, T. P., 1963, Flow-layering in alpine periodotite-gabbro complexes: *Mineralogical Society of America Special Paper 1*, p. 55-61.

Uebel, P.-J., 1977, Internal structure of pegmatites, its origin and nomenclature: *Neues Jahrbuch für* Mineralogie Abhandlungen, v. 131, p. 83-113.

Wager, L. R., and Brown, G. M., 1967, *Layered igneous rocks*: San Francisco, W. H. Freeman and Co., 588 p.

White, W. H., and six others, 1981, Character and origin of Climax-type molybdenum deposits, p. 270-316 *in* Skinner, B. J., editor, *Seventy-fifth anniversary volume*: El Paso, Economic Geology Publishing Co.

Willis, D. B., 1977, A kinematic model of preferred orientation: *Geological Society of America Bulletin*, v. 88, p. 883-894.

Wright, A. E., and Bowes, D. R., 1968, Formation of explosion-breccias: *Bulletin Volcanologique*, v. 32, p. 15-32.

# ■ Field Studies of Metamorphic Rocks

### 15-1. Protoliths of Metamorphic Rocks

The classical genetic kinds of metamorphism—thermal, dynamic, high-pressure, metasomatic—rarely affect rocks separately. They may act together; they may act in turn during one period of metamorphism; or they may act in various combinations during widely separated periods. Age relations are thus crucial in studies of metamorphic rocks, and they are most easily resolved where specific minerals or textures can be related to structures that give clear evidence of relative ages. Age relations are thus described in most sections of this chapter and are brought together in the last section in an abbreviated checklist. Structures and structural methods are described in Chapter 12, and metamorphic rocks in Chapter 4.

A second major aspect of field studies entails mapping metamorphic zones, igneous bodies, metasedimentary and metavolcanic stratigraphic units, and relations of metamorphic rocks to major structures. Such mapping may identify geographic or broadly structural varieties of metamorphism: contact metamorphism, regional metamorphism, and so on.

In every type of study, interpretation depends partly on knowing the protolith of each metamorphic rock. Highly metamorphosed or deformed rocks may have to be mapped as lithodemic units (Section 5-3), but metamorphic rocks should, if possible, be treated as metamorphosed sedimentary or volcanic formations or lesser units. In ideal cases, metamorphosed units can be traced to other areas where they are unmetamorphosed or slightly metamorphosed. Stratigraphic sequence may be unique enough to correlate metamorphosed units positively with unmetamorphosed units.

If stratigraphic correlations are not possible, protoliths of undeformed or moderately deformed rocks can usually be deduced from relict primary features, such as those described in Chapters 4, 9, 10, 13, and 14. Growth of metamorphic minerals typically blurs these features or obscures them by darkening the entire rock, but relics can often be seen on smooth, clean, weathered surfaces or on wet surfaces. Metamorphosed sedimentary beds that were originally size-graded from sand or silt upward to clay may become inversely size graded, from finer quartz-rich rock at the base to rock with large grains of aluminum-rich silicates at the top. Igneous intrusions can usually be recognized by their cross-cutting relations and by moderately affected igneous texture in their central or least-sheared parts. Feldspar phenocrysts are especially likely to remain intact during both dynamic and thermal metamorphism.

Fossils are often preserved in low-grade metasediments and even in high-grade contact metamorphic aureoles. In phyllite and slate they can be found by splitting the rock along the cleavage at places where it is parallel to bedding, such as in the limbs of isoclinal folds. Fragments of echinoids in limestone tend to resist recrystallization and deformation because each is a large single crystal. Large corals and mollusks in sandstone may retain their original shapes even though they recrystallize to calcite aggregates or are replaced by wollastonite or garnet.

If metamorphism has destroyed all primary features, original lithology can often be estimated from the proportions of metamorphic minerals in each rock, assuming the common situation of isochemical metamorphism (Fig. 15-1). In addition, a high proportion of Fe to Mg is indicated by the presence of chloritoid, staurolite, and abundant black biotite in aluminous rocks, which were originally iron-rich claystone. Deep red rather than tan to green garnet in calcsilicate rock suggests a ferruginous limestone protolith or, more commonly, a metasomatized limestone (skarn). Protoliths rich in potassium (illitic shale, micaceous siltstone, and igneous rock ranging in composition from syenite to granite) are suggested by abundant muscovite, biotite, or potassium feldspar; however, potassium is sometimes added

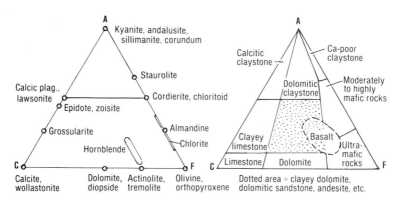

**Fig. 15-1.** Diagrams comparing compositions of some common metamorphic minerals *(left)* with common rock compositions *(right)*. Compositions are based only on the three components $Al_2O_3$ (A), CaO (C), and FeO + MgO (F). To use the diagrams, estimate the proportions among the minerals shown; locate that proportion on the mineral diagram; and find the same location on the rock diagram. For example, a rock containing about equal amounts of andalusite, cordierite, and plagioclase would plot near the center of the triangle among those three minerals, a point lying in the field of dolomitic claystone in the rock diagram. Minerals with compositions not represented on the diagrams, such as quartz, may also be present. Note, too, that some rock compositions are not represented on the right-hand diagram. Rules for plotting mineral and rock compositions exactly have been given by Turner (1981, chapt. 5) and Winkler (1979, chapt. 5)

metasomatically (Section 15-7). Metamorphic rocks containing more than about 40% quartz are likely to have been sandstone or silicic igneous rock, two cases that can generally be differentiated by relict structures. Protoliths deficient in silica, as dolomite and peridotite, are implied by olivine, spinel, periclase, or brucite. Lateritic metaclaystone or metasomatically desilicated metashale are suggested by corundum, by highly micaceous quartz-free rocks, and by quartz-free rocks with abundant $Al_2SiO_5$ polymorphs or abundant garnet.

When the protolith has been narrowed to a few possibilities, it may be possible to interpret it further by lithologic association and by thickness and shape of layers. A nearly pure quartzite, for example, might be interpreted as metamorphosed quartzose sandstone rather than metachert on the basis of thick layers intercalated with aluminous schist. Primary associations of rocks are described in Chapters 4, 9, 10, 13 and 14.

### 15-2. Metamorphic Mineral Reactions

Although laboratory studies are generally needed to identify minerals and reaction relations completely (e.g., to assign rocks to metamorphic facies), field study can locate scarce but crucial variants that would not be sampled otherwise and can resolve age relations which would be obscure in thin sections. This section suggests ways of recognizing minerals in fine-grained rocks and describes relations that indicate mineral reactions.

Many of the more important metamorphic minerals in fine-grained rocks occur at least locally as euhedral or subhedral porphyroblasts, many of them being distinctively colored. Examples are garnet, kyanite, andalusite, staurolite, chloritoid, ankerite, epidote, zoisite, lawsonite, glaucophane, biotite, muscovite, and hornblende. These minerals are usually apparent in local coarse segregations, if nowhere else, perhaps associated with veins (Section 15-5).

Several other important minerals characteristically form anhedral porphyroblasts sieved by fine inclusions. Cordierite generally forms ellipsoidal porphyroblasts in fine-grained rocks and coarse ellipsoidal aggregates in gneiss. It can be recognized on weathered surfaces as etched-out pits coated with limonite. Scapolite often forms anhedral porphyroblasts in fine-grained rocks, and can be recognized by its two mutually perpendicular cleavages.

Geometric relations among porphyroblasts, their inclusions, and groundmass fabric may indicate whether the large grains grew before, during, or after deformation (Fig. 15-2).

Changes in color or luster in fine-grained rocks may suggest specific mineral reactions. Greenish-gray phyllite, presumably containing abundant chlorite and muscovite, becomes slightly reddish-gray and darker in tone when biotite forms abundantly. Reddish-gray color also suggests abundant biotite in hornfels. Gray-green to yellow-green metabasalt turns dark gray to black when chlorite (or actinolite) and epidote react to form abundant hornblende.

Arrested mineral reactions that can often be recognized without a hand lens are: (1) partial reaction of chert bodies and limestone to form wollastonite; (2) partial reaction of chert in dolomite to form tremolite and calcite, or diopside with or without wollastonite; (3) conversion of epidote-rich nodules (which commonly develop during low-grade metamorphism of basaltic or andesitic lavas and pyroclastic rocks) to deep red andradite and quartz; (4) partial reaction of andraditic bodies to black hornblende; (5) calcitic concretions in metashales that may then be partly reacted to rocks rich in plagioclase and biotite; (7) calcitic concretions in metashale reacted to garnetiferous rocks that are partly reacted to hornblende-rich envelopes; (6) dolomitic concretions that are first metamorphosed to rocks rich in tremolite or actinolite and then partly to hornblende or pyroxene; (8) gneisses or other siliceous metamorphic rocks containing fragments of ultramafic rocks reacted to zoned bodies with hornblende-rich rims; and (9) metalimestone and metashale that interact along contacts to a reaction skarn, presumably in the presence of abundant pore fluid (Section 15-7). The zoned inclusions of Fig. 14-8 illustrate some of these reaction relations.

Using a hand lens, arrested reactions can sometimes be recognized by rims or relict cores of mineral grains, or by aggregates pseudomorphing mineral grains that had distinctive shapes. Some common reactions are: (1) garnet, staurolite, or cordierite to biotite; (2) garnet or clinopyroxene to hornblende; (3) ilmenite to sphene; (4) garnet and clinopyroxene (together) to glaucophane; (5) garnet, staurolite, or cordierite to chlorite; (6) aluminosilicates to white mica; (7) plagioclase to epidote; (8) orthopyroxene or olivine to serpentine, perhaps with talc or brucite; (9) quartz and calcite to wollastonite; (10) quartz and dolomite to diopside; (11) clinopyroxene to actinolite; and (12) actinolite to hornblende (as dark rims). Garnets may express compositional variations during metamorphism by color zoning. Garnets with tan or greenish (grossularite) cores and dark red (andradite) rims reflect a

**Fig. 15-2.** *A*. Inclusion of a fine early schistose fabric, indicating the porphyroblast grew before coarser grains of the groundmass. *B*. Included sigmoidal fabric indicating porphyroblast grew during gradual rotation, presumably due to simple shear within the groundmass. *C*. Partially replaced relics parallel to groundmass fabric, indicating post-tectonic growth of the porphyroblast.

change in the oxidation state of iron. The cores of zoned garnets may also be replaced preferentially, the rim surviving as a partial spherical layer (atoll) around secondary minerals.

Finally, mineral reactions are implied where a rock believed to have been originally uniform shows geographic variations in mineral content, as where a uniform lithologic unit is traced toward an igneous body. Such changes may be abrupt, gradual, or sporadic, and generally require mapping specific mineralogic variations, as described in the section that follows.

### 15-3. Metamorphic Zones Based on Minerals or Textures

Metamorphic zones mark the geographic extent of certain new minerals or textures in rocks of specific original lithology. Their boundaries are generally mapped at the first appearance of a metamorphic mineral or texture. Mapping thus entails locating all outcrops where this change is first noted, and finally connecting the mapped points with a line. A progression of mapped zones shows the extent of certain metamorphic conditions or processes and is of great value in determining the cause of metamorphism; for example, where the zones are concentric around an intrusion.

First appearances of metamorphic minerals or textures differ among various protoliths; therefore, zones must be based strictly on rocks of the same original lithology. Ideally, the kind of rock chosen for mapping should be easily recognizable, widely distributed in the area, and sensitive to changes in metamorphic conditions. Zone-mapping will be most successful if stratigraphic rock units are mapped throughout the area first, during which time their metamorphic variations are studied in a general way. Thin sections made after this first stratigraphic mapping will be very valuable in the zone-mapping, but many of the kinds of zones described below can be mapped at least approximately with a hand lens.

*Mineral zones* used widely in areas of schist are based on carbonate-free metaclaystones mafic enough to develop garnet and staurolite at appropriate metamorphic grades. The progression of index minerals that forms in many cases is: (1) chlorite and muscovite (formed from detrital clays); (2) biotite; (3) almandine garnet; (4) staurolite; (5) kyanite; (6) sillimanite; and (7) potassium feldspar (or disappearance of muscovite). Atherton (1965) has noted that the first appearances of garnet and staurolite are particularly valuable because they take place in spite of moderate variations in rock composition, whereas the appearance of biotite depends appreciably on rock composition. The change from clays to chlorite and muscovite may be marked by a transition from gray slate to phyllite or to pale gray or green-gray schist. Where biotite has formed abundantly, greenish rocks change to dark gray schist with a reddish tone. Garnet, staurolite, and kyanite commonly form euhedral porphyroblasts that are easily visible in certain layers or segregations. Sillimanite generally forms silky patches of fine needles (fibrolite) in its first appearance or is associated with biotite as fine lineated clusters approximately parallel to mica cleavage. Potassium feldspar is typically segregated into folia or patches with quartz, and can be distinguished from untwinned plagioclase by its lesser degree of alteration or by staining (Section 2-3). Disappearance of muscovite may be impractical to use as a zone boundary because the mineral often develops during retrograde metamorphism.

A somewhat different progression of minerals in schistose metaclaystone has been interpreted to indicate lower metamorphic pressures than the one just described (Miyashiro, 1973, p. 167). The lower-grade index minerals are the same as those just described, but andalusite forms at about the same position as staurolite; kyanite is scarce or absent; and cordierite appears at higher grade. The highest grade zone is thus characterized by association of cordierite, sillimanite, and potassium feldspar, locally with andalusite.

Mafic igneous rocks associated with schists can be separated into low-grade and intermediate-grade zones where hornblende first appears, and into a high-grade zone at the first appearance of pyroxene. Garnet may or may not form at intermediate grade. Epidote is usually consumed in the reaction to form hornblende but in unusually calcareous rocks it may last into a higher grade, or it may be retrograde. Mafic rocks formed from dolomitic mudstone may develop diopside at low to moderate grades of metamorphism.

Calcareous rocks are often difficult to zone because of their compositional variations; however, garnet and (or) hornblende typically appear in them at positions approximately equivalent to the garnet zone of metaclaystones, and clinopyroxene (diopside or salite) at positions equivalent to the sillimanite zone (Tanner, 1976). Meionitic scapolite may develop from calcic plagioclase in deep-seated high-grade rocks.

Siliceous dolomite, or dolomite with silica added metasomatically, is progressively metamorphosed to associations that might be mapped as zones: (1) talc-calcite; (2) tremolite-calcite; (3) diopside-calcite; and (4) forsterite-calcite.

Metaclaystones in contact aureoles commonly show this progression of mineral appearances: (1) biotite, (2) andalusite, (3) cordierite, (4) sillimanite, and (5) potassium feldspar. The first three appearances may be so closely spaced as to appear superimposed at the outer boundary of some aureoles (Compton, 1960). Staurolite may appear together with andalusite, and it may disappear when cordierite develops. Kyanite forms in the inner parts of some aureoles, typically but not necessarily before andalusite (Pitcher and Berger, 1972, p. 325). Contact metamorphosed mafic rocks show mineral appearances much like those described above for schists; however, garnet is much less likely to form and hypersthene may appear in the innermost zone of high-level aureoles. Contact metamorphism of impure calcareous rocks may lead either to hornfels and schist with minerals much like those already described for calcareous rocks, or to coarse calc-silicate rocks that are zoned mainly due to metasomatism (see *skarn*, Section 15-7).

**Textural zones** can be defined by progressive coarsening (grain growth) of metamorphic minerals. The progression in dynamothermally metamorphosed claystones and in fine-grained rocks of some other compositions is: (1) slate; (2) phyllite; (3) schist; and (4) segregated coarse schist or gneiss (Section 15-5). These textural changes are typically accompanied by mineral changes like those just described; in some cases, however, the rocks can all be in one mineral zone. Turner (1941) and Bishop (1972) mapped low-grade (chlorite-zone) metagraywackes in four textural zones that might be adapted to other kinds of rocks: (1) textures dominantly original and rock nonfoliated; relict grains fuzzy but obvious; (2) textural relics of original grains somewhat flattened or elongated, and rock semischistose due to strain and recrystallization in matrix; (3) rocks essentially fine schists with scarce textural relics; and (4) rocks gneissose, or schistose with thin segregated lenses and layers.

Contact aureoles can also be zoned on the basis of textural progressions. The outer boundaries of aureoles in metaclaystones, for example, have been mapped at the first appearance of tiny porphyroblasts (which may appear only as small knots or dimples in foliation), and an inner zone has been mapped where hornfels or schist are coarse enough for biotite grains to be easily visible (Compton, 1960, p. 1401). In basic rocks, an outer zone has been mapped at the first appearance of tough hornfels, and an inner zone where hornblende is coarse enough to be easily visible (Compton, 1955, p. 37).

### 15-4. Metasomatism

Almost all rocks gain and lose substances during metamorphism, although in most cases these changes involve only volatile components, as $H_2O$, $Cl_2$, $F_2$, $CO_2$, S, and organic compounds. Numerous cases of moderate to extreme silicate metasomatism are known, however, and such changes should be anticipated in rocks that have undergone hydrothermal alteration (Section 15-7), or appear to be segregated (Section 15-5). Metasomatism may also be expected between adjacent rocks that contain materials likely to react at metamorphic temperatures. Two examples are marble and aluminous metasediments that react to form skarn along mutual contacts (Einaudi and others, 1981), and peridotite that reacts with diabase or other sources of calcium to form rodingite (Hietanen, 1981, p. 22).

Field relations indicating metasomatic origin for rock bodies include the relations shown in Fig. 15-3 as well as the following: (1) broadly gradational boundaries; (2) mineralogic changes in structure of known original com-

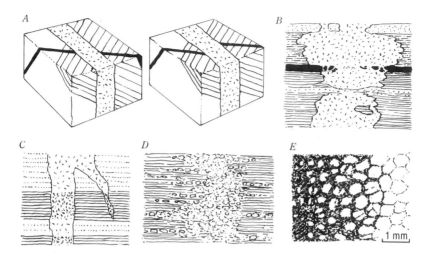

**Fig. 15-3.** Structural relations indicating replacement. *A.* Layering and a thin dike (*black*) can be projected through the dike in question (*left*) rather than being displaced in a dilational sense (*right*). *B.* Irregular boundaries, septa crossing the body, included relics (skialiths), and enlargements of the body where it crosses certain kinds of rocks or permeable zones. *C.* Compositional variations contiguous with compositional layers outside the body, or with contacts between contrasting rocks. *D.* Boundaries of body, often consisting of crystal strings, extending outward along cleavage or fractures. *E.* Progressive diminution of mineral grains (*light*) in a younger matrix.

position, as calcite ooliths now consisting of pyrite; (3) physical effects of fluids, such as unusually coarse grain or numerous mineral segregations; and (4) abundant minerals containing OH, F, Cl, $CO_2$, S, or B, such as micas, scapolite, topaz, idocrase, humites, tourmaline, and axinite.

The nature and amounts of substances exchanged during metasomatism may be calculated if the original composition of the altered rocks can be determined. In cases where the original rock is observably uniform and the metasomatic bodies are small and distinct, a few carefully chosen samples may give fairly reliable results (Fig. 15-4). It is difficult, on the other hand, to test for metasomatism across entire metamorphic zones and thus typically over distances of kilometers. The original composition of a rock unit may well have varied over such distances, and can thus be determined only with some degree of uncertainty. This degree can be reduced in some cases by sampling the original rock in several parts of a metamorphosed area, and thus obtaining average compositions and degrees of variance for texturally or mineralogically similar rocks (Compton, 1960, p. 1407). If possible, the sampling should be planned after first studying the statistics of variance

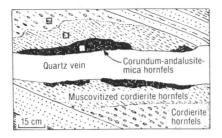

**Fig. 15-4.** Uniform layer of cordierite hornfels (metashale) altered locally to black corundum-bearing hornfels and muscovitized hornfels along a quartz vein. Chemical analyses of samples from sites such as those indicated by the small squares show an $SiO_2$ content of 65% in the uniform hornfels, 62.5% in the muscovitized hornfels, and 41.5% in the black hornfels. In the same three rocks, $K_2O$ is 2.5%, 5%, and 6%. For the complete analyses, see Compton (1960, p. 1390, nos. 5A, B, and C).

or by seeking help from geologists who have used statistics in devising sampling strategies.

Calculations of gains and losses are generally based on the rule that volume remains unchanged during metasomatism (see, e.g., Lindgren 1933, p. 92). This rule does not necessarily hold for every case, however, and therefore should be checked in the field as completely as possible. Expansion or contraction can in some cases be detected by measuring thicknesses of rock units in the area affected and outside of it. Changes in volume may also be indicated by patterns of fractures, by genetic varieties of fractures, and by displacements on minor faults (Sections 12-5, 12-6, and 14-7). Open fractures and other cavities that have been filled (indicating shrinkage) may be recognized by their mineral fabrics and layering (Fig.

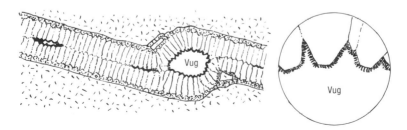

**Fig. 15-5.** Features suggesting progressive filling of an opening fracture: symmetical mineral zoning, geometric form of the vein, comb structure (fabric), and encrusted mineral terminations in vugs (*right*).

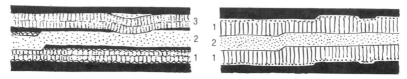

**Fig. 15-6.** Repeated dilation of wall rocks (black) indicated by geometic relations, by septa, by cross-cutting relations, and by symmetrical mineral zones. *Left,* veins reopened along one margin, emplaced in order of the numbers. *Right,* vein emplaced in medial fracture of earlier vein.

15-5). Evidence of dilation (Fig. 15-3*A*) and of repeated dilation (Fig. 15-6) may be helpful.

The actual sequence of alterations must be determined in order to interpret metasomatism. The two zones illustrated in Figure 15-4, for example, could have been formed simultaneously by a fluid that migrated outward from a channel (now the vein) and became modified physically or chemically as it penetrated the rocks. Possibly, however, the outer zone formed first and then a different fluid formed the inner zone; or perhaps the inner zone formed first and the outer zone was formed by a fluid that moved deeper into the rock at a later time. The inner zone is so changed texturally that it could well have gone through several states of alteration. Such possibilities can be resolved only by finding enough examples to see all stages in the formation of the zones. These examples must generally include small veins as well as large ones, with some crossing others so that an age sequence can be established. An example of a widely distributed and thoroughly studied array of veins and alteration zones is that at Butte, Montana, described and interpreted by Meyer and Hemley (1967, and in references cited by them). These zones and structural relations are described briefly in Section 15-7.

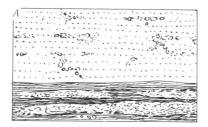
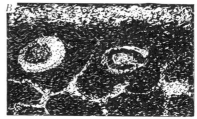

**Fig. 15-7.** *A.* Feldspathic segregations in massive metasandstone (*above*) and along foliation in mica schist. *B.* Epidote-rich segregations in metabasalt, partly as nodules and web veins.

## 15-5. Segregated Metamorphic Rocks; Gneisses

The rocks described in this section may be of any metamorphic grade but are similar in that all have been differentiated metamorphically into two megascopic domains of contrasting composition. Rocks in which one domain is of granitic or other igneous-appearing material are described together with other migmatites in Section 15-6. In all segregated rocks, the two contrasting domains may be more or less equal in volume but typically one predominates. This domain contains lenses, layers, rods, veins, pods, or irregular bodies of the less voluminous domain. Some domains are no more than clusters or strings of porphyroblasts. In thoroughly segregated rocks, the principal minerals in one set of domains are likely to number only two or three species and to differ from an equally simple assemblage in the other set. An important additional relation is that small numbers of grains of each mineral assemblage occur in the other assemblage.

Segregation (metamorphic differentiation) may be caused by chemical potentials arising without deformation, or may be due partly or entirely to deformation. Segregation with little if any deformation is suggested by lack of schistose texture and by the shapes of the segregated bodies (Fig. 15-7). Features inherited from the original rocks or from an earlier metamorphic stage are undeformed. In mafic metalavas, amygdular and brecciated zones are commonly replaced by epidote and quartz or carbonates before the other parts of the rock. Glassy rims of pillows tend to become differentiated from the central parts of pillows (Vallance, 1965). Smith (1968) described metalavas in which low-grade segregation without deformation was expressed by yellow-green bodies, containing epidote and quartz, that crossed primary igneous structures.

Segregation by penetrative deformation has been demonstrated in mylonite and recrystallized mylonite that have developed compositional layering

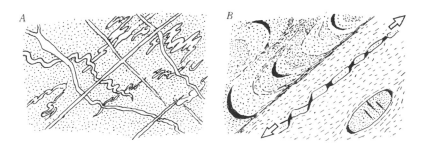

**Fig. 15-8.** *A.* Succession of deformed veins, the more folded being cut by progressively less folded ones. *B.* Segregations (*black*) at fold hinges, between boudins, and at the ends of a concretion in metaclaystone, suggesting the extension indicated by the arrows.

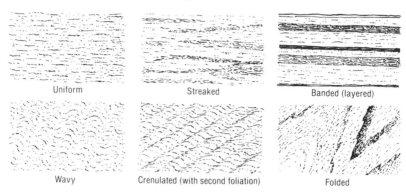

**Fig. 15-9.** Structural varieties of gneiss. Gneisses are commonly migmatitic and may thus show superimposed structural varieties such as those in Fig. 15-12.

in originally uniform rocks, the layers typically being 0.5 to 10 mm thick (Fig. 4-19*H*). Prinz and Poldervaart (1964) described mafic dikes consisting of pyroxene and plagioclase that were hydrated and segregated into thin hornblende-rich and plagioclase-rich domains during mylonitization.

Segregation is common during dynamothermal metamorphism, and its relative age may be indicated by degree of deformation (Fig. 15-8*A*) or by segregated pods and rods formed at crests or troughs of folds, between boudins, or next to bodies that remained intact during deformation (Fig. 15-8*B*). Small-scale segregation along appressed limbs of folds is characteristic of crenulation cleavage (Fig. 12-17*A*). Quartz, epidote, and carbonates are likely to be segregated into pore fluids during dynamothermal metamorphism of schist, and may be deposited as veins in nearby fractured (brittle) rock, such as quartzite or massive metalava.

Relations suggesting metamorphic differentiation rather than metasomatism include: (1) gradations from undifferentiated to differentiated rocks (Turner, 1941); (2) two domains, each with a simple mineral assemb-

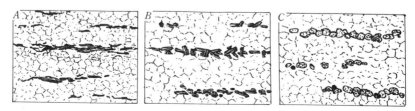

**Fig. 15-10.** Textural varieties of gneiss. See text for interpretations.

lage, together making up the composition of the original rock; (3) small numbers of mineral grains of one domain in the other domain; (4) domains relatively small in size and more or less evenly distributed; (5) lack of alteration envelopes in the wallrocks of veins (Section 15-7); and (6) age relations showing that the two domains developed simultaneously.

**Gneisses** are foliated and (or) lineated rocks having uniformly distributed domains which are so small as to impart a textural aspect when viewed from a few meters away. Examined with a hand lens, however, the domains can be seen to consist of groups of certain mineral grains. Most, or all, of the rock is granoblastic and cleaves only crudely parallel to the foliation. Figure 15-9 illustrates some structural varieties.

Gneiss may form in the same ways as other foliated segregated rocks, and criteria for recognizing a specific origin are similar to those already noted. In addition, gneiss formed by deformation acting throughout grain growth is characterized by strongly lineated and often plicated domains, with micas, amphiboles, and other platy or linear grains lying parallel to foliation and lineation (Fig. 15-10*A*). In gneiss formed mainly by grain growth after deformation, platy and elongate minerals are oriented more or less randomly (Fig. 15-10*B*), or grains that are usually platy or prismatic may be nearly equidimensional (Fig. 15-10*C*). Protomylonite may have a gneissose aspect (Fig. 15-11*A*), as may mylonite in which groundmass grains have grown to visible sizes (Fig. 15-11*B*). The progressive development of gneiss from sheared ultramafic rocks and gabbro has been described by Moore (1973).

Gneiss developed from igneous rock (*orthogneiss*) may be difficult to distinguish from gneissose metasedimentary rock (*paragneiss*) if the igneous rock has been segregated into layers. A sedimentary protolith is indicated by intercalated quartzite, marble, or other carbonate-bearing rocks, and mapping or reconnaissance may disclose gradations or suggest correlation

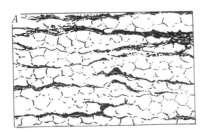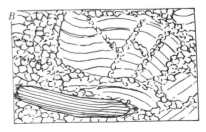

**Fig. 15-11.** *A*. Gneissose protomylonite, the dark domains consisting of mylonitic materials. *B*. Mylonitic marble with bent phlogopite (*below*) and deformed calcite porphyroclast crossed by trains of recrystallized (annealed) calcite, which also forms a matrix to the larger relics.

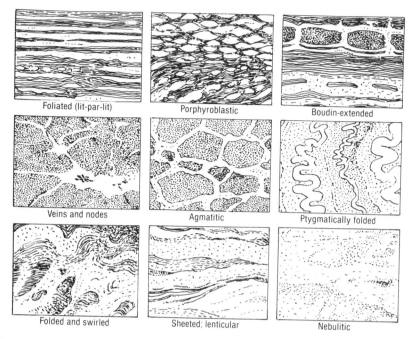

**Fig. 15-12.** Structural varieties of migmatite, with leucosome shown by the lighter patterns.

with less metamorphosed strata. Mapping of orthogneiss may disclose transgressive intrusive relations or lateral gradations into more uniform rocks with relics of igneous texture. Many of the outcrop-size structures described in Chapter 14 may be partially preserved in orthogneiss.

## 15-6. Migmatites

These composite rocks consist of two domains: (1) metamorphic rock and (2) igneous or igneous-appearing rock. The latter domain is most commonly quartzofeldspathic but may be dioritic and even gabbroic. It is typically the lighter colored of the two, and is thus called the *leucosome*. Structural varie-

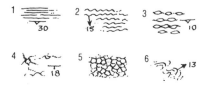

**Fig. 15-13.** Map symbols for structural varieties of migmatite: (1) foliated; (2) crenulated; (3) augen-bearing; (4) veins and nodes; (5) agmatitic, and (6) folded, showing orientation and typical shapes in horizontal section.

ties of migmatites are illustrated in Fig. 15-12. Migmatites differ from igneous rocks with inclusions in that the metamorphic domain is dominant or is the host domain. Exceptions are nebulites, which might also be called schlieric granite except that they are contiguous with other migmatites.

Migmatites can often be interpreted genetically, although more than one process is likely to have operated at a given site. Intruded leucosome will show dilative relations (Fig. 15-3A) and perhaps other features typical of igneous rocks (Chapter 14). Criteria for replacive bodies are presented in Section 15-4. Segregation is indicated by the relations described in Section 15-5 and by: (1) mafic or cordierite-rich haloes next to the leucosome; (2) granitic leucosome most abundant in metashales and quartzofeldspathic rocks; (3) lack of granitic leucosome in quartzite and marble; and (4) leucosome compositions that vary with the compositions of the host rocks; for example, granite leucosome with abundant potassium feldspar in schists containing abundant mica or potassium feldspar versus hornblende tonalite leucosome in amphibolite. Segregated migmatite may also grade to other segregated rocks (Section 15-5). Migmatites generated by circulation of vapor or hydrothermal fluid are likely to be rich in OH-bearing minerals relative to unaffected rocks, an example being the alteration of granulite facies rocks during a second metamorphism in northwest Scotland (Sutton and Watson, 1951). A number of papers describing the origin and interpretation of migmatites are included in Atherton and Gribble (1983).

Migmatites often show structural relations implying relative ages of events. Examples are folded bodies of leucosome cut by unfolded ones; metasomatic migmatites consistently cut by intrusive leucosome; migmatite cut by dilative basic dikes that are later migmatized; and migmatite cut off at an unconformity that is later crossed by additional leucosome.

Mapping the distribution of migmatites provides a means of interpreting them relative to major intrusions, folds, unconformities, and fracture zones. Mapping is generally keyed to abundance of leucosome at each locality, which can be plotted as an approximate percentage or by colored lines, dots, or patches spaced to indicate proportion of leucosome. The symbols may also show the structural type of migmatite at each locality (Fig. 15-13).

## 15-7. Hydrothermal Alteration

Hydrothermal metamorphism, or alteration, is caused by circulation of a hot fluid through rocks. The fluid may be a gas, a liquid, or a mixture of the two, and generally carries a variety of substances in solution. Metasomatic changes are typical and may be dramatic; however, some compositional changes are no more than moderate additions of volatile substances. The fluid is related to nearby igneous intrusions in many cases, being either magmatic water exsolved at a late magmatic stage or meteoric groundwater that has been heated by the pluton and circulated geothermally (Section 14-8). In some cases the fluid is groundwater that has been heated because of its deep position in the earth and mobilized by broad gravity-related potentials (Hanor, 1979).

Hydrothermal fluids tend to alter rocks selectively, some of the more susceptible rocks being limestone, dolomite, and mafic or glassy igneous rocks. Alteration also tends to be most intense along permeable zones such as open fractures, breccias, uncemented coarse clastic rocks, and rocks that are readily hydrofractured. In rocks of moderate permeability, alteration is likely to be strongest near contacts with less permeable rocks, or at fault gouge that acted as a barrier to circulation. Because permeability is increased by fracturing, the most brittle rocks in a given association are often the most veined and altered.

Some hydrothermally altered rocks contain only one abundant new mineral, and they are classified and named accordingly: *silicified, dolomitized, chloritized, albitized,* and so on. Two distinctive kinds of hydrothermally metamorphosed rocks named long ago by miners are *skarn*, which is described later in this section, and *greisen*, a coarse muscovite-rich rock that forms in and around granite plutons and generally also contains quartz, alkali feldspars, and one or more of topaz, fluorite, and tourmaline. Some additional kinds of alteration in and near plutons of silicic to intermediate composition have been classified into the following general types (Meyer and Hemley, 1967):

*Potassic:* plagioclase and mafic minerals altered to potassium feldspar and(or) biotite; typically a late magmatic alteration (Section 14-8).

*Propylitic:* plagioclase and mafic minerals altered to epidote and chlorite; these fine-grained green rocks are generally peripheral to rocks altered at higher temperatures (e.g., by potassic alteration).

*Sericitic:* feldspars altered to sericite; generally later than potassic alteration and typically of same age as principal ore sulfides; this alteration and the two following are generally postmagmatic (Section 14-8).

*Intermediate argillic:* plagioclase largely altered to clay minerals; potassium feldspar partly altered to clays; sericite may be present.

*Advanced argillic:* feldspars totally altered to clays and other fine-grained aluminous minerals (alunite, diaspore, pyrophyllite, andalusite, etc.).

Varieties of altered rocks may vary from these ideal types, and the names just listed are used somewhat loosely. Therefore textures, mineralogic rock compositions, and the nature of rocks prior to alteration should be recorded as fully as possible.

Alterations may be selective of original mineral species, as noted above, or may be pervasive (as in most cases of advanced argillic alteration). Alterations are sometimes restricted to the margins of veins, fractures, or igneous bodies, and in these cases are commonly zoned, as will be described below. **Alterations associated with veins.** Veins are deposited by hydrothermal fluids moving along fractures. They are classified as dilative or replacive (Figs. 15-3, 15-5, and 15-6) and are described according to their thickness and extent, their shape, the minerals in them, and the textural and structural arrangements of these minerals.Important structural details are breccia bodies, folds, faults, stylolites, and reopened (multiple) vein fillings (Fig. 15-6). Age relations are determined at places where veins cross or where their alteration envelopes overlap. Strikes and dips taken over large areas may help in classifying veins by age and perhaps point to the causes of fracturing (Sections 12-5, 12-6, and 14-7).

Unless alteration becomes pervasive in veined rock bodies, individual veins are typically enveloped in symmetrical zones of wall-rock alteration. These zones may be uniform and single or may be composite. Composite zones must be studied carefully to determine whether or not the different parts developed simultaneously or in an age sequence. Simultaneous composite zoning, which is probably the most common type, was first proven at Butte, Montana, where an initial phase of high-temperature potassic alteration was followed by a protracted period of hydrothermal alteration during which large numbers of veins that were opened at various times were all enveloped by the same kinds of composite zones (Sales and Meyer, 1948; Meyer and Hemley, 1967). The Butte study also documented progressive spatial changes in vein thickness and mineralogic composition that are coupled to changes in wall-rock alteration (Fig. 15-14). These distributions point to the general source of hydrothermal fluid and its course through the overlying rock.

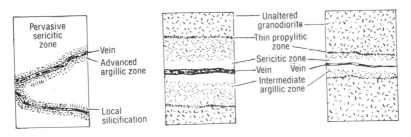

**Fig. 15-14.** Alteration envelopes on quartz veins in the central (*left*), intermediate (*center*), and peripheral parts of the main-stage ore zone at Butte, Montana. Ore minerals in the veins are (*left*) pyrite, covellite, digenite, djurleite, and enargite, with or without chalcocite and bornite; (*center*) pyrite, sphalerite, chalcopyrite, bornite, chalcocite, and tennantite, with or without enargite; and (*right*) pyrite, with or without rhodochrosite, rhodonite, sphalerite, chalcopyrite, and bornite. Based on p. 189 of Meyer and Hemley (1967).

**Skarn** is calcsilicate rock of coarse and often uneven grain size and is commonly zoned mineralogically. Both properties correlate with a hydrothermal origin. Studies of textures and stable isotopes (Taylor and O'Neil, 1977) indicate that skarn develops later than isochemical contact metamorphism. It is thought to form in two stages, first by high-temperature metasomatism by magmatic fluids, and then by hydrolyzing reactions between the early skarn and increasingly meteoric geothermal fluids. Metasomatism is proven by skarn zones lying across original bedding or metamorphic structures, or by calcsilicate rocks lying in pure marble.

Skarn is of particular value in field studies because of its coarse grain and because certain varieties may suggest depth of metamorphism. Einaudi and others (1981) noted that skarns alongside mesozonal (moderately deep-seated) plutons tend to have a garnet-pyroxene zone near the igneous body and a zone of wollastonite, often with idocrase, next to the marble. Scheelite is a characteristic mineral, and the low oxidation states typical of these deep-seated rocks cause garnet to be grossularitic (often pale colored) and pyroxene to be hedenbergitic. These early stage minerals are commonly altered in later stages to biotite, hornblende, actinolite, and epidote.

Skarns next to shallow plutons tend to be finer grained than mesozonal skarn, to become more altered and veined in late stages, and to contain abundant sulfides of Cu, Mo, Zn, and Pb. Garnet is more abundant in the inner zone of shallow skarns, and a typically high oxidation state makes the garnet more andraditic, clinopyroxene more diopsidic, and magnetite and hematite common accessories (Einaudi and others, 1981). Garnet is generally greenish (often more grossularitic) away from the igneous body, and an outermost zone is characterized by wollastonite with accessory garnet, idocrase, and clinopyroxene. When the pluton is sericitized and argillized, the early skarn minerals are altered to abundant tremolite, actinolite, iron-rich smectite, and lesser amounts of calcite, siderite, quartz, opal, iron oxides, pyrite, talc, epidote, and chlorite. Shallow skarns are typically fractured repeatedly, so that late-stage minerals may form alteration envelopes along thin veins. In cases where fluids carry Zn and Pb, these metals are deposited as sulfides in skarns formed in permeable zones far from the intrusion. Early stage zones in these outer skarns contain garnet with magnetite and chalcopyrite near the fluid channel, and clinopyroxene with sphalerite next to marble.

## 15-8. Age of Metamorphism; Sequence of Metamorphic Events

A numerical age of metamorphism can be determined by dating igneous bodies intruded during metamorphism, as indicated by fabrics and structures parallel to (and locally continuous with) those in the metamorphic rocks. Dating metamorphic minerals will give a minimum age of metamorphism. A maximum geologic age is indicated by the youngest rock unit affected by metamorphism, and a minimum geologic age by the oldest sedimentary unit deposited on the metamorphic rocks or containing detritus eroded from them.

*Age of metamorphism relative to deformation* is indicated by relations among metamorphic minerals and deformational fabrics and structures. *Pretectonic* metamorphic minerals may be recognized by: (1) being bent, kinked, broken, pulled-apart, or sheared-out into fine aggregates; (2) being replaced by younger minerals that take part in a planar or linear fabric; or (3) growing parallel to cleavages that are deformed. *Syntectonic* metamorphism is indicated by: (1) platy minerals lying parallel to cleavage in folds; (2) lineate minerals oriented parallel to hinges of folds; (3) platy or lineate minerals oriented parallel to zones of simple shear; (4) porphyroblasts (typically of staurolite, biotite, garnet, or chlorite) that include sigmoidal or spiraled trains of small platy or lineate minerals that also are in the groundmass (Fig. 15-2B); and (5) porphyroblasts that include fabrics folded less intensely than in the adjoining groundmass. *Posttectonic* metamorphism is indicated by: (1) micas, amphiboles, or other platy or lineate minerals lying across the rock fabric or cleavage (Fig. 15-2C); (2) large grains that have included crenulations or have grown over small faults or other dynamic features; and (3) aggregates of minerals with random orientation that have replaced grains in schistose rocks.

*Age relations among metamorphic events* have been described or illustrated in this chapter and are brought together here in condensed lists. Age relations are particularly visible and reliable where they can be observed in structures, and these include: (1) folded or otherwise deformed veins or foliations associated with unfolded metamorphic veins that have the same general orientations; (2) metamorphic veins cut by later metamorphic veins; (3) folded metamorphic foliations and lineations (Section 12-3); (4) metamorphic features cut by dilative dikes that have been metamorphosed; (5) alteration zones cutting across foliated rocks of regional extent; (6) undeformed replacive rocks or strings of porphyroblasts guided by foliations formed by penetrative deformation; (7) clasts of metamorphic rocks in metamorphosed conglomerate or breccia; (8) shear zones of fine-grained metamorphic rocks cutting coarser rocks; (9) relict slices of coarse-grained metamorphic rocks in mylonite; (10) weakly foliated rocks cut by zones of strongly foliated rocks; (11) shear zones displacing earlier shear zones; and (12) metamorphic veins with structures indicating two or more periods of opening.

Metamorphic age relations seen within rock fabrics include: (1) porphyroblasts that have grown across schistosity; (2) diversely oriented sets of lineate minerals; (3) diversely oriented sets of platy minerals; (4) porphyroclasts of metamorphic minerals in recrystallized mylonite; and (5) coarse

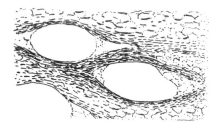

**Fig. 15-15.** Texturally composite augen (here, large rounded single crystals of feldspar with associated finer aggregates) suggesting shear and partial recrystallization of porphyroblasts.

undeformed mica or chlorite growing across cleavages.

At the scale of individual mineral grains or small groups of grains, these relations suggest sequence in metamorphism: (1) bent, kinked, broken, or separated (pulled apart) crystals; (2) large bent carbonate crystals with traces of twin lamellae; (3) augen consisting of rounded single crystals and polycrystalline "tails" (Fig. 15-15); (4) porphyroclasts with margins recrystallized along with the groundmass (Fig. 4-19*I*); (5) pseudomorphs of metamorphic minerals; (6) minerals with reaction rims; (7) zoning in porphyroblasts indicated by differences in color; (8) small grains growing in larger ones along fractures; and (9) relics of minerals known to be unstable in a given mineral assemblage.

Age relations may also be developed by geologic mapping, especially of metamorphic zones: (1) relict contact aureoles around metamorphosed intrusions; (2) zones of hydrothermal alteration lying across metamorphic zones of regional extent; (3) broad zones of segregated rocks superimposed on fine-grained schists; (4) zones of contact aureoles lying across regionally extensive zone-sequences; and (5) bands of mylonite or broad zones of schistose rocks cutting through other metamorphic zones. Two other important relations have been discussed by Read (1957, p. 263ff): (1) metamorphic mineral zones may be formed in inverted structural order (high-grade zones over low-grade zones); therefore, inverted zones are not evidence of postmetamorphic deformation; and (2) metamorphic zonations from low to higher metamorphic grade are not evidence, in themselves, that the higher grade rocks went through the stages represented by the lower grade rocks. In fact, high grade rocks with relict primary features, formed alongside lower grade rocks in which the features have been destroyed, imply that the higher grade rocks did not go through the state represented by the lower grade rocks.

**References Cited**

Atherton, M. P., 1965, The chemical significance of isograds, p. 169-202 *in* Pitcher, W. S., and Flinn, G. W., editors, *Controls of metamorphism*: Edinburgh and London, Oliver and Boyd.

Atherton, M. P., and Gribble, C. D., editors, 1983, *Migmatites, melting and metamorphism*: Nantwich, England, Shiva Publishing Ltd., 326 p.

Bishop, D. G., 1972, Progressive metamorphism from prehnite-pumpellyite to greenschist facies in the Dansey Pass area, Otago, New Zealand: *Geological Society of America Bulletin*, v. 83, p. 3177-3198.

Compton, R. R., 1955, Trondhjemite batholith near Bidwell Bar, California: *Geological Society of America Bulletin*, v. 66, p. 9-44.

Compton, R. R., 1960, Contact metamorphism in Santa Rosa Range, Nevada: *Geological Society of America Bulletin*, v. 71, p. 1383-1416.

Einaudi, M. T., Meinert, L. D., and Newberry, R. J., 1981, Skarn deposits, p. 317-391 *in* Skinner, B. J., editor, *Seventy-fifth anniversary volume*: El Paso, Economic Geology Publishing Co.

Hanor, J. S., 1979, The sedimentary genesis of hydrothermal fluids, p. 137-172 *in* Barnes, H. L., editor, *Geochemistry of hydrothermal ore deposits*, 2nd edition, New York, John Wiley & Sons.

Hietanen, A., 1981, *Geology west of the Melones fault between the Feather and North Yuba Rivers*: U.S. Geological Survey Professional Paper 1226-A, 35 p.

Lindgren, W., 1933, *Mineral deposits*, 4th edition: New York, McGraw-Hill Book Co., 930 p.

Meyer, C., and Hemley, J. J., 1967, Wall rock alteration, p. 166-235 *in* Barnes, H. L., editor, *Geochemistry of hydrothermal ore deposits*. New York, Holt, Rinehart and Winston.

Miyashiro, A., 1973, *Metamorphism and metamorphic belts*: New York, John Wiley & Sons, 492 p.

Moore, A. C., 1973, Studies of igneous and tectonic textures and layering in the rocks of the Gosse Pile intrusion, central Australia: *Journal of Petrology*, v. 14, p. 49-80.

Pitcher, W. S., and Berger, A. R., 1972, *The geology of Donegal: a study of granite emplacement and unroofing*: New York, Wiley-Interscience, 435 p.

Prinz, M., and Poldervaart, A., 1964, Layered mylonite from Beartooth Mountains, Montana: *Geological Society of America Bulletin*, v. 75, p. 741-744.

Read, H. H., 1957, *The granite controversy*: London, Thomas Murby, 430 p.

Sales, R. H., and Meyer, C., 1948, Wall rock alteration at Butte, Montana: *American Institute of Mining Engineers Transactions*, v. 178, p. 9-35.

Smith, R. E., 1968, Redistribution of major elements in the alteration of some basic lavas during burial metamorphism: *Journal of Petrology*, v. 9, p. 191-219.

Sutton, J., and Watson, J., 1951, The pre-Torridonian metamorphic history of the Loch Torridon and Scourie areas in the North-West Highlands, and its bearing on the chronological classification of the Lewisian: *Quarterly Journal of the Geological Society of London*, v. 106, p. 241-307.

Tanner, P.W.G., 1976, The progressive metamorphism of thin calcareous bands from the Moinian rocks of N. W. Scotland: *Journal of Petrology*, v. 17, p. 100-134.

Taylor, B. E., and O'Neil, J. R., 1977, Stable isotope studies of metasomatic Ca-Fe-Al-Si skarns and associated metamorphic and igneous rocks, Osgood Mountains, Nevada: *Contributions to Minerology and Petrology*, v. 63, p. 1-49.

Turner, F. J., 1941, The development of pseudo-stratification by metamorphic differentiation in the schists of Otago, New Zealand: *American Journal of Science*, v. 239, p. 1-16.

Turner, F. J., 1981, *Metamorphic petrology; mineralogical, field, and tectonic aspects*, 2nd edition: New York, McGraw-Hill Book Co., 524 p.

Vallance, T. G., 1965, On the chemistry of pillow lavas and the origin of spilites: *Mineralogical Magazine*, v. 34, p. 471-481.

Winkler, H.G.F., 1979, *Petrogenesis of metamorphic rocks*, 5th edition: New York, Springer-Verlag, 348 p.

# ■ Preparing Illustrations a
# Writing Reports

## 16-1. From Field Study to Report Writing

On returning from the field, the following preparations for report writing can be done at once:

1. Have films developed and printed.

2. Unpack rock samples and lay them out in order by age, by collecting date, or by geographic position.

3. Get help from paleontologists to identify fossils, make geologic age assignments, and assess depositional facies.

4. Read through the field notes carefully, making lists of (a) samples that should be prepared for petrographic or other laboratory analyses, and (b) data and ideas that are significant in light of the full field study.

5. Read all field summaries and unit descriptions, listing or indexing significant items.

6. Study pertinent literature and geologic maps, taking notes on: (a) items that will help in describing the regional setting of the study; (b) data that can be added directly to the main results of the study; and (c) disagreements in unit names or other specifics that will require further research or correspondence with the authors or suitable experts.

7. Begin final assembly and checking of the geologic map, cross sections, and columnar sections, as described in Section 16-2.

*Composing a plan for the report* should not be difficult if the purpose of the project was clear (Section 1-3) and if evidence and interpretations have been reviewed periodically during the field season (Section 1-6). The purpose of the project predetermines the general nature of the report as well as its readers, who must be considered during all stages of organization and writing. Reports for nongeologists require careful organization and use of only the most common geologic terms. Geologist readers vary in degree of specialization, field experience, and access to a library, so the rule, again, is to write as simply and clearly as possible. This rule applies especially to reports intended for a broad spectrum of geologists and for the permanent scientific record, and this is the general case assumed in this chapter. Reports on specialized studies that will be published in journals must be so brief as to require thorough planning and tight writing, but they should be directed to a general audience as much as possible (Cochran and others, 1974). Reports making specific recommendations, such as for drilling sites or engineered structures, have the thrust of selling ideas, often to specific readers, and must be organized and worded accordingly.

Outlines and further suggestions for several kinds of reports are presented in Section 16-4. No preexisting outline, however, provides a basis for actually beginning a specific report, whereas the following questions may:

1. In view of the project's purpose and a final review of the literature, what are the more valuable conclusions resulting from the study?

2. What data and relations led to these conclusions?

3. Were the methods of the study unusual enough so that they should be described? Regardless of their originality, will a description be needed to convince the reader of their validity and precision?

In order to adopt a direct writing style at the outset, respond to these questions as if they were being asked by a reader. For question 2, for example, list the data and relations in the order that will be most convincing to the reader. Write as though you were talking to the reader, or perhaps as though you were talking informally to several geologists. Go over what you have written several times, on several days, and at times when you are not rushed.

When you are satisfied with your answers to the three questions, recompose the answers into a summary of the project. Give the purpose first, then the methods, then the chief findings, and finally the conclusions. This summary will be a brief version of your report and might be condensed later into the report's abstract. The purpose and conclusions may be brought together to make the first paragraph of the report itself. Before organizing further sections, however, complete and study the map and sections as well as other illustrations that will contribute directly to the statements of your summary. You may want to modify your summary after doing these things.

### 16-2. Major Illustrations First

Maps, cross sections, and columnar sections convey most of the data in geologic reports. If they are prepared before the report is written, the report is likely: (1) to be more accurate than otherwise; (2) not to repeat information shown clearly on these illustrations; and (3) include descriptions of data and relations not apparent in the illustrations. If pencil drafts of the illustrations have not been completed in the field (Sections 1-7, 6-5, and 11-9), they should be completed before summarizing the field study, as described above.

The final illustrations should meet the needs of the project yet require as little time and expense as possible. If the map and cross sections are placed on one plate, readers can visualize the geology in three dimensions conveniently. The plate should be as small as consistent with clarity, and Fig.

16-1 suggests a layout. Use of color on the final map is desirable and practical when only a few copies will be needed, whereas colored maps and sections are costly to reproduce and are usually published only by government agencies. When planning a map for publication, the publisher's requirements must be determined at the outset.

A major question is whether the geologic map will need a topographic base or whether a planimetric base of drainage and roads will be adequate. An inexpensive way to produce a combined topographic and geologic map is to draft the geology in black ink on a green chronoflex topographic base (Section 6-1) or on a screened black reproduction of the topographic base. When printed by most of the methods described below, the black ink lines will reproduce distinctly darker than the green or screened lines, so that topography can be distinguished from geology. If mapping was done on plane table sheets, and the contours and other geographic features will obscure the geology if all are shown in black, the geography and contours can be traced first in black and reproduced as a screened transparent positive on which the geology is inked. Use of a standard topographic map as a base is limited by the fact that it will be illegible when reduced more than 50%.

**Geographic names** must be accurate, and all names used in the text of a report should appear on at least one map accompanying the report. The Domestic Names Committee of the U.S. Board on Geographic Names supervises the standardization of geographic names in the United States, and only names authorized by this committee appear on published maps and charts of the U.S. Geological Survey, the U.S. Forest Service, and the National Ocean Survey. These sources thus serve as standards for use on new geologic maps and in reports. The basic policy of the committee is to

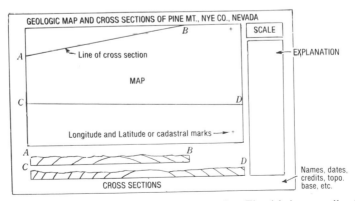

**Fig. 16-1.** Layout of map, cross sections, and explanation. The title is generally placed along the upper or lower margin or in a box in one of the corners. See figures and text that follow for details.

follow present-day local usage of names. If locally named features that would be useful on a map are not named on the published maps and charts mentioned above, it is advisable to write the committee about the use of the new name. Before concocting names for completely unnamed features, it is recommended to read the section on geographic names in Bishop, Eckel, and others (1978) before proposing a new name to the committee.

**Engineering geologic maps** may cover such a wide variety of rock properties and other information that more than one map may be needed. *Documentation maps* show locations of all measurements, samples, or observations as well as the data obtained. *Derived maps* present information extrapolated over relevant areas, generally by colors, patterns, or contours. *Structure contour maps* are derived maps showing elevations on a unit contact or other surface of interest, and *isopach maps* are contoured so as to show thicknesses of a unit or of overburden. Additional kinds of derived maps show various grades of a rock property, or presence or absence of a property or a critical unit (Section 5-7). Where a surficial deposit covers another deposit of interest, the colors or patterns for the two units are interstriped, or other patterns or colors are used to indicate the unit sequence. Complex sequences can be indicated by brief logs at data sites (Fig. 5-10). Still another kind of derived map shows areal distributions of units and structures at some specific depth or at some elevation above a datum. Varnes (1974) has described and illustrated several kinds of engineering geologic maps, and examples of actual map series are those by Easterbrook (1973, 1976) and Péwé and Bell (1976).

**Duplication processes** range greatly in cost, convenience, and nature of products, and may limit the size of the final plate. The information given here is general and should be checked thoroughly at local blueprint companies. For large plates, such as geologic maps, the least-expensive prints are made from a transparent or translucent original by the diazo process (and similar processes), which give black, blue, or brown lines on white paper or transparent Mylar base. The paper copies often yellow with age and are not strong enough for repeated handling; however, "card stock" gives white-based durable prints. The Xerox Large Document Printer produces copies up to 24 in. wide on paper of good quality and permits reduction from either an opaque or transparent original.

The various "blueprint" processes are generally more expensive because they produce a negative print, and therefore a transparent negative must be prepared first. They give durable prints on high quality paper, with maximum size limited by the size of the vacuum frame available.

Generally the most expensive process is that of having the plate photographed and copies printed from the negative. This procedure permits precise enlargement or reduction and can give large copies (e.g., 42 in. wide) on high quality paper or Mylar.

If copies 18 x 24 in. or smaller will be adequate, *Photomechanical transfer* (PMT) provides excellent copies on a variety of materials at moderate cost. Small illustrations can also be reproduced inexpensively by Xerox or Color Xerox, with enlargement or reduction commonly available.

**Drawing the map** usually consists of taping a transparent sheet over a penciled compilation and tracing the geologic features in black ink. The features are inked in an order such that features inked later are broken so as not to obscure features already inked (Fig. 16-2). A suggested order is: (1) locations of fossil collections; (2) structure symbols without numbers; (3) faults; (4) contacts; (5) cross-section lines; (6) numbers for structure symbols; (7) numbers for fossil locations; and (8) letter symbols for rock units. If the tracing is not made on a chronoflex or screened map base, these items are inked in order before item 8: (1) culture; (2) drainage; (3) geographic names; (4) contours; and (5) lines of longitude and latitude or lines of cadastral surveys (in most cases these are shown as ticks in the margin rather than across the entire map).

If the units will be patterned in black and white, the patterns are added after the map has been checked and cleaned. Patterns, like colors, are used to emphasize structural relations and to point up locations of certain units. It is generally desirable to keep patterns light except for units that need emphasis and carry no structural symbols (Fig. 16-2). Commonly, alternate units (or more) need not be patterned, because unit symbols identify them adequately. Prepared stick-on patterns (Craftint, Zipatone) are always uniform but are difficult to apply to small irregular areas and to trim around structure symbols and geographic names. It may thus be quicker and easier to hand-pattern all units except those that form large, simple areas. Maps of foliated rocks are more expressive if patterned parallel to the strike of foliation (Fig. 16-2). If structure symbols and numbers prove difficult to draw, stick-on and rub-on copy is available commercially.

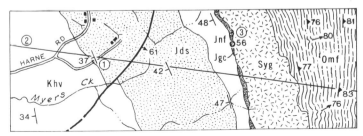

**Fig. 16-2.** Part of a geologic map, showing features broken for crucial data, as at 1, where section line and road are broken for a structure symbol. Point 2 indicates the section line and point 3 a fossil locality. The patterning of schist, on the right, is drawn parallel to local strike of the foliation.

When the map appears to be complete, a print should be made and colored in order to catch errors and omissions. The map may then be compared point by point with field sheets as a final check.

**Explanations** of geologic maps and sections serve three crucial purposes: (1) identifying the map units and possibly describing them; (2) showing their geologic ages or their age relations; and (3) identifying all geologic structures, surficial features, and special engineering geology symbols. Prior to the late 1960s, the first two purposes were traditionally met by a single column of unit boxes, each identified by its map symbol, unit name, and perhaps a brief description, and with geochronologic and chronostratigraphic units shown by brackets on the sides (Fig. 16-3). In cases where several units of the same age had to be shown side by side, and especially in cases where allochthonous assemblages required separate columns, the explanations were large and some were difficult to follow. The U.S. Geological Survey resolved these difficulties by an explanation in two parts: (1) an upper part consisting of small unit-boxes arranged to show age relations at a glance (Fig. 16-4); and (2) a lower part composed of larger boxes each accompanied by the unit name and generally a brief description of the unit (Fig. 16-4).

In both kinds of explanations, the structure symbols are placed below the unit explanation and are ordered more or less as in Appendix 7. They are followed by symbols for mapped surficial features, boreholes, and so on. Explanations are sometimes shortened by omitting the more common structural symbols, as strike and dip of bedding and standard lines for contacts and faults; however, symbols are not standardized internationally, so

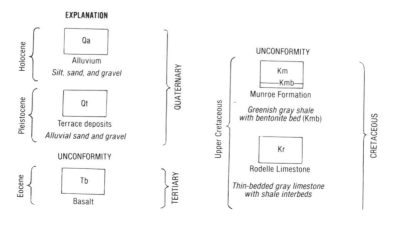

**Fig. 16-3.** Part of an explanation for a geologic map.

the readership of the report should be considered. Appendix 7 includes some optional symbols, and the definition of any symbol should be changed if a more specific or limited definition will give the map more meaning.

**The title** is usually lettered across the top or base of the map (Fig. 16-1), in a box at any corner, or above the explanation. It includes these items (numbered in Fig. 16-5): (1) name of the map, which should be in the same geographic terms as the title of the report; (2) the map's general location; (3) the name of the geologist and his or her affiliation, if any; and (4) the date when the map was completed.

Other items included in Fig. 16-5 and usually placed in the lower margin of the map or elsewhere are (5) a bar-scale in metric and English units; (6) the contour interval; (7) arrows showing the directions of true and magnetic north and the amount of the magnetic declination; (8) a location map; (9) mention of field personnel, dates of the mapping, or credit to persons who drafted the map or contributed material to it; (10) if parts of the area were mapped by different persons, a small map to locate the parts; and (11) identification of the base map. Finally, ticks and labels for longitude and latitude, for townships and ranges, or for other cadastral lines are placed in the

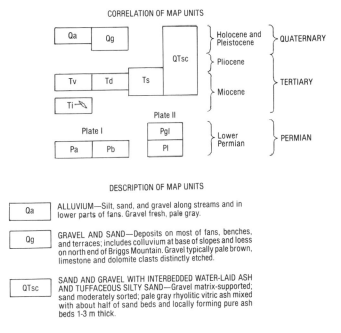

**Fig. 16-4.** Explanation for a geologic map, with the upper part showing age relations among units and the lower part descriptions of the units (the lower part is a fragment.)

margins close to the edges of the map (see a U.S. Geological Survey quadrangle map).

**Cross sections** are prepared in pencil draft as described in Section 6-5.
When the geologic map is completed and checked, they are compared with it
and corrected as necessary by placing each transparent section strip along
its section line or by using a set of dividers to make sure that all contacts
and faults are in exactly the right position relative to the ends of the section.
The map should then be studied thoroughly to be sure that the sections
include all geologic features which do not cross a section line but project into
the sections at depth.

The sections are then traced with ink. Sections on a separate plate must
have a bar scale, a title, and an explanation or a reference to the explanation
of the map plate. If the sections will be colored, lines parallel to bedding or
foliation are all that need be added to the units. If the sections will not be
colored, the units can be indicated by patterns like those used on the map or
by simplified lithologic patterns drawn parallel to bedding (Appendix 8),
whichever makes their structural relations clearer.

*Engineering geologic cross sections* and sections of surficial deposits commonly are constructed with exaggerated vertical scales in order that thicknesses and depths to units can be scaled directly from the illustrations. The

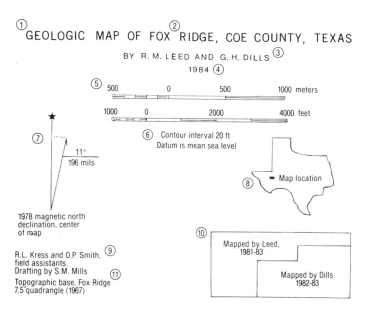

**Fig. 16-5.** Title and accessory information for a geologic map. The numbered items are
identified in the text.

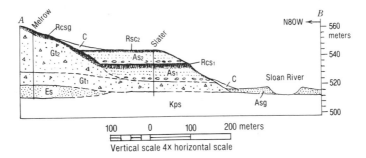

**Fig. 16-6.** Cross section with vertical scale exaggerated to show dimensions of surficial deposits, based partly on two wells (vertical lines). *Rcsg*, soil of clayey sandy gravel; *Rcs₁* and *Rcs₂*, soils of clayey sand; *C*, colluvium; *Asg*, alluvial sandy gravel; *As₁* and *As₂*, sands of alluvial terraces; *Gt₁* and *Gt₂*, successive tills; *Es*, eolian sand; and *Kps,* Cretaceous formation.

resulting exaggeration of surface slopes may be useful in pointing up certain landforms (Fig. 16-6). Although dips become exaggerated also, this is usually not a problem with surficial units.

*Columnar sections* are constructed in pencil as described in Section 11-9. They can be redrafted into several kinds of illustrations:

1. A page-sized column, typically generalized and with lithologic descriptions, to present an overview of the stratigraphy of an area. This illustration might be used in the part of the report that introduces the stratigraphic sequence (Section 16-4).

2. Detailed columns, generally without lithologic descriptions, of specific members, formations, or sequences of lithologically related units. They

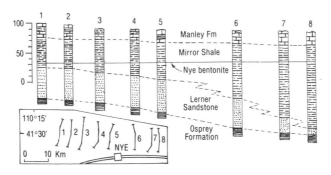

**Fig. 16-7.** Serial columnar sections documenting a marine transgression, with location map showing lines along which sections were measured.

might be used to illustrate cyclic or rhythmic sequences, series of unconformities, evidence for transgression or regression, sites of fossil collections
or other important samples, or serve in place of descriptive text. Examples
are given by Sando and others (1975, pl. 1).

3. Two or more columns illustrating lateral variations in lithology or age,
or showing correlations. Lithologic descriptions are generally omitted, and
the sections are aligned on a biozone boundary or key bed that is drawn as a
horizontal line through all of them, or they may be plotted with their bases
at the same level (Fig. 16-7). Correlation lines must be labeled if they are
based on data other than lithology. Sando and others (1975) have given
several examples. Any columnar sections drafted on separate plates must
be accompanied by all of the items described in Section 11-9.

### 16-3. Photographs, Drawings, and Diagrams

When the main subjects of the report have been determined (Section 16-1),
field sketches and photographs can be selected to present points in evidence
or clarify complex relations. This section describes some examples and
suggests ways of making them effective.

**Photographs** that show convincing relations are scarce and of exceptional value as illustrations. Writers tend to see more in their photographs
than readers do, and therefore the effectiveness of photographs should be
tested on other persons. Color photographs are generally superior for illustrating scenes and earth materials, whereas black and white photographs
are superior for structures and other features that must be shown crisply
or with a strong sense of the third dimension. Glossy prints are required for
published reports, but matte-surfaced prints are generally more attractive
and far easier to retouch. Retouching removes dust spots and interfering
highlights and shadows. To retouch, mix black ink and china white paint in
a white dish with a pointed sable brush, and apply the paint a small spot at
a time. Photographs can be clarified by adding unit symbols, arrows, and
obscure contact lines, or by tracing a line drawing from the photograph and
mounting the two side by side. If no hammer or other recognizable object is
in the photograph, the dimensions of an obvious feature should be given in
the caption.

Vertical aerial photographs often make valuable illustrations, and they
can be reduced to small prints of high quality by first obtaining transparent
negatives, which are generally available from suppliers of aerial photographs. Topographic forms on vertical photographs will appear natural
rather than inverted if they are oriented so as to appear lighted from the
upper left. Geographic orientation can be described in the caption or shown
by a north arrow on the print.

**Drawings** are used for a variety of illustrations: (1) objects shown naturally in three dimensions; (2) simplified views or cutaway sections in two dimensions; (3) small maps and cross sections; and (4) generalized diagrams, charts, and graphs. The illustrations in this book show examples of most of these varieties.

*Three-dimensional drawings* are generally shaded as though lighted from the upper left. Shading can also give a sense of texture if applied as dots (stippling) for grainy surfaces, parallel lines for smooth or lineated surfaces, and irregularly applied short lines for rough surfaces. Dark objects can be textured effectively by inking over the entire object on scratch board (cardboard with a thin coating of white clay) and scratching away part of the ink with a knife, as in Fig. 14-1. Dark backgrounds emphasize the shapes of light objects (Fig. 4-3). Smooth all-value shading is done with ink wash, powdered carbon, or pencil, and must be reproduced photographically.

*Perspective drawings* can be constructed offhand by making horizontal lines converge toward a horizon line that is erased when the drawing is completed (Fig. 16-8A). For objects with large vertical dimensions, vertical lines must also converge upward or downward (Fig. 16-8B). Preparation of accurately dimensioned perspective drawings has been described by Hoelscher and Springer (1956) and Lobeck (1958). *Isometric drawings* are easier to construct than dimensioned perspective drawings and the reader can make measurements on them by using their three axial scales, which are identical (Fig. 16-9).

*Page-size or smaller maps and sections* are often essential to illustrate structural relations, certain geologic aspects of an area, or regional relations. They should be oriented to read like the text of the report rather than being placed sideways. The location map may also illustrate geologic features mentioned in the introduction to the report or in the description of the regional setting (Fig. 16-10). An effective method of illustrating correlations among different sets of data is to make small duplicate outline maps of the area, plot a different set of data on each map, and mount the maps together, as four-to-the-page, so that readers can compare them easily. Outline maps of a pluton, for example, might be used to show fracture systems, dikes, and

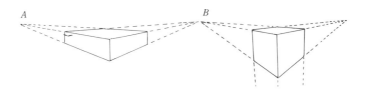

**Fig. 16-8.** Perspective diagrams based on convergence of parallel lines toward two points (*A*) and three points (*B*).

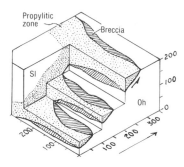

**Fig. 16-9.** Isometric diagram of a breccia body (*lined pattern*) along a fault, developed by mine workings and drill holes. Isometric drawings may be constructed from maps and subsurface data by: (1) making measurements along N-S, E-W, and vertical lines, and (2) laying out these distances on the three mutually perpendicular axes of the diagram, each of which is parallel to one of the scaled lines in the figure.

a variety of alterations. Series of small maps can also present stages in paleogeography, tectonic history, or development of igneous centers. Additional kinds of maps are described briefly under *Engineering geologic maps* in Section 16-2. Diagrammatic cross sections are often used to summarize stratigraphic relations among surficial deposits and soils, or to show depositional models interpreted from stratigraphic data.

*Diagrams* commonly used are stereographic plots of structural data, charts showing stratigraphic correlations among a number of units (as Sando and others, 1975, p. 50); and charts summarizing past uses of strati-

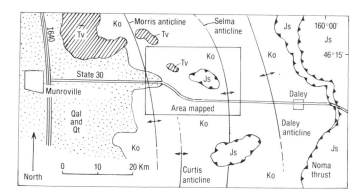

**Fig. 16-10.** Map used in text of a report to show location of the area, access to it, and major geologic structures and rock units in the vicinity.

graphic names compared to their use by the author (as in Sando and others, 1975, p. 6). Charts summarizing stratigraphic units and their corresponding events are especially valuable for reports on Quaternary history (as Mullineaux, 1970, p. 73). Diagrams comparing thicknesses of lithostratigraphic units to geologic age-ranges have been described by Meyer and Dickinson (1984). For showing horizontal orientations of currents, wind-rose graphs are recommended over sector graphs because the latter give visual bias to the more abundant orientations (Fig. 16-11). A variety of other graphs are described and illustrated diagrammatically by Bishop and others (1978, p. 70), and Tufte (1983) has described the basis, design, and preparation of a great variety of diagrams.

## 16-4. Designing the Report

Every scientific report consists of four basic elements that are arranged logically as follows:

1. What was sought or intended in making the study.
2. What was actually done.
3. The data or relations that were found or measured.
4. Conclusions that can be inferred from the data and from all other information available.

The principal subjects in each category can be determined by composing a preliminary summary, as suggested in Section 16-1. The summary provides a central thrust or design for organizing additional topics and details. Features relevant to the design are organized with care, down to the last detail, whereas irrelevant although interesting ones are omitted or given subordinate treatment. A detailed outline for one report can never serve for another, and thus the outlines presented here are of the broadest sort. Most

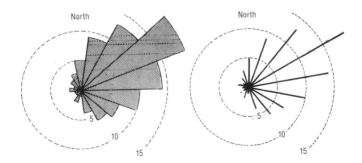

**Fig. 16-11.** Current data accumulated in 20° intervals of azimuth and depicted by a sector diagram (*left*) and wind-rose diagram. The numbered circles indicate numbers of measurements.

apply to descriptive rather than specialized reports, and are organized along traditional lines in parallel with the sequence noted above.

It is recommended that every report start with a statement of the project's purpose and the principal conclusions, thus giving readers a sense of direction as well as an interest in what is coming. This introduction does not contravene the abstract, which consists mainly of the actual findings of the study (Section 16-6).

*Reports on general geology* are traditionally organized as follows:

1. *Introduction.* (a) Statement of purpose and description of main conclusions; (b) location of area, as by referring to a small map that can also show geographic and geologic data needed to describe the purpose and conclusions (Fig. 16-10); (c) access and geography, if essential to others finding the area; (d) summary of previous work or annotated list of more important publications (only where truly pertinent to the project); and (e) acknowledgment of help or information received (or this can be placed at the end of the report).

2. *Methods.* (a) Description of any that are new or so unusual that the data of the report cannot be judged otherwise; (b) dates of the field work and other studies; and (c) mention of kinds of laboratory work, such as petrography or micropaleontology.

3. *Regional setting.* Geologic relations and histories needed as a framework for the data of the area studied; typical examples are regional structures, regionally significant stratigraphic sequences, broadly interrelated igneous bodies, and regional patterns of metamorphism.

4. *Rock units (Stratigraphy* if all units are lithostratigraphic). This major section may be introduced by a general description of the rock sequence and its genetic interrelations, and consists otherwise of systematic descriptions of the units in order of decreasing age, typically including: (a) mention of the principal kinds of rocks, the thickness or shape of the unit, and perhaps its geographic distribution; (b) brief explanation of unit name, as by reference to its original definition and any revisions; (c) complete description of rocks (Section 3-3) and their lateral variations (descriptions need not duplicate information on stratigraphic columnar sections); (d) description of unit contacts; (e) fossils or isotopic determinations and probable geologic age of the unit; and (f) interpreted origin.

5. *Structure.* An introduction may clarify geographic and age relations in complex areas; descriptions of folds, faults, joints, brecciated rock, intrusive relations, and mélanges may be put in whatever order will make their interrelations clearest; metamorphic minerals, fabrics, and structures should be described together with related tectonic structures (Section 15-8); page-size or smaller maps, sections, and diagrams may be essential.

6. *Geologic history.* Interpreted events, citing their age relations and geologic ages, ordered chronologically; references to others' data and ideas are

typically included; diagrammatic maps or sections may help in presenting paleogeography or stages in tectonic evolution.

**Reports with emphasis on biostratigraphy** may be organized like the foregoing outline except for:

1. In *Methods*, (a) discussion of the kinds of fossils used; (b) their interrelations as faunas or floras; (c) methods of collecting and separating them; and (d) discussion of the validity of the methods.

2. For each stratigraphic unit, and generally after the full rock description: (a) lists of specific fossils collected; (b) zonations or other correlations; and (c) ecology and environment.

3. The final interpretation will typically require correlation charts, serial columnar sections, diagrams of depositional models, or paleographic maps.

**Reports with emphasis on structure** will typically include:

1. In *Methods*, (a) quantitative procedures and theory behind them; (b) tests for the validity of the data; and (c) determinations of numerical ages.

2. In *Structure*, any information on (a) apparent movement relations, (b) kinds and amounts of strain; (c) metamorphic and diagenetic changes associated with deformation; (d) relations to igneous bodies; and (e) actual age relations as distinct from inferred age relations.

3. *Geologic history* will treat one or more of many possible interpretive topics, depending on the specific study; stereographic diagrams and maps showing stages of tectonic development are two of many possible diagrammatic figures.

**Reports with emphasis on ore deposits** may include:

1. In *Methods*, (a) mapping methods and (b) sampling strategies and procedures, with reference to theory.

2. In *Regional setting*, distribution of ore deposits and mineralized areas.

3. In *Rock units*, emphasis on units that are related to ore and mineralization.

4. An additional section on *Ores and mineralization* including: (a) distribution relative to rock units, fracture systems, or other structures; (b) size and shape; (c) mineral assemblages, with age relations; (d) geometric and age relations between ores, wall-rock alteration, and other metamorphic features; (e) age and origin; (f) maps and descriptions of mines and partly developed areas; and (g) suggestions for prospecting and other exploration.

**Reports with emphasis on engineering geology** are typically specialized because of the wide range of subjects (Section 5-7; and see Slosson, 1984). A broad study might include:

1. Emphasis on *Methods*, including (a) sources of all subsurface information, referring to standard tests to save space but making sure reader can

tell what was done; and (b) new methods and a discussion of their validity. Throughout the report, inferred information must be distinguished clearly from observed or tested data.

2. Emphasis in *Rock units* on surficial deposits, soils, and engineering properties (Section 5-7) with age and origin of rocks treated briefly unless they are important to engineering decisions. Abundance and nature of exposures are important and should be documented by outcrop maps (Section 5-6). Where possible, descriptions of surficial units should include their correspondence to specific landforms.

3. Emphasis in *Structure* on faults, degree and orientation of fracturing, and discontinuities (as contacts) that might lead to sliding.

4. A section on *Geomorphic relations* as they apply to surface stability, erosion problems, and site selections.

5. A section on *Hydrology* with a complete description of surface water systems and groundwater (specific topics are suggested in the subsection *Environmental assessments*, Section 5-7).

6. A section on *Geologic processes* or *Historic events* describing specific erosion, sliding, earthquakes, ground collapse, subsidence, creep, and other on-going processes(Section 5-7).

7. A section on *Economic geology* describing availability of construction materials.

8. *Engineering geology*, a principal section giving specific practical judgments or recommendations derived from all the foregoing, such as foundation conditions per area, siting of septic and refuse systems, slope stabilities per use or area, restrictions on quarrying and fill, and so on.

Maps, borehole logs, diagrams, and cross sections should carry as much of these descriptions as possible. Engineers' needs and vocabulary should guide the writing in all cases (Rose, 1965).

**Reports consisting mainly of recommendations** are typically specific in scope and not intended for general geologic readers nor for the scientific record. They are usually prepared for a limited readership and are commonly in-organization reports, often essentially memoranda. They will work best for the writer if: (1) the purpose of the study and thus the report is clearly understood by both the writer and the reader; (2) the writer takes the reader's interest, schedule, and idiosyncrasies into account; and (3) the writing is as direct and convincing as possible, with details presented in lists, tables, maps, and sections of an appendix. The report should always begin with a statement of purpose and a brief description of the recommendations and their principal logic or basis. The remainder of the report develops a fuller argument, and can be organized much as the outlines given above. Tichy (1966) and Rose (1965) have given valuable suggestions for preparing this kind of report.

**16-5. The Writing Itself**

Because many geologic relations combine elements of time, place, and degree, clear geologic writing requires extra effort. Some general suggestions are: (1) deal with only one topic at a time; (2) think carefully about what you want to say; and (3) say it as simply and directly as possible. Brevity is desirable but not as important as saying exactly what you mean. After writing a sentence, study it to see if each word carries the meaning accurately. Perhaps the statement can be made more direct by deleting some adjectives, adverbs, or articles.

Use terms that will be understood by geologists generally, and avoid saying anything that will interrupt the reader's progress. In-words (jargon) and vague new terms tend to have this effect. Many readers will also pause at sexist terms, such as the *he* without the accompanying *she*. Above all, do not let your opinions show in descriptive writing. Your readers will pause, become wary, and begin to think thoughts other than the ones you intended.

Unless the meaning is obvious, try not to interchange words expressing place with those expressing time, prevalence, or abundance. *Locally, here and there, nowhere, everywhere* are place words; *often, sometimes, frequently, never, always* are time-words; and *usually, typically, generally, invariably, ordinarily,* and *commonly* are words expressing prevalence. Specific terms of abundance, such as *many, most, few, several, abundant,* and *scarce* should be used where possible instead of vague terms, such as *considerable, appreciable, adequate,* and *some.*

Statements can generally be made more direct by removing any variations of *occur, exist, are present, are found,* and *are observed.* Thus change *Faults found on the west side of* to *Faults on the west side of;* change *Occurrences of hornfels on Stevens Hill indicate that a pluton is present at some depth beneath* to *Hornfels on Stevens Hill indicates an underlying pluton;* and change *Observations of cross-bedding in the sandstones indicate currents* to *The cross-bedded sandstones indicate currents.* Tichy (1966, p. 123) compiled a valuable list of similarly diffuse verbs and verbal constructions.

Avoid explaining too much (teaching), and use long, pretentious words only when their meaning is just right. Long words based on Latin or Greek are numerous in geology and can in some cases be replaced by simpler ones, such as *stratification* by *bedding, highly indurated* by *tough, dessicated* by *dry; lithologic unit* by *rock unit, arenaceous* by *sandy, portion* by *part, melanocratic* by *dark, transformation* by *change,* and *homogeneous* by *uniform.* Nongeologic words and phrases that result in false elegance or wordiness can be corrected by using the list prepared by Tichy (1966, p. 137).

*Data, strata, criteria,* and *media* are the plural forms of *datum, stratum, criterion,* and *medium.* And note the use of each of the italicized words:

The study *comprised* current measurements, mapping, and petrography.
The formation is *composed* of (or *consists* of) sandstone and shale.
The three intrusions *constitute* a genetic series.
The evidence *implies* these events—.
The geologist *inferred* the event from these relations—.

The section "Review of English I" in Bishop and others (1978) presents many additional suggestions for word usage and correct writing.

Degrees of precision and accuracy are important in geologic descriptions and can often be covered without long statements. Doubt in identification or correlation may be expressed by inserting a question mark, as *metamorphosed Daley(?) Shale; Tertiary(?) intrusions,* and *rhyolite(?) vitric tuff.* Ranges and averages of quantitative data can be written as *5-17 m (average, 12 m).* Significance of numbers is shown by rounding off to the order of reliability; for example, a stratigraphic thickness accumulated by measuring ten units to the nearest meter might sum to 2717 m, but would be given in the text as *about 2720 m* or *about 2700 m.* Customarily, numbers followed by a plus or minus quantity, such as *113 ± 5,* indicate reliability with approximately a 90% degree of confidence (two standard deviations).

Use the metric system generally, but include English equivalents when appropriate for readers who will need them.

Rock names are described in Chapter 4 and geographic names in Section 16-2. Fossil names are covered by rules of biologic taxonomy that were presented succinctly by Schenk and McMasters (1956). Guidelines for use of fossil names and for expressing degrees of uncertainty in identifications are also described by Bishop and others (1978, p. 181). Geologic terms can be checked for spelling and meaning by reference to Bates and Jackson (1980, 1984).

Use of stratigraphic names and names of unstratified rock units should follow the stratigraphic code (North American Commission on Stratigraphic Nomenclature, 1983). Modifications of the code will be made from time to time, and references to them are available from the Chairman of the Commission on Stratigraphic Nomenclature (c/o American Association of Petroleum Geologists, Box 979, Tulsa, OK 74101). Correct current usage of specific unit names can be obtained from the U.S. Geological Survey, as described in Section 5-3.

**Style,** as used here, is the overall aspect of writing that guides the reader's flow of thought, of appreciation, and of feeling. When style is right, the reader gets exactly what the writer intended to convey. Moreover, the reader is unaware of getting it, and thus reads on easily to the next topic and the next. The meanings of words and phrases are a major element of style, as are their spoken sound and the degree of emphasis and rhythm they impart to phrases and sentences. Lengths of sentences and paragraphs, parallelism within paragraphs, and departures from normal syntax are further elements of style. To learn about these specifics and at the same time get a sense of first-rate style, read Strunk and White (1959) from cover to cover. Read it carefully and in one sitting if possible (it's a little book).

**Revision** is typically essential but of little value until a complete draft of the report can be set aside for several weeks. Other tasks can be completed during this period, as final drafting of figures and writing figure captions. When it is time for revising, try reading the report aloud or have someone read it to you. Another way to sense major problems with awkward or heavy writing is to first read something that is well written; for example, reread Strunk and White (1959). Comparing the figures and figure captions with the text will almost always lead to simplification. Let the figures carry as much of the presentation as possible, and avoid giving the same information in text and captions.

## 16-6. Specific Parts of the Report; Format

**Acknowledgments** are traditionally placed either in the introduction or at the end of the report. They credit help from individuals and organizations not cited elsewhere in the report. Each credit should be simple and specific and need not state gratitude (which will be implied).

**References cited** in the report are listed alphabetically by author in a separate section placed immediately after the last part of the text. The exact style of citations varies from decade to decade and journal to journal, so that recent journal issues should be consulted for preparing lists that will be published. Publications of the U.S. Geological Survey or the Geological Society of America may be used as guides for reports that will not be published; however, fuller references (e.g., those in this book) are never incorrect. The basic rule is to include enough information so that readers can find the items in libraries easily. Names of journals should not be abbreviated unless the terms will be absolutely clear.

Citations in the text are given by the author's name and the year of publication. Page numbers should be included if the reader will need them to find the relevant material, such as topics in long documents that are not subheaded or indexed. Citations for unpublished information should give the person's full name, an affiliation if appropriate, and the nature and date of the communication; for example, *The fossils were identified by William P. Nash of the Illinois State Survey (written communication, 1974).*

Many abstracts prepared for talks are inaccurate or incomplete because they are written long before studies are completed. Thus always correspond with authors before using and citing these sources.

**Lists and tables** may save space or may present data more clearly than it could be presented in the text. Lists are organized in one or two columns, and the items should have parallel construction if they are statements. The stratigraphic log in Section 11-9 is an example. Items in tables are divided by horizontal and vertical lines or by enough space to imply these lines. Tables typically present data sets or classifications based on two kinds of variance (Table 4-2). Tables may be ideal for comparing the characterizing properties and features of a sequence of rock units.

Lists are generally introduced directly by the text and are within the text, whereas tables have titles and are presented much like figures. References to sources and other pertinent information in tables are generally placed at the base of the table. Bishop and others (1978, p. 82) have given additional suggestions and several examples for composing moderately complex tables.

**The abstract** was well defined by Landes (1966) as "a condensation and concentration of the *essential information* in the paper" (his italics). A good time to prepare an abstract is just after the first revision of the report, when the material is well in mind but the report is not quite finished. The preliminary summary described in Section 16-1 may be a useful first version of an abstract but it may have to be condensed, because most abstracts should be no more than two typed pages long, and those for published papers shorter than a page. A general suggestion, however, is not to be too concerned about length at the outset. Say all that seems necessary and condense it later.

Overall, the abstract should give the study's purpose, methods, and principal results. The results are its main subject and should comprise data and relations as well as the conclusions they imply. The brevity required causes some writers to give up and write such statements as *Structural and textural relations between the granitic bodies and the metamorphic rocks indicate that metamorphism and intrusion were synchronous*, when they could have said *Concentric patterns of mapped mineral zones and metamorphic textures in early granitic dikes indicate metamorphism and intrusion were synchronous*. The idea is to make the abstract as informative and specific as possible. It is the only part of the report that most people will manage.

Points need not be in the same order as they are in the text although that is preferable. Ideal coverage will vary from report to report; however, it is generally preferable to deal firmly with important topics rather than to cover all topics.

As in the main text, write to the reader. A sign of success will be wanting to modify the text when the abstract is finished.

**The format** of a report comprises: (1) typing it double-spaced throughout, with margins of at least 1 in. on all sides; (2) using headings and subheadings to guide the reader where necessary, and making their rank obvious; (3) leaving an extra space after each headed section; (4) numbering pages throughout, including illustrations bound with the text; (5) omitting footnotes, if necessary by modifying the text; and (6) inserting figures and tables where they will be most useful to the reader. *Plates* are oversized illustrations that are folded and inserted in an envelope glued firmly inside the back cover of the report or placed in a separate, fully titled cover. Plates should either be folded so that the plate number can be seen without unfolding them, or the number should be lettered on the outside of the folded plate.

**Front matter** generally consists of a title page, an abstract, a table of contents, and a list of illustrations, in that order. In addition to the title of the report, the title page includes the name of the author, his or her affiliation during the project, and the date when the report was completed (or published). The abstract has already been described. The table of contents is a paginated list of headings and subheadings, but perhaps without those of lowest rank, such as the repetitive subheads under each rock unit. The rank of headings is indicated by indenting each successively lesser rank at last five spaces to the right. The list of illustrations gives numbers, captions, and pages of text figures, and then numbers and names of plates. Long captions should be reduced in the list, as from *Outcrop of Farley Conglomerate viewed from the east, showing cross sections of filled stream channels* to *Outcrop of Farley Conglomerate* or *Channel structures in Farley Conglomerate*, depending on the chief message.

### References Cited

Bates, R. L., and Jackson, J. A., editors, 1980, *Glossary of Geology*, 2nd edition: Falls Church, VA, American Geological Institute, 749 p.

Bates, R. L., and Jackson, J. A., editors, 1984, *Dictionary of geological terms*: Garden City, NY, Anchor Press/Doubleday, 571 p.

Bishop, E. E., Eckel, E. B., and others; Eric, J. H., coordinator, 1978, *Suggestions to authors of the reports of the United States Geological Survey*, 6th edition: Washington, D.C., U.S. Government Printing Office, 273 p.

Cochran, W., Fenner, P., and Hill, M., editors, 1979, *Geowriting; a guide to writing, editing, and printing in earth science*, 3rd edition: Falls Church, VA, American Geological Institute, 80 p.

Easterbrook, D. J., 1973, *Map showing percolation rates of earth materials in western Whatcom County, Washington*: U.S. Geological Survey Miscellaneous Geologic Investigations Map I-854-A.

Easterbrook, D. J., 1976, *Map showing engineering characteristics of geologic materials, western Whatcom County, Washington*: U.S. Geological Survey Miscellaneous Investigations Series Map I-854-D.

Hoelscher, R. P., and Springer, C. H., 1956, *Engineering drawing and geometry*: New York, John Wiley & Sons, 520 p.

Landes, K. K., 1966, A scrutiny of the abstract, II: *American Association of Petroleum Geologists Bulletin*, v. 50, p. 1992.

Lobeck, A. K., 1958, *Block diagrams and other graphic methods used in geology and geography*, 2nd edition: Amherst, MA, Emerson-Trussell Book Co., 212 p.

Meyer, L., and Dickinson, W. R., 1984, Time-thickness diagrams: Simultaneous display of lithostratigraphic thickness and chronostratigraphic age: *Geology*, v. 12, p. 7.

Mullineaux, D. R., 1970, *Geology of the Renton, Auburn, and Black Diamond quadrangles, King County, Washington*: U.S. Geological Survey Professional Paper 672, 92 p.

North American Commission on Stratigraphic Nomenclature, 1983, North American stratigraphic code: *American Association of Petroleum Geologists Bulletin*, v. 67, p. 841-875.

Péwé, T. L., and Bell, J. W., 1976, *Map showing foundation conditions in the Fairbanks D-2 SW quadrangle, Alaska*: U.S. Geological Survey Miscellaneous Investigations Series Map I-829-E.

Rose, D., 1965, A civil engineer reads a geology report: *Geotimes*, v. 10, no. 1, p. 9-12.

Sando, W. J., MacKenzie, G., Jr., and Dutro, J. T., Jr., 1975, *Stratigraphy and geologic history of the Amsden Formation (Mississippian and Pennsylvanian) of Wyoming*: U.S. Geological Survey Professional Paper 848-A, 83 p.

Schenk, E. T., and McMasters, J. H., 1956, *Procedure in taxonomy*, 3rd edition (revised by A. M. Keen and S. W. Muller): Stanford, CA, Stanford University Press, 149 p.

Slosson, J. E., 1984, Genesis and evolution of guidelines for geologic reports: *Association of Engineering Geologists Bulletin*, v. 21, p. 295-316.

Strunk, W., Jr., with revisions, an introduction, and a new chapter on writing by E. B. White, 1959, *The elements of style*, New York, Macmillan Co., 71 p.

Tichy, H. J., 1966, *Effective writing for engineers, managers, scientists*: New York, John Wiley & Sons, 337 p.

Tufte, E. R., 1983, *The visual display of quantitative information*: Cheshire, CT, Graphics Press, 197 p.

Varnes, D. J., 1974, *The logic of geologic maps, with reference to their interpretation and use for engineering purposes*: U.S. Geological Survey Professional Paper 837, 48 p.

# ■ Appendixes

## APPENDIX 1. Equipment and Supplies for Geologic Field Work

Most of the items in this check list are described in Chapter 2 and in Section 6-2, or Section 7-1. For a list of equipment for plane table mapping, see Table 8-1.

Adhesive tape
Aerial photographs and indexes
Altimeter
Auger, soil
Binoculars
Calculator, pocket
Camera, tripod, film, etc.
Canteens
Cement, cellulose
Chemicals for staining rocks
Cold chisel
Color pencils
Colored tape or paint for marking localities
Compass, Brunton or other
Clinometer (or Brunton compass)
Drawing board
Erasers
Field case for maps and photographs
Field glasses
First aid kit
Flashlight
Gloves
Gold pan
Grain-size card
Hammer, geologist's
Hand lens
Hand level (or Brunton compass)
Hydrochloric acid, dilute
Ink, waterproof; black, brown, blue, red, and green
Insect repellent
Jacob staff
Knapsack
Lettering set
Looseleaf binder
Magnet
Maps, topographic, geologic
Microscope, binocular

Mineral hardness set
Moil
Needles, for marking photographs, etc.
Notebooks, field
Paper, lined
Paper, quadrille
Paper, scratch
Pen, drop circle
Pen holders
Pen points
Pen, ruling
Pens, ballpoint
Pens, inkflow, for photographs
Pencils, 3B to 9H
Pencil pocket clips
Pencil pointer (file or sandpaper)
Pick or mattock
Plastic drafting film, matte-surfaced
Pocket knife
Proportional dividers
Protracters
Rain gear
Rangefinder
Reference library
Safety goggles
Sample bags
Scale, plotting, 6 in.
Shovel
Stereographic net
Stereoscopes
Straightedge, steel
Tables, mathematical
Tally counter
Tape, 6-ft
Tape, 100-ft
Triangles, drawing
T-square
Typewriter
Watch

## APPENDIX 2. Abbreviations of Geologic Terms

Abbreviations for nouns are capitalized to distinguish them from adjectives (see, for example, *dolomite* and *dolomitic*). For a more extensive list, see Mitchell, J.G., and Maher, J.C., 1957, Suggested abbreviations for lithologic descriptions: *Bulletin of the American Association of Petroleum Geologists*, vol. 41, p. 2103-2107.

| | | | |
|---|---|---|---|
| abundant | abnt | clinopyroxene | Cpx |
| acicular | acic | cobble | Cbl |
| actinolite | Act | conglomerate | Cgl |
| aggregate | Aggr | contact | Ctc |
| albite | Ab | cordierite | Cord |
| amorphous | amor | corundum | Cor |
| amount | Amt | cross-bedded | xbdd |
| amphibole | Amph | cross-bedding | Xbdg |
| amphibolite | Ampht | cross-laminated | xlam |
| andalusite | Andal | cross section | X sect |
| angle | ∠ | crystal | Xl |
| angular | ang | crystalline | xln |
| andesite | And | diameter | Diam |
| anhedral | anhed | different | diff |
| anhydrite | Anhy | diopside | Diop |
| approximate | approx | disseminated | dissem |
| arenaceous | aren | dolomite | Dol |
| argillaceous | arg | dolomitic | dol |
| argillite | Arg | elevation | Elev |
| arkosic | ark | equivalent | equiv |
| arsenopyrite | Ars | evaporite | Evap |
| asphaltic | asph | exposure | Exp |
| average | Ave | feldspathic | feld |
| bedded | bdd | foliated | fol |
| bedding | Bdng | foliation | Fol |
| bentonite | Bent | foraminifer | Foram |
| biotite | Bio | formation | Fm |
| bituminous | bit | fragmental | frag |
| boulder | Bldr | glauconite | Glauct |
| brachiopod | Brach | granite | Gr |
| breccia | Bx | granodiorite | Grd |
| calcareous | calc | granular | gran |
| carbonaceous | carb | graptolite | Grap |
| cavernous | cav | graywacke | Gwke |
| cement | Cmt | greenstone | Grnst |
| chalcedony | Chal | gypsiferous | gyp |
| chalcopyrite | Cp | hematitic | hem |
| chlorite | Chl | horizontal | horiz |
| claystone | Clst | hornblende | Hbl |
| cleavage | Clv | hornfels | Hfls |

| | | | |
|---|---|---|---|
| hypidiomorphic | hypid | pyritic | py |
| igneous | ign | pyroxene | Px |
| ignimbrite | Ignm | pyroxenite | Pxt |
| ilmenite | Ilm | pyrrhotite | Pyrr |
| inclusion | Incl | quartz | Qz |
| interbedded | intbdd | quartzite | Qzt |
| intrusion | Intr | radiolarian | Rad |
| irregular | ireg | reconnaissance | Recon |
| joint | Jnt | regular | reg |
| kaolinite | Kaol | rhyolite | Rhy |
| K-feldspar | Kspar | rocks | Rx |
| laminated | lam | rounded | rndd |
| limestone | Ls | sandstone | Ss |
| limonite | Lim | saturated | sat |
| lithologic | lith | secondary | sec |
| magnetite | Mag | sediment | Sed |
| maximum | Max | sedimentary | sed |
| member | Mbr | serpentine | Spt |
| metamorphic | met | siliceous | sil |
| microline | Micr | siltstone | Sltst |
| montmorillonite | Mont | soluble | sol |
| mudstone | Mdst | sphalerite | Sphal |
| muscovite | Musc | station | Sta |
| nepheline | Neph | staurolite | Staur |
| nodular | nod | structure | Struc |
| olivine | Ol | stratigraphic | strat |
| orthopyroxene | Opx | surficial | surf |
| orthoclase | Orth | tabular | tab |
| outcrop | Otcp | temperature | T |
| pebble | Pbl | topographic | topo |
| pegmatite | Peg | tourmaline | Tourm |
| peridotite | Perid | tremolite | Trem |
| permeability | Perm | unconformity | Uncf |
| phenocryst | Pheno | variegated | vrtg |
| phlogopite | Phlog | vegetation | Veg |
| phosphatic | phos | vertebrate | Vrtb |
| plagioclase | Plag | volcanic | volc |
| point | Pt | volume | Vol |
| porphyritic | porph | wollastonite | Woll |
| probable | prob | xenolith | Xen |

# APPENDIX 3. Percentage Diagrams For Estimating Compositon By Volume*

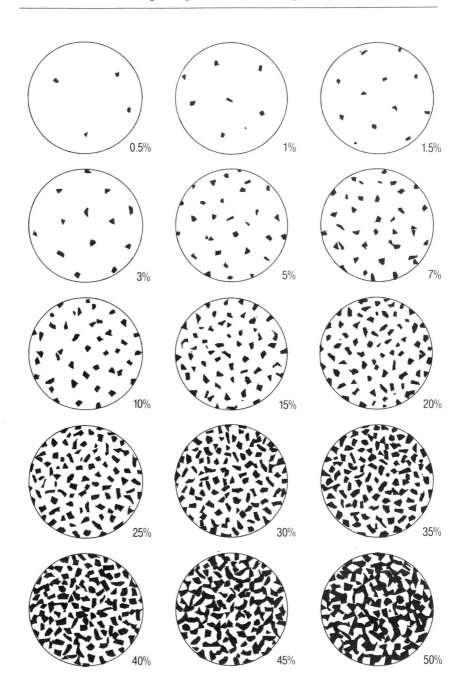

*To convert the results to weight percentages, multiply each volume percentage by the specific gravity of that mineral and recalculate the resulting numbers so that they sum to 100.

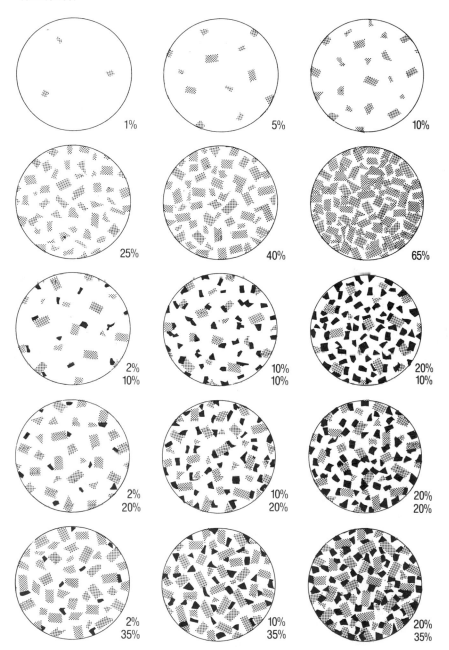

**APPENDIX 4: Strength (Coherence) and Hardness of Rocks and Sediments**

A material's *strength* is its resistance to breaking under compression, and its *hardness* is its resistance to scratching or grinding. Strength varies with degree of coherence and, for most clay-rich materials, with moisture content. The scales presented here are not standards but may serve for systematic comparisons. The hammer tests should be made with a 2-lb hammer on pieces about 4 in. (10 cm) thick placed on a solid, hard surface, and tests with the hands should be made on pieces about 1.5 in. (4 cm) thick. The pieces must not have incipient fractures, and therefore several should be tested. The sound tests, which are only accessory, should be made on solid outcrops or on fragments thicker than 1 ft (0.3m).

A. SCALE OF STRENGTH
1. *Loose.* Sediment flows when dry and thus cannot be sampled in aggregate.
2. *Very friable.* Sediment crumbles so easily that pieces are difficult to collect intact.
3. *Friable.* Sediment crumbles under light pressure in the hands.
4. *Somewhat friable.* Rock breaks in the hands under moderate pressure.
5. *Firm, slightly friable.* Rock breaks with difficulty in the hands but readily by hammer blows, sounding a dull "whop."
6. *Weak, nonfriable.* Rock cannot be broken in the hands but breaks under light hammer blows, sounding "whop."
7. *Moderately strong.* Rock breaks under moderate hammer blows, sounding "whap."
8. *Strong.* Rock breaks under hard hammer blows, sounding "whack," with a ring.
9. *Very strong.* Rock difficult to break with hammer, sounding "boink."
10. *Unusually strong.* Rock impossible to break with hammer, sounding like steel.

B. SCALE OF ROCK HARDNESS
1. *Soft.* All rocks weaker than 5 on the scale above.
2. *Moderately hard.* Slightly friable or nonfriable rocks consisting mainly of soft minerals, as carbonates, sulfates, micas, and clays.
3. *Hard.* Nonfriable rocks consisting almost entirely of minerals with hardnesses of 4, 5, or 6 on the Mohs scale, and quartz-rich rocks with strength of 6 or 7.
4. *Very hard.* Rocks stronger than 7 on the scale above and consisting mainly of minerals harder than 6 on the Mohs scale.

## APPENDIX 5.
### Township-section Cadastral System of the U.S. Bureau of Reclamation

The dimensions given in the diagrams below are ideal (for example, some townships are somewhat more or less than 6 mi on a side). Most section corners and some ¼ corners can be found in the field by a vertical pipe with a scribed brass head, or by a concrete or stone marker. As examples of location notations, the 40-acre plot (lower right) is the SE¼ of the SE¼ of Sec. 12, T2S, R1W, and point A is 394 ft south and 552 ft east of the N ¼ corner of Sec. 12, T2S, R1W.

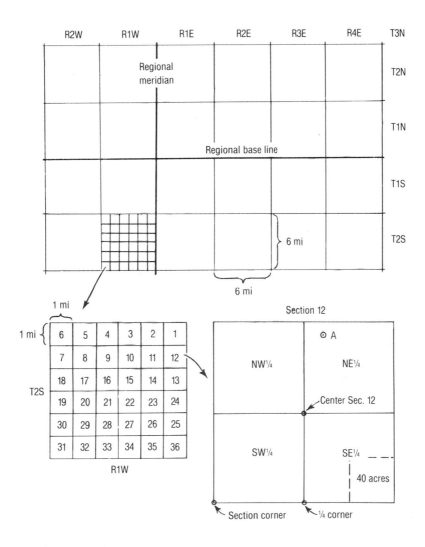

**APPENDIX 6:**
**Use of Charts for Standardizing Colors of Sediments and Rocks**

---

Color is an important aspect of materials, and it must be standardized for two reasons: (1) a given color will often be named differently by different persons, and (2) a given color will appear differently when seen next to other colors (e.g., gray appears bluish when seen next to orange or brown earth colors). Both of these problems can be resolved by use of charts that provide a color notation based on standard color chips, and that permit comparisons against a neutral gray background. The most widely accepted system of color notation in the United States is the Munsell color system, and it is the basis of the *Munsell Soil Color Charts,* which provide a detailed coverage of colors characteristic of soils, sediments, and most rocks. These charts are bound in a hard loose-leaf cover and are available from the Munsell Color Company, 2441 N. Calvert St., Baltimore, MD 21218. A somewhat broader but less detailed coverage is given in the *Rock-Color Chart,* available from The Geological Society of America, P.O. Box 9410, Boulder, CO 80301. The descriptions and instructions given below specify the *Munsell Soil Color Charts* but can be applied readily to the *Rock-Color Chart.*

Each chart is based on a primary color or on a specific mixture of primary colors, designated by a symbol at the top of the chart. As examples, the hue of the chart with the symbol 5R is spectrum red; the chart 5Y is spectrum yellow; and the chart 5YR spectrum orange. Additional charts are based on intermediate hues, such as 7.5 YR, which is a yellowish orange. For each chart, the basic hue is mixed with near-white in five or six graded steps arranged across the top of the chart, from near-white on the left to the fullest *chroma* (amount of the basic hue) appearing in most soils and rocks (see the diagram). Black is then mixed with each of these chromatic divisions to give seven grades *(values)* forming vertical sets of chips that range from mixtures with nearly pure white to mixtures with nearly pure black (pure white would have a value of 10 and black a value of 0). Colors thus become lighter in value from bottom to top and more intensely colored (chromatic) from left to right. No chart has a full array of chips because colors not represented in natural materials are omitted.

Small sediment or rock samples are placed under the apertures in order to find a color match. The color is then recorded as the letter-and-number of the chart plus the numbers of the value and chroma. For example, a rock matching the chip X in the diagram would be color 5YR 6/4. Intermediate colors can usually be estimated to the nearest half-and-half mixture, and are designated accordingly (as 5YR 6.5/4). Color names are given on the page facing each chart but are not subdivided as fully as the color chips and their notation.

5YR

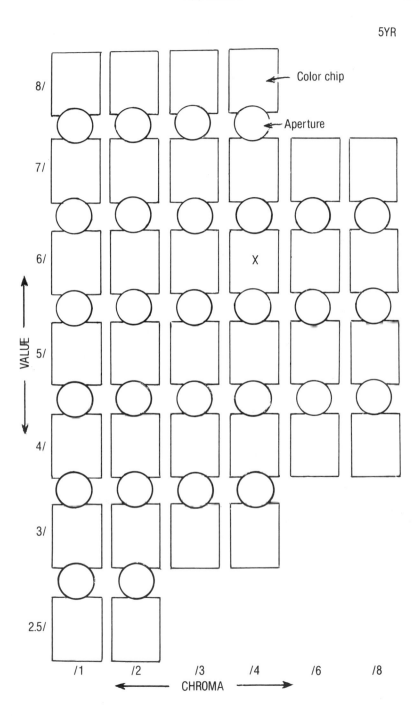

## APPENDIX 7: Symbols for Geologic Maps

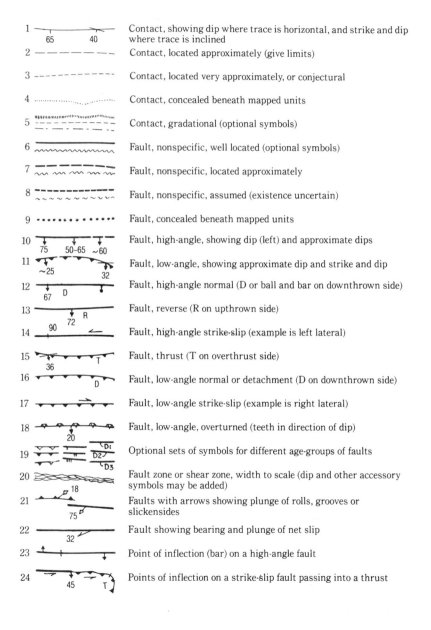

1. Contact, showing dip where trace is horizontal, and strike and dip where trace is inclined

2. Contact, located approximately (give limits)

3. Contact, located very approximately, or conjectural

4. Contact, concealed beneath mapped units

5. Contact, gradational (optional symbols)

6. Fault, nonspecific, well located (optional symbols)

7. Fault, nonspecific, located approximately

8. Fault, nonspecific, assumed (existence uncertain)

9. Fault, concealed beneath mapped units

10. Fault, high-angle, showing dip (left) and approximate dips

11. Fault, low-angle, showing approximate dip and strike and dip

12. Fault, high-angle normal (D or ball and bar on downthrown side)

13. Fault, reverse (R on upthrown side)

14. Fault, high-angle strike-slip (example is left lateral)

15. Fault, thrust (T on overthrust side)

16. Fault, low-angle normal or detachment (D on downthrown side)

17. Fault, low-angle strike-slip (example is right lateral)

18. Fault, low-angle, overturned (teeth in direction of dip)

19. Optional sets of symbols for different age-groups of faults

20. Fault zone or shear zone, width to scale (dip and other accessory symbols may be added)

21. Faults with arrows showing plunge of rolls, grooves or slickensides

22. Fault showing bearing and plunge of net slip

23. Point of inflection (bar) on a high-angle fault

24. Points of inflection on a strike-slip fault passing into a thrust

| | | |
|---|---|---|
| 25 | | Fault intruded by a dike |
| 26 | | Faults associated with veins |
| 27 | | Anticline, showing trace and plunge of hinge or crest line (specify) |
| 28 | | Syncline (as above), showing dip of axial surface or trough surface |
| 29 | | Folds (as above), located approximately |
| 30 | | Folds, conjectural |
| 31 | | Folds beneath mapped units |
| 32 | | Asymmetric folds with steeper limbs dipping north (optional symbols) |
| 33 | | Anticline (top) and syncline, overturned |
| 34 | | Antiformal (inverted) syncline |
| 35 | | Synformal (inverted) anticline |
| 36 | | Antiform (top) and synform (stratigraphic sequence unknown) |
| 37 | | Separate dome (left) and basin |
| 38 | | Culmination (left) and depression |
| 39 | | Small anticline and syncline, showing shapes in horizontal section |
| 40 | | Vertically plunging anticline and syncline |
| 41 | | Monocline, south-facing, showing traces of axial surfaces |
| 42 | | Steeply plunging monocline or flexure, showing trace in horizontal section and plunge of hinges |
| 43 | | Plunge of hinge lines of small folds, showing shapes in horizontal section |
| 44 | | Strike and dip of beds or bedding |
| 45 | | Strike and dip of overturned beds |
| 46 | | Strike and dip of beds where stratigraphic tops are known from primary features |
| 47 | | Strike and dip of vertical beds or bedding (dot is on side known to be stratigraphically the top) |
| 48 | | Horizontal beds or bedding (as above) |

49    $\overset{40}{-'-}$   —ı—    Approximate (typically estimated) strike and dip of beds

50   ~90   $\overset{30}{\cdot}$   ~30    Strike of beds exact but dip approximate

51   $\overset{12}{}$    $\overset{12}{}$    Trace of single bed, showing dip where trace is horizontal and where it is inclined

52   $\overset{15}{}$   $\overset{15}{}$   $\overset{15}{}$    Strike and dip of foliation (optional symbols)

53 —◇—   —◆—   ⊢—    Strike of vertical foliation

54 —◇—   —✦—   ⊢⊤    Horizontal foliation

55   $\overset{12}{}$   $\overset{12}{}$   $\overset{12}{}$    Strike and dip of bedding and parallel foliation

56   $\overset{10}{}$    $\overset{5}{}$    Strike and dip of joints (left) and dikes (optional symbols)

57     Vertical joints (left) and dikes

58     Horizontal joints (left) and dikes

59 •••••••    $\overset{\uparrow}{\text{qz-py}}$    Strike and dip of veins (optional symbols)

60 •••••••    qz-py    Vertical veins

61 •••••    qz∥py    Horizontal veins

62    ——→ 35    Bearing (trend) and plunge of lineation

63    ◇ᴸ    ◄——►    Vertical and horizontal lineations

64    ≻—► 20    Bearing and plunge of cleavage-bedding intersection

65    ✕—►   ◇—►    Bearing and plunge of cleavage-cleavage intersections

66    ●—►   $\overset{bio}{}$ ◻—►    Bearings of pebble, mineral, etc. lineations

67    $\overset{40}{}$   —◇—►    Bearing of lineations in plane of foliation

68    $\overset{15}{}$ ◄△►   ◄—◆—►    Horizontal lineation in plane of foliation

69    —◆◇    Vertical lineation in plane of vertical foliation

70    ——→   —//⧸→   ⧸⧸⧸→   ∞—→    Bearing of current from primary features; from upper left: general; from cross-bedding; from flute casts; from imbrication

| | | |
|---|---|---|
| 71 | | Bearing of wind direction from dune forms (left) and cross-bedding |
| 72 | | Bearing of ice flow from striations (left) and orientation of striations |
| 73 | | Bearing of ice flow from drumlins |
| 74 | | Bearing of ice flow from crag and tail forms |
| 75 | | Spring |
| 76 | | Thermal spring |
| 77 | | Mineral spring |
| 78 | ▲ | Asphaltic deposit |
| 79 | BIT | Bituminous deposit |
| 80 | | Sand, gravel, clay, or placer pit |
| 81 | | Mine, quarry, or open pit |
| 82 | | Shafts: vertical, inclined, and abandoned |
| 83 | | Adit, open (left) and inaccessible |
| 84 | X | Trench (left) and prospect |
| 85 | ● ○ ◌ | Water wells: flowing, nonflowing, and dry |
| 86 | ● | Oil well (left) and gas well |
| 87 | | Well drilled for oil or gas, dry |
| 88 | | Wells with shows of oil (left) and gas |
| 89 | | Oil or gas well, abandoned (left) and shut in |
| 90 | ⊙ | Drilling well or well location |
| 91 | | Glory hole, open pit, or quarry, to scale |
| 92 | | Dump or fill, to scale |

## APPENDIX 8:
## Lithologic Patterns for Stratigraphic Columns and Cross Sections

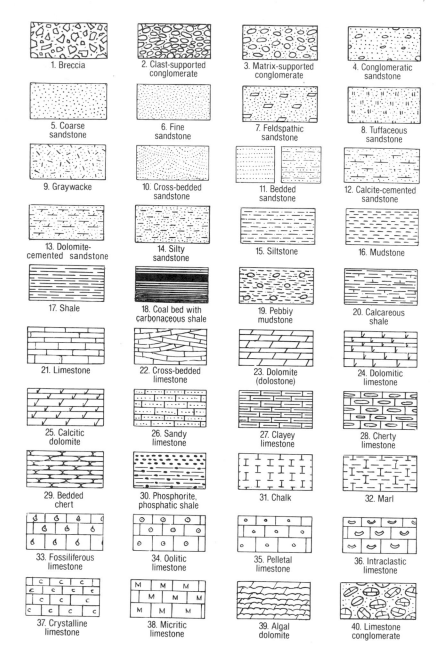

1. Breccia
2. Clast-supported conglomerate
3. Matrix-supported conglomerate
4. Conglomeratic sandstone
5. Coarse sandstone
6. Fine sandstone
7. Feldspathic sandstone
8. Tuffaceous sandstone
9. Graywacke
10. Cross-bedded sandstone
11. Bedded sandstone
12. Calcite-cemented sandstone
13. Dolomite-cemented sandstone
14. Silty sandstone
15. Siltstone
16. Mudstone
17. Shale
18. Coal bed with carbonaceous shale
19. Pebbly mudstone
20. Calcareous shale
21. Limestone
22. Cross-bedded limestone
23. Dolomite (dolostone)
24. Dolomitic limestone
25. Calcitic dolomite
26. Sandy limestone
27. Clayey limestone
28. Cherty limestone
29. Bedded chert
30. Phosphorite, phosphatic shale
31. Chalk
32. Marl
33. Fossiliferous limestone
34. Oolitic limestone
35. Pelletal limestone
36. Intraclastic limestone
37. Crystalline limestone
38. Micritic limestone
39. Algal dolomite
40. Limestone conglomerate

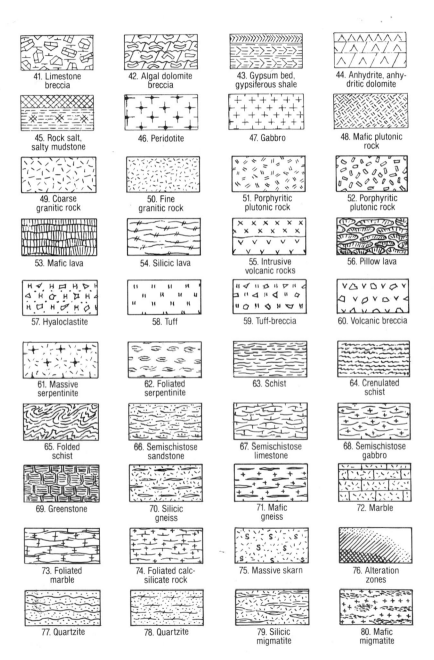

41. Limestone breccia

42. Algal dolomite breccia

43. Gypsum bed, gypsiferous shale

44. Anhydrite, anhydritic dolomite

45. Rock salt, salty mudstone

46. Peridotite

47. Gabbro

48. Mafic plutonic rock

49. Coarse granitic rock

50. Fine granitic rock

51. Porphyritic plutonic rock

52. Porphyritic plutonic rock

53. Mafic lava

54. Silicic lava

55. Intrusive volcanic rocks

56. Pillow lava

57. Hyaloclastite

58. Tuff

59. Tuff-breccia

60. Volcanic breccia

61. Massive serpentinite

62. Foliated serpentinite

63. Schist

64. Crenulated schist

65. Folded schist

66. Semischistose sandstone

67. Semischistose limestone

68. Semischistose gabbro

69. Greenstone

70. Silicic gneiss

71. Mafic gneiss

72. Marble

73. Foliated marble

74. Foliated calc-silicate rock

75. Massive skarn

76. Alteration zones

77. Quartzite

78. Quartzite

79. Silicic migmatite

80. Mafic migmatite

## APPENDIX 9:
## Fossil and Structure Symbols for Columnar Sections and Field Notes*

| Symbol | Name | Symbol | Name |
|---|---|---|---|
| | Algae | | Tree trunk fallen |
| | Algal mats | | Trilobites |
| | Ammonites | | Vertebrates |
| | Belemnites | | Wood |
| | Brachiopods | | Beds distinct |
| | Bryozoans | | Beds obscure |
| | Corals, solitary | | Unbedded |
| | Corals, colonial | | Graded beds |
| | Crinoids | | Planar cross-bedding |
| | Echinoderms | | Trough cross-bedding |
| | Echinoids | | Ripple structures |
| | Fish bones | | Cut and fill |
| | Fish scales | | Load casts |
| | Foraminifers, general | | Scour casts |
| | Foraminifers, large | | Convolution |
| | Fossils | | Slumped beds |
| | Fossils abundant | | Paleosol |
| | Fossils sparse | | Mud cracks |
| | Gastropods | | Salt molds |
| | Graptolites | | Burrows |
| | Leaves | | Pellets |
| | Ostracodes | | Oolites |
| | Pelecypods | | Pisolites |
| | Root molds | | Intraclasts |
| | Spicules | | Stylolite |
| | Stromatolites | | Concretion |
| | Tree trunk in place | | Calcitic concretion |

*Chiefly after the *Standard Legend* of the Royal Dutch/Shell Group of Companies (Shell International Petroleum Maatschappij B. V., The Hague, July 1977)

## APPENDIX 10: Major Geochronologic and Chronostratigraphic Units in Use by the U.S. Geological Survey[1]

| Eon or Eonothem | Era or Erathem | Period or System | | Epoch or Series | | | Age estimates of boundaries in millions of years[2] |
|---|---|---|---|---|---|---|---|
| Phanerozoic | Cenozoic (Cz) | Quaternary (Q) | | Holocene | | | 0.010 |
| | | | | Pleistocene | | | 2   (1.7–2.2) |
| | | Tertiary (T) | Neogene Subperiod or Subsystem (N) | Pliocene | | | 5   (4.9–5.3) |
| | | | | Miocene | | | 24   (23–26) |
| | | | Paleogene Subperiod or Subsystem (Pɛ) | Oligocene | | | 38   (34–38) |
| | | | | Eocene | | | 55   (54–56) |
| | | | | Paleocene | | | 63   (63–66) |
| | Mesozoic (Mz) | Cretaceous (K) | | Late Early | Upper Lower | | 96   (95–97) |
| | | | | | | | 138   (135–141) |
| | | Jurassic (J) | | Late Middle Early | Upper Middle Lower | | 205   (200–215) |
| | | Triassic (Ŧ) | | Late Middle Early | Upper Middle Lower | | ~240 |
| | Paleozoic (Pz) | Permian (P) | | Late Early | Upper Lower | | 290   (290–305) |
| | | Carbon-iferous Periods or Systems (C) | Pennsylvanian (IP) | Late Middle Early | Upper Middle Lower | | ~330 |
| | | | Mississippian (M) | Late Early | Upper | | 360   (360–365) |
| | | Devonian (D) | | Late Middle Early | Upper Middle Lower | | 410   (405–415) |
| | | Silurian (S) | | Late Middle Early | Upper Middle Lower | | 435   (435–440) |
| | | Ordovician (O) | | Late Middle Early | Upper Middle Lower | | 500   (495–510) |
| | | Cambrian (Ꞓ) | | Late Middle Early | Upper Middle Lower | | ~570[2] |
| Precambrian (pꞒ)[3] | Proterozoic (P) | Late Proterozoic[4] (Z) | | | | | 900 |
| | | Middle Proterozoic[4] (Y) | | | | | 1600 |
| | | Early Proterozoic[4] (X) | | | | | 2500 |
| | Archean (A) | Late Archean[4] (W) | | | | | 3000 |
| | | Middle Archean[4] (V) | | | | | 3400 |
| | | Early Archean[4] (U) | | | | | 3800? |
| | pre-Archean (pA)—an informal time term without specific rank. | | | | | | 4550 |

1. Format modified slightly from Sohl, N.L., and Wright, W.B. (1980, *Changes in stratigraphic nomenclature by the U.S. Geological Survey, 1979:* U.S. Geological Survey Bulletin 1502-A, p. A1-A3.), with Precambrian units from Harrison, J.E., and Peterman, Z.E. (1980, A preliminary proposal for the Precambrian of the United States and Mexico: *Geological Society of America Bulletin,* v. 91, p. 1128-1133). See these articles for sources of the original data.
2. Ranges reflect uncertainties of isotopic and biostratigraphic age assignments. Ages of boundaries not closely bracketed by data shown by ~.
3. A time term without specific rank.
4. Time terms only.

## APPENDIX 11: Natural Trigonometric Functions, at 0.5° Intervals

| ∠↓ | Sine | Cos. | Tan. | Cot. | | ∠↓ | Sine | Cos. | Tan. | Cot. | |
|---|---|---|---|---|---|---|---|---|---|---|---|
| 0° | .0000 | 1.000 | .0000 | infin. | 90° | 23° | .3907 | .9205 | .4245 | 2.356 | 67° |
|  | .0087 | .9999 | .0087 | 114.6 | |  | .3987 | .9171 | .4348 | 2.300 | |
| 1° | .0175 | .9998 | .0175 | 57.29 | 89° | 24° | .4067 | .9135 | .4452 | 2.246 | 66° |
|  | .0262 | .9997 | .0262 | 38.19 | |  | .4167 | .9100 | .4557 | 2.194 | |
| 2° | .0349 | .9994 | .0349 | 28.64 | 88° | 25° | .4226 | .9063 | .4663 | 2.145 | 65° |
|  | .0436 | .9990 | .0437 | 22.90 | |  | .4305 | .9026 | .4770 | 2.097 | |
| 3° | .0523 | .9986 | .0524 | 19.08 | 87° | 26° | .4384 | .8988 | .4877 | 2.050 | 64° |
|  | .0610 | .9981 | .0612 | 16.35 | |  | .4462 | .8949 | .4986 | 2.006 | |
| 4° | .0698 | .9976 | .0699 | 14.30 | 86° | 27° | .4540 | .8910 | .5095 | 1.963 | 63° |
|  | .0785 | .9969 | .0787 | 12.71 | |  | .4617 | .8870 | .5206 | 1.921 | |
| 5° | .0872 | .9962 | .0875 | 11.43 | 85° | 28° | .4695 | .8829 | .5317 | 1.881 | 62° |
|  | .0958 | .9954 | .0963 | 10.39 | |  | .4772 | .8788 | .5430 | 1.842 | |
| 6° | .1045 | .9945 | .1051 | 9.514 | 84° | 29° | .4841 | .8746 | .5543 | 1.804 | 61° |
|  | .1132 | .9936 | .1139 | 8.777 | |  | .4924 | .8704 | .5658 | 1.767 | |
| 7° | .1219 | .9925 | .1228 | 8.144 | 83° | 30° | .5000 | .8660 | .5774 | 1.732 | 60° |
|  | .1305 | .9914 | .1317 | 7.596 | |  | .5075 | .8616 | .5890 | 1.698 | |
| 8° | .1392 | .9903 | .1405 | 7.115 | 82° | 31° | .5150 | .8572 | .6009 | 1.664 | 59° |
|  | .1478 | .9890 | .1490 | 6.691 | |  | .5225 | .8526 | .6128 | 1.632 | |
| 9° | .1564 | .9877 | .1584 | 6.314 | 81° | 32° | .5299 | .8480 | .6249 | 1.603 | 58° |
|  | .1650 | .9863 | .1673 | 5.976 | |  | .5373 | .8434 | .6371 | 1.570 | |
| 10° | .1736 | .9848 | .1763 | 5.671 | 80° | 33° | .5446 | .8387 | .6494 | 1.540 | 57° |
|  | .1822 | .9833 | .1853 | 5.396 | |  | .5519 | .8339 | .6619 | 1.511 | |
| 11° | .1908 | .9816 | .1944 | 5.145 | 79° | 34° | .5592 | .8290 | .6745 | 1.483 | 56° |
|  | .1994 | .9799 | .2035 | 4.915 | |  | .5664 | .8241 | .6873 | 1.455 | |
| 12° | .2079 | .9781 | .2126 | 4.705 | 78° | 35° | .5736 | .8192 | .7002 | 1.428 | 55° |
|  | .2164 | .9763 | .2217 | 4.511 | |  | .5807 | .8142 | .7133 | 1.402 | |
| 13° | .2250 | .9744 | .2309 | 4.331 | 77° | 36° | .5878 | .8090 | .7265 | 1.376 | 54° |
|  | .2334 | .9724 | .2401 | 4.165 | |  | .5948 | .8039 | .7400 | 1.351 | |
| 14° | .2419 | .9703 | .2493 | 4.011 | 76° | 37° | .6018 | .7986 | .7536 | 1.327 | 53° |
|  | .2504 | .9681 | .2586 | 3.867 | |  | .6088 | .7934 | .7673 | 1.303 | |
| 15° | .2588 | .9659 | .2679 | 3.732 | 75° | 38° | .6157 | .7880 | .7813 | 1.280 | 52° |
|  | .2672 | .9636 | .2773 | 3.606 | |  | .6225 | .7826 | .7954 | 1.257 | |
| 16° | .2756 | .9613 | .2867 | 3.487 | 74° | 39° | .6293 | .7771 | .8098 | 1.235 | 51° |
|  | .2840 | .9588 | .2962 | 3.376 | |  | .6361 | .7716 | .8243 | 1.213 | |
| 17° | .2924 | .9563 | .3057 | 3.271 | 73° | 40° | .6428 | .7660 | .8391 | 1.192 | 50° |
|  | .3007 | .9537 | .3153 | 3.172 | |  | .6494 | .7604 | .8541 | 1.171 | |
| 18° | .3090 | .9511 | .3249 | 3.078 | 72° | 41° | .6561 | .7547 | .8693 | 1.150 | 49° |
|  | .3173 | .9483 | .3346 | 2.989 | |  | .6626 | .7490 | .8847 | 1.130 | |
| 19° | .3256 | .9455 | .3443 | 2.904 | 71° | 42° | .6691 | .7431 | .9004 | 1.106 | 48° |
|  | .3338 | .9426 | .3541 | 2.824 | |  | .6756 | .7373 | .9163 | 1.091 | |
| 20° | .3420 | .9397 | .3640 | 2.747 | 70° | 43° | .6820 | .7314 | .9325 | 1.072 | 47° |
|  | .3502 | .9367 | .3739 | 2.675 | |  | .6883 | .7254 | .9490 | 1.054 | |
| 21° | .3584 | .9336 | .3839 | 2.605 | 69° | 44° | .6947 | .7193 | .9657 | 1.036 | 46° |
|  | .3665 | .9304 | .3939 | 2.539 | |  | .7009 | .7133 | .9827 | 1.018 | |
| 22° | .3746 | .9272 | .4040 | 2.475 | 68° | 45° | .7071 | .7071 | 1.000 | 1.000 | 45° |
|  | .3827 | .9239 | .4142 | 2.414 | |  | .7133 | .7009 | 1.018 | .9827 | |
|  | Cos. | Sin. | Cot. | Tan. | ∠↑ |  | Cos. | Sine | Cot. | Tan. | ∠↑ |

## APPENDIX 12: Equivalence Among Common English and Metric Units

Underlined numbers are exact; others are rounded to the nearest numeral on the far right.

1 m = 1000 mm; 100 cm; 10 dm; 0.1 dkm; 0.01 hectometer; 0.001 km

1 km = 1000 m; 0.621 mi; 3281 ft

1 m = 39.37 in.; 3.281 ft; 1.094 yd; 0.000624 mi

1 cm = 10 mm; 0.394 in.

1 in. = 2.54 cm; 25.4 mm; 0.0833 ft; 0.02778 yd

1 ft = 0.3048 m; 12 in.; 0.061 rods; 1/6 fathom; 0.0001894 mi

1 yd = 0.9144 m; 3 ft; 0.1818 rods; 0.0005682 mi

1 rod = 5.0292 m; 198 in.; 16.5 ft; 5.5 yd

1 mi = 1609 m; 1.609 km; 5280 ft; 1760 yd; 320 rods

1 international nautical mile = .999 U.S. nautical mile; 1.151 mi; 1852 m

1 sq km = 100 hectares; 247.1 acres; 0.386 sq mi

1 hectare = 10,000 sq m; 2.471 acres; 11,960 sq yd

1 sq m = 10.76 sq ft; 1.196 sq yd

1 sq mi = 259 hectares; 2.59 sq km; 640 acres

1 acre = 4048 sq m; 0.405 hectare; 43,560 sq ft; 4840 sq yd

1 sq ft = 929.0 sq cm; 144 sq in.

1 liter = 1 cu dm; 1000 cc; 61.02 cu in.; 1.057 liq quarts; 0.264 gallon

1 cu m = 1000 liters; 1.308 cu yd; 35.32 cu ft; 264.2 gallons

1 cu ft = 7.481 gal; 28.3 liters; 0.0283 cu m; 0.0370 cu yd

1 acre ft = 1233 cu m; 43,560 cu ft

1 U.S. liq ounce = 29.57 cc; 1.805 cu in.

1 U.S. liq quart = 0.964 liter; 57.75 cu in.; 0.833 British quart

1 U.S. gallon = 3.785 liters; 231 cu in.; 0.833 British gal; 0.0238 bbl pet

1 kilogram = 1000 gr; 2.205 lb

1 gram = 0.035 ounce; 15.43 grains

1 ton metric (megagram) = 1000 kilograms; 2205 lb; 1.102 net ton; 0.984 gross ton

1 lb avoir. = 453.6 grams; 7000 grains; 16 oz; 1.215 lb troy

1 ounce (avoir.) = 437.5 grains; 28.35 grams

1 ton, net = 0.907 metric ton; 2000 lb avoir.; 0.893 gross ton

$1°F = 5/9°C$; T in °F = (T in °C × 9/5) + 32; T in °C = (T in °F − 32)5/9

## APPENDIX 13: Table for Interconversion of True Dip and Apparent Dip

The true dip of a planar feature is seen in vertical sections oriented perpendicular to the strike of the feature. Vertical sections oriented otherwise show apparent dip. All beds have horizontal apparent dips in any vertical section parallel to their strike, and the apparent dip increases as the acute angle between the vertical section and the strike increases, approaching the true dip as the angle between the section and the strike approach 90°. The values of apparent dip given below correspond to the true dips shown at the left of the table and to the angles between strike and the line of the vertical section shown at the top of the table and in the diagram. The values of apparent dip are rounded to the nearest 0.5° because dips are rarely measured or plotted more precisely.

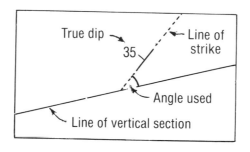

| True dip | ACUTE ANGLE BETWEEN STRIKE AND LINE OF VERTICAL SECTION | | | | | | | | | | | | | | | |
|---|---|---|---|---|---|---|---|---|---|---|---|---|---|---|---|---|
| | 2.5 | 5 | 10 | 15 | 20 | 25 | 30 | 35 | 40 | 45 | 50 | 55 | 60 | 65 | 70 | 80 |
| 5 | 0.0 | 0.5 | 1.0 | 1.5 | 2.0 | 2.0 | 2.5 | 3.0 | 3.0 | 3.5 | 4.0 | 4.0 | 4.5 | 4.5 | 5.0 | 5.0 |
| 10 | 0.5 | 1.0 | 2.0 | 2.5 | 3.5 | 4.0 | 5.0 | 6.0 | 6.5 | 7.0 | 8.0 | 8.0 | 8.5 | 9.0 | 9.5 | 10.0 |
| 15 | 1.0 | 1.5 | 3.0 | 4.0 | 5.0 | 6.5 | 8.0 | 9.0 | 10.0 | 11.0 | 11.5 | 12.5 | 13.0 | 13.5 | 14.0 | 15.0 |
| 20 | 1.0 | 2.0 | 3.5 | 5.5 | 7.0 | 9.0 | 10.0 | 12.0 | 13.0 | 14.5 | 15.5 | 16.5 | 17.5 | 18.0 | 19.0 | 20.0 |
| 25 | 1.0 | 2.0 | 4.5 | 7.0 | 9.0 | 11.0 | 13.0 | 15.0 | 17.0 | 18.0 | 20.0 | 21.0 | 22.0 | 23.0 | 24.0 | 25.0 |
| 30 | 1.5 | 3.0 | 6.0 | 8.0 | 11.0 | 14.0 | 16.0 | 18.5 | 20.5 | 22.0 | 24.0 | 25.0 | 26.5 | 27.5 | 28.5 | 29.5 |
| 35 | 2.0 | 3.5 | 7.0 | 10.5 | 13.5 | 16.5 | 19.5 | 22.0 | 24.0 | 26.5 | 28.0 | 30.0 | 31.0 | 32.5 | 33.5 | 35.5 |
| 40 | 2.0 | 4.0 | 8.0 | 12.0 | 16.0 | 19.5 | 23.0 | 26.0 | 28.5 | 30.5 | 33.0 | 34.0 | 36.0 | 37.0 | 38.5 | 39.5 |
| 45 | 2.5 | 5.0 | 10.0 | 14.5 | 19.0 | 23.0 | 26.5 | 30.0 | 33.0 | 35.0 | 37.0 | 39.0 | 41.0 | 42.0 | 43.0 | 44.5 |
| 50 | 3.0 | 6.0 | 11.5 | 17.0 | 22.0 | 27.0 | 31.0 | 34.5 | 37.5 | 40.0 | 42.5 | 44.0 | 46.0 | 47.0 | 48.0 | 49.5 |
| 55 | 4.0 | 7.0 | 14.0 | 20.0 | 26.0 | 31.0 | 35.5 | 39.5 | 42.5 | 45.0 | 47.5 | 49.5 | 51.0 | 52.5 | 53.5 | 54.5 |
| 60 | 4.5 | 8.5 | 16.5 | 24.0 | 30.5 | 36.0 | 41.0 | 45.0 | 48.0 | 51.0 | 53.0 | 55.0 | 56.0 | 57.5 | 58.5 | 59.5 |
| 65 | 5.5 | 10.5 | 20.5 | 29.0 | 36.0 | 42.0 | 47.0 | 51.0 | 54.0 | 56.5 | 58.5 | 60.0 | 62.0 | 63.0 | 63.5 | 64.5 |
| 70 | 6.5 | 13.0 | 25.5 | 35.0 | 43.0 | 49.0 | 54.0 | 57.5 | 60.5 | 63.0 | 64.5 | 66.0 | 67.0 | 68.0 | 69.0 | 69.5 |
| 75 | 9.0 | 18.0 | 33.0 | 44.0 | 52.0 | 57.5 | 62.0 | 65.0 | 67.5 | 69.0 | 70.5 | 72.0 | 73.0 | 73.5 | 74.0 | 75.0 |
| 80 | 13.5 | 26.5 | 44.5 | 56.0 | 63.0 | 67.5 | 70.5 | 73.0 | 74.5 | 76.0 | 77.0 | 78.0 | 78.5 | 79.0 | 79.5 | 80.0 |
| 85 | 26.0 | 45.0 | 63.5 | 71.5 | 75.5 | 78.0 | 80.0 | 81.5 | 82.0 | 83.0 | 83.5 | 84.0 | 84.0 | 84.5 | 84.5 | 85.0 |

## APPENDIX 14. Equal-area (Schmidt) Stereographic Net

To prepare the net for use: (1) cut it out and glue it to a smooth-surfaced flat board, such as one made of ⅛-inch masonite; (2) make a small hole through the board exactly at the center of the net; (3) insert a thumbtack through the hole from the back of the board, and glue or tape it firmly in place; (4) cut a rectangular sheet of plastic drafting film or tracing paper that is somewhat larger than the net; (5) stick a small piece of drafting tape on the back of the sheet at its center; (6) press the sheet down over the board so that the tack comes through the piece of tape and the sheet lies flat on the net; (7) mark and label pencil ticks on the sheet at the N and S poles of the net.

Data may now be plotted on the net, and Fig. 3-11 describes how to plot a planar feature and a lineation that lies parallel to the feature. The same figure describes each step in converting a lineation measured by its pitch into a lineation based on bearing (trend) and plunge. In addition, Fig. 9-12 describes the steps in rotating inclined linear features and planar features so that they are horizontal, and Fig. 9-14 describes rotations used to unfold or unfault linear and planar features. Turner and Weiss (see the references for Chapter 12) have described basic as well as more complex procedures with a stereographic net; and basic procedures are also described in some structure texts.

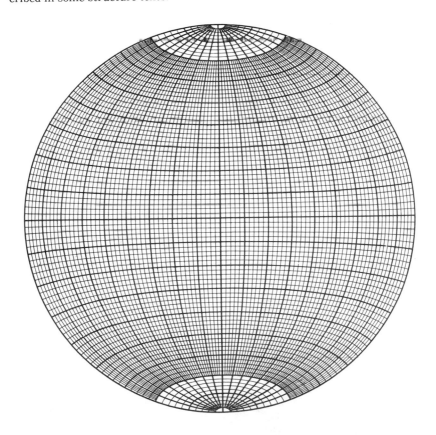

# ▪ Index

Page numbers in *italic type* refer to figures or tables.